Materials for the Engineering Technician

Materials for the Engineering Technician

THIRD EDITION

Raymond A. Higgins

BSc (Birm), C. Eng., FIM

Former Senior Lecturer in Materials Science, The College of Commerce and Technology, West Bromwich; some time Chief Metallurgist, Messrs Aston Chain and Hook Co., Ltd., Birmingham; and Examiner in Metallurgy to The Institution of Production Engineers, The City and Guilds of London Institute, The Union of Lancashire and Cheshire Institutes, and The Union of Educational Institutes

A member of the Hodder Headline Group
LONDON • SYDNEY • AUCKLAND
Copublished in North, Central and South America by
John Wiley & Sons Inc., New York • Toronto

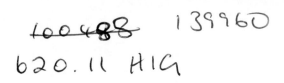
First published in Great Britain in 1972
Second edition 1987
Third edition published in Great Britain by Arnold,
a member of the Hodder Headline Group
338 Euston Road, London NW1 3BH

Co-published in North, Central and South America by
John Wiley & Sons Inc., 605 Third Avenue,
New York, NY 10158-0012

British Library Cataloguing in Publication Data
A catalogue record for this book is available from the British Library

Library of Congress Cataloging-in-Publication Data
A catalog record for this book is available from the Library of Congress

ISBN 0 340 67654 X
ISBN 0 470 23626 4 (Wiley)

4 5 6 7 8 9 10

Typeset in 10.5/11pt Times by
Mackreth Media Services, Hemel Hempstead
Printed and bound in Great Britain by
J W Arrowsmith Ltd, Bristol

To
the memory of my father

VICTOR HIGGINS
1882–1942
Watchmaker

*As a small boy, standing by his workbench,
I learned how to harden steel and brass.*

Contents

Preface to the first edition

My grandfather used to take trout from the River Tame, which runs alongside our new college here in West Bromwich. During my boyhood, fish no longer lived there, though occasionally an adventurous water-vole would make an exploratory drive. Today the Tame is a filthy open sewer, which will support neither animal nor vegetable life, and even the lapwings have forsaken its surrounding water-meadows. The clear-water tributaries of the Tame, where as a boy I gathered watercress for my grandmother – and tarried to fish for sticklebacks – are now submerged by the acres of concrete which constitute the link between the M5 and M6 motorways. On its way to the sea, the Tame spews industrial poison of the Black Country into the unsuspecting Trent, doing little to maintain the ecological stability of that river.

We Midlanders are not alone in achieving this kind of environmental despoliation. It seems that the Americans have succeeded in poisoning considerable areas of their Great Lakes. At this rate of 'progress', we must seriously consider the ultimate pollution of the sea, which, far from being 'cruel', provides Man with a great deal of his food. More important still, much of his supply of oxygen is generated by the vast forests of marine kelp which is generally referred to, somewhat unkindly, as 'seaweed'.

We would do well to consider the extent to which a more effective use of materials might alleviate the pollution of our environment. Suppose, for example, a motor car were constructed of materials such that it would last much longer. Presumably, since fewer motor cars would then be produced, the total amount of environmental pollution associated with their production would also be reduced. Further, since their scrap value would be higher, less of them would be left around to disfigure – or pollute – our countryside. Obviously, if such a course were followed, the whole philosophy of consumer production would need to be rethought, since a reduced total production would mean underemployment per worker (judged by present standards, rather than those which will obtain during the twenty-first century). However, these are problems to be solved by the planner and the politician, rather than by the mechanical engineer.

Pollution of the environment is, however, only one factor in the use of materials that the engineer must consider. The supplies of basic raw materials are by no means inexhaustible, and, by the end of the present century, reserves of metals like copper and tungsten will have run out, if we continue to gobble them up at present rates. In the more distant future it may well become a punishable offence to allow iron in any of its forms to rust away and become so dispersed amongst its surroundings that it cannot conveniently be reclaimed. The motor car of the future will need to be made so that it does not corrode. Such an automobile, made to last say forty years, would inevitably cost a lot of money, but, when worn out, would be returned to the factory for 'dismemberment', so that almost 100 per cent of its basic materials would be reclaimed. Only in this way will looting and pollution of the environment be reduced to tolerable proportions, particularly if the world population continues to increase even at only a fraction of its present rate.

It follows that a greater appreciation of the properties of all available materials is necessary to the mechanical engineer, and, although this book has been written with the mechanical engineering technician specifically in mind, it is hoped that it will provide a useful introduction to materials science for others engaged in engineering production. The book has been so written that only a most elementary knowledge of general science is necessary in order that it can be read with advantage. As a point of interest, the original idea of the work was conceived when the author was blizzard-bound in his caravan one Easter mountaineering holiday in Skye. It was finished in high summer in the Haute Savoie, when plagues of vicious insects in the alpine meadows made it equally unpleasant to venture forth.

Since writing the above paragraphs, I have learned that plans to clean up our River Tame are now 'under consideration', so we may get some action – eventually. In the meantime, the luckless river continues to pursue its fetid course outside the window of the lecture room where I teach. At 9.00 a.m. on Monday morning, the water is clear enough for me to discern an old motor tyre, presumably thrown there by small boys. By midday, the water has assumed a particularly disagreeable appearance, and the tyre has disappeared in a 'soup' of nacreous hue. As the working week proceeds, the river changes in colour from time to time, as the products of our 'effluent society' are discharged into it. Overhead, a hopeful kestrel still hunts this poisoned land; I wish him luck.

R. A. HIGGINS
The College of Commerce and Technology
West Bromwich, Staffs

Preface to the third edition

In this edition the section devoted to metals and alloys has been updated but the chapters dealing with most of the non-metallic materials have been completely rewritten. This has been made necessary by developments in syllabus content but also by numerous advances in the fields of ceramics and composite materials as well as in the increased industrial uses of some polymers.

It has become the fashion recently to include sections in course syllabi, and in textbooks, on 'selection of materials and processes'. Consequently, I have retained and expanded the chapter dealing with that subject in this book. However, one would have thought that such an aim would be one of the principal themes running throughout any book on 'materials science' and that it would of necessity discuss typical uses of any manufactured alloy described. During more than forty years of writing and revising textbooks on metallurgy and materials science I have always included detailed tables listing the *compositions*, *properties* and *uses* of most groups of engineering materials. I would suspect that such tables are of more value to the student confronted by a project in which he has to choose a material which may satisfy a given set of requirements than to be provided with a few 'case studies' which may only lead him into a series of 'blind alleys'. The only case study I recount is that of the type of high-temperature alloy required for the turbine blades of jet engines. Engineers at that time were unable to 'buy one off the shelf' but, having decided the properties required of such a material, persuaded the metallurgists to 'get on and make it!' The student however will be expected to choose a material from those available. It is hoped that the tables in this book will help him to do so – and in many cases to quote its relevant British Standards Specification Number.

In the preface to the first edition of this book, written some quarter of a century ago, I complained of the damage which was currently being done to the environment. In particular I mentioned the effects of a hundred years of industrial development on our local River Tame. Since then there has been a considerable improvement in the condition of that river and, whilst as far as I can see no fish have yet returned, some simple vegetation is now flourishing and a few mallard are able to disport themselves on the still rather murky waters without their feet falling off. Sadly, I fear that much of this improvement is due to the demise of large sections of Black Country industry during the post-war years. The old 'Johnson's Iron Works' of West Bromwich where, as a boy, I watched enthralled – and petrified – as large white-hot 'snakes' rushed from between the rolls of an ancient Belgian 'looping-mill' to squirm and writhe across the shop floor before being grabbed by tongs and deftly guided into the next roll grooves, is now of course no more. As I motor towards the local countryside I pass hectares of devastation where once great steel works stood – land to be used for what? – yet another shopping complex? Surely we must earn wealth before we can spend it. Or have we abandoned our wish to become again a great industrial nation?

However, as a small crumb of comfort, I understand that a large sum of money has recently been allocated to clean up and rehabilitate our River Tame and my beloved Sandwell Valley. So perhaps eventually it may be possible to see and hear the abundance of wildlife which as a boy some seventy years ago I enjoyed on country walks there with my father.

At the time of producing the first edition of this book I was also apprehensive that we were running short world-wide of usable metallic ores. I had not at that time foreseen another world recession or the sudden end of the Cold War – both events which have led to decreased demand for raw materials. For whatever reasons – internal political contriving or military confrontation from without – the recession will end and we must consider then the effects renewed demand will have in relation to the availability of raw materials.

Having spent several years of my early working life in the copper-alloys industry I have a certain affection for that rosy-tinted metal and whilst a number of cheaper substitutes for its more ductile alloys have been developed I can see little prospect of its being replaced as a conductor of electricity for the majority of its industrial and domestic uses. Consequently, in the future, we will have to use yet 'leaner' ores for the extraction of copper which in turn will involve higher production costs – and larger mounds of rejected spoil, which brings us back to the question of environmental pollution, though in this instance not in our own 'backyard'. But enough! I have rabbited on for too long. In any case I am told that very few readers so much as glance at the preface of a textbook. I must confess that I am one of them!

R. A. HIGGINS
Walsall, West Midlands

Prologue: The cost of a nail

I found this old nail – some 135 mm long – whilst poking about among the debris from restoration work in the medieval hill-top village of San Gimignano, Tuscany. In the Middle Ages such nails were made by hand, a blacksmith using a hammer and an anvil, in this case containing a suitable slot so that he could form the nail head. His only additional equipment

would be a charcoal fire assisted by goat-skin bellows to enable him to heat the piece of iron from which he forged and pointed the nail. Because of the way in which they were made medieval nails were roughly square in cross-section and differed little from those with which Christ was crucified more than a thousand years earlier.

Such products were of course very labour intensive and, even allowing for the relatively low wages and long hours worked by craftsmen in those days, nails would have been quite costly items. Producing the necessary lumps of iron from the original ore was also an expensive process adding further to the cost of the nail.

Following the Industrial Revolution methods used to manufacture nails have naturally become increasingly mechanized. In modern processes steel rod – or wire – is fed into a machine where a die automatically forges the head whilst almost simultaneously cutting through the stock rod at suitable angles to form the point of the resulting nail. The process is completed in a fraction of a second and the rod, or wire, immediately travels forward to produce the next nail. Since a single operator may tend several machines the immediate labour cost of producing a nail is very low compared with medieval times. However the situation is not quite as simple as that – things rarely are!

Obviously a modern nail-making machine is a complex and fairly expensive piece of equipment which must be serviced frequently if it is to show a suitable profit before it finally becomes 'clapped out' (how often has one purchased nails whose points proved to be blunt due to lack of maintenance of the shearing blades?). There are of course many other overheads as well as running costs which must be taken into account in assessing the cost of producing a modern nail. One thing is certain: technological advances during and since the Industrial Revolution have reduced the unit cost reckoned in man-hours in producing a nail. That unit, the man-hour, is of course the only reliable one we can use, assuming that men are always working under similar financial pressures. Monetary units, whether medieval groats or ducats, current pounds, marks or dollars or future 'Euros' are subject to fluctuations in *real* value due to the vagaries of inflation which are contrived by politicians and outside the control of us engineers.

CHAPTER 1

Atoms, molecules and crystals

1.10 Atoms are very tiny particles indeed. A pin head contains about 350 000 000 000 000 000 of them – give or take a few billions. All truly solid materials consists of atoms which are arranged in some pattern peculiar to that material. These atoms are held together by forces of attraction which have their origin in electrical charges within each atom. Altogether there are ninety-two different types of atom which occur naturally though during the 'nuclear age' scientists have succeeded in producing some new ones, for example plutonium. Of the naturally occurring atoms the smallest and simplest is that of hydrogen; whilst the largest, some two-hundred and thirty-eight times as massive, is that of uranium.

A chemical *element* contains atoms all of one type. Therefore there are ninety-two different 'natural' chemical elements of which over seventy are metals. Some of these metallic elements are extremely rare, whilst others are useless to the engineer either by virtue of poor mechanical properties or because they are chemically very reactive. Consequently less than twenty of them (appendix) are in common use in engineering alloys.

1.11 Of the non-metallic elements carbon is perhaps the one which forms the basis of most engineering materials since it constitutes the 'backbones' of all plastics. Moreover it can be used in strong fibre form and is an essential constituent of all common heat-treatable steels. Chemically similar to carbon – indeed a member of the same chemical 'family' – the element silicon has become famous in the form of the 'silicon chip', but along with oxygen (as silicon dioxide or *silica*) it is the basis of many refractory building materials. Oxygen and silicon are by far the most common elements in the Earth's crust and account for some 75 per cent of it in the form of clays, sands, stones and rocks like granite.

The chemical bonding of atoms

1.20 In 1805 the British chemist John Dalton proposed that the atom was the smallest particle in which matter could exist and was indivisible. Here he was repeating a theory suggested by the Ancient Greeks some two thousand years earlier. We now know that there are many *subatomic* particles in the Universe but only three of them need be mentioned in this book. They are the main 'building blocks' of all atoms and are the *electron*, the *proton* and the *neutron*.

The proton and the neutron are roughly equal in mass whilst the electron is only about one two-thousandth of the mass of the other two.[1] However, it is the electrical charges carried by these particles which are more important. The electron carries a unit charge (1.6×10^{-19} coulombs) of *negative* electricity whilst the proton carries an equal but opposite charge of *positive* electricity. The neutron, as its name suggests, carries no resultant electrical charge and in all atoms other than the radioactive ones the neutron can be regarded as 'dead weight' in the nucleus. It affects only the 'atomic weight' or, more properly, the *atomic mass* of the element, for which reason we need consider the neutron no further here.

All of the relatively massive protons and neutrons are concentrated in the core – or *nucleus* – of the atom, whilst the electrons are arranged in a series of 'orbits' or 'shells' around the nucleus. The numbers of protons and electrons in any stable atom are equal and, since the charges carried by protons and electrons are equal but opposite, any stable atom carries no resultant charge. It is electrically neutral.

Each electron shell contains a specific number of electrons. The first shell, nearest to the nucleus, contains only two electrons, the second eight, the third eighteen, and so on. Since the number of protons in the nucleus governs the total number of electrons in all shells around it, it follows that there are generally insufficient electrons to complete the final outer shell (Fig. 1.1). At the same time there is electrical attraction between the outer-shell electrons of one atom and the protons in the nuclei of neighbouring atoms and it is these forces of attraction which cause both chemical combination and physical changes such as crystallization to take place.

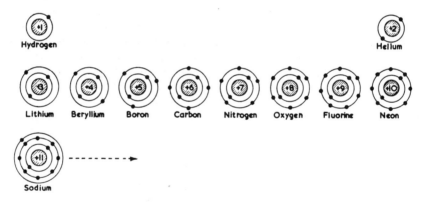

Fig. 1.1   *This shows the electron structure of the first eleven elements in order of 'atomic number', i.e. the number of protons in the nucleus. Although these diagrams show electrons as being single discrete particles in fixed orbits this interpretation should not be taken too literally. It is better to visualize an electron as a sort of mist of electricity surrounding the nucleus!*

As a result of such reactions strong bonds are formed between atoms so that the atoms are left with completed outer electron shells. This is achieved by atoms either losing, gaining or sharing electrons.

1.21   The electrovalent bond is the attractive force produced by reactions which take place between metals and non-metals. Thus the extremely reactive metal sodium combines with the equally reactive non-metallic gas chlorine, to form crystals of sodium chloride (common table salt). Since the atomic structures involved are simpler we will instead consider the similar

---

[1]   The mass of the electron is 9.11×10^{-31} kg.

combination which occurs between the metal litlium (now an essential element in the batteries used in electronic watches, automatic cameras and the like) and the extremely reactive non-metallic gas fluorine.[2]

An atom of lithium contains only three protons in its nucleus and, therefore, three electrons in orbit around that nucleus. Two of these electrons complete the first shell, leaving a lone electron in the second shell. The atom of fluorine on the other hand contains nine protons in its nucleus and, consequently, nine electrons in orbit around it. Two of these fill the first shell so that there are no less than seven electrons in the second shell.

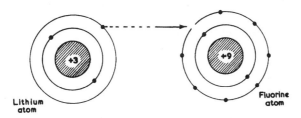

Fig. 1.2

Since the nuclear charge of the lithium atom is relatively small its attraction for the lone electron in the outer orbit is weak and so, if a lithium atom and a fluorine atom approach each other, the lone electron of the lithium atom is snatched away so that it joins the outer electron shell of the fluorine atom. This leaves the lithium particle with a complete first shell (2 electrons) and at the same time the electron which it has lost goes to complete the second shell (8 electrons) of the fluorine particle. Because the lithium particle has lost a negatively charged electron it now has a resultant positive charge; whilst the fluorine particle having gained an electron now has a resultant negative charge. These charged particles are no longer simply atoms – the lithium particle is now less than an atom whilst the fluorine particle is more than an atom. Charged particles of this type are called *ions* (Fig. 1.3). Metals always form positively charged ions whilst non-metals form negatively charged ions.

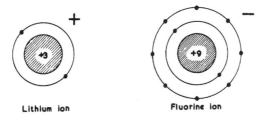

Fig. 1.3

As these lithium and fluorine ions carry opposite charges they will attract each other. However, whilst unlike charges attract, like charges repel so that lithium and fluorine ions arrange themselves in a geometrical pattern in which each fluorine ion is surrounded by six lithium ions as its nearest

---

[2] The gas fluorine is one of the same group or family of chemical elements – the halogens (or salt-makers) – as chlorine, a reactive poisonous gas used on the Western Front in the First World War. Fluorine is the most chemically reactive of all non-metals and will even attack glass.

neighbours whilst each lithium ion is surrounded by six fluorine ions (Fig. 1.4). This compound, lithium fluoride, forms a relatively simple cubic type of *crystal structure*. Other such *salts* may form more complex crystal patterns depending upon the relative sizes of the ions involved and the ratio of the electrical charges carried by each type of ion. Here we have dealt with atoms which lose (or gain) only one electron; those which lose (or gain) two electrons will produce ions carrying twice the electrical charge.

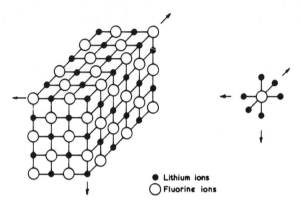

● Lithium ions
○ Fluorine ions

Fig. 1.4   *The crystal structure of lithium fluoride.*

1.22   The metallic bond Most metals have one, two or at most three electrons in the outermost shell of the atom. These outer-shell electrons are loosely held to the atomic nucleus and as a metallic vapour condenses and subsequently solidifies these outer-shell electrons are surrendered to a sort of common pool and are virtually shared between all atoms in the solid metal. Since the resultant metallic ions are all positively charged they will repel each other and so arrange themselves in some form of regular (crystal) pattern in which they are firmly held in position by the attractive force between them and this permeating 'cloud' of negatively charged electrons (Fig. 1.5).

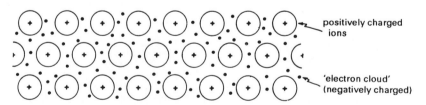

positively charged
ions

'electron cloud'
(negatively charged)

Fig. 1.5   *A simple interpretation of the metallic bond concept. The positively charged ions repel each other but are held in place by the attractive force supplied by the negatively charged 'electron cloud'.*

The metallic bond theory of metals explains many of the main characteristics of metallic elements:

- All metals are good conductors of electricity. What we call a 'current of electricity' is in fact a stream of moving electrons. Since electrons constituting the 'electron cloud' are free to move within the body of the metal it follows that a metal *conducts electricity*. If we force in some electrons at one end of a metal wire (by using the 'pressure' supplied by a 'head' of electrons, or an electrical *potential difference* as it is termed) an equal number of electrons will be forced out at the other end of the wire provided we give them another conductor into which they can escape.

- Metals are good conductors of heat. The application of heat to a piece of metal causes electrons to vibrate more actively and these vibrations can be passed on quickly from one electron to another within the electron cloud which is continuous within the body of metal.
- Most metals are ductile because layers of ions can be made to slide over each other by the application of a shearing force. At the same time metals are strong because the attractive force provided by the electron cloud opposes the slipping apart of those layers of ions.
- Metals are lustrous in appearance since the free, vibrating surface electrons fling back units of light as these fall upon the surface of the metal.

1.23   The covalent bond is formed between atoms of those non-metallic elements in which, for various reasons, there is a strong attractive force between the nucleus and the outer-shell electrons. Hence instead of a *transfer* of electrons from one atom to the other there is a *sharing* of electrons between two atoms thus binding them together.

The hydrogen atom is the most simple of all atoms. It consists of a single electron in orbit around a nucleus consisting of a single proton. Two atoms of hydrogen will therefore combine to form a *molecule* of hydrogen and in this way the two electrons are shared between the two atoms thus *completing the electron shell for each atom* by this process of sharing the two electrons (Fig. 1.6). Thus a mass of hydrogen gas does not contain single atoms but rather a conglomeration of rapidly moving *diatomic*[3] molecules.

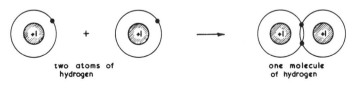

two atoms of
hydrogen

one molecule
of hydrogen

Fig. 1.6

Similarly an atom of carbon which has four electrons in its outer shell (the second electron shell) (Fig. 1.1) can combine with *four* atoms of hydrogen. The resultant unit containing a tightly bound association of one carbon atom with four hydrogen atoms constitutes a molecule of the gas methane, the principal component of the natural gas obtained currently from the North Sea bed and the most simple of all of the many thousands of carbon-hydrogen compounds, or *organic compounds* as they are called, which are derived from living matter whether plant or animal.

In the methane molecule (Fig. 1.7) the carbon atom shares four electrons contributed by the four hydrogen atoms and so the outer shell of eight electrons is completed. Similarly each hydrogen atom shares one of the electrons provided by the carbon atom and so its outer shell of two is completed. The methane molecule is thus a stable unit.

---

[3] Containing two atoms.

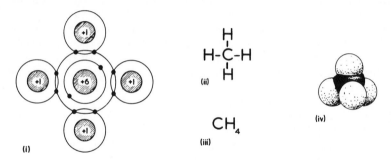

Fig. 1.7   *Four ways of representing a molecule of the gas methane: (i) drawn to represent the nature of the covalent bond; (ii) the usual way by which chemists indicate covalent bonds (–) in the* structural formula *of a molecule; (iii) the chemist's method of representing the* empirical formula *of a compound, in which the proportions of each type of atom are indicated; (iv) the probable shape of a methane molecule. The four hydrogen nuclei, which being positively charged repel each other, arrange themselves as far from each other as is possible, i.e. in a tetrahedral pattern around the carbon atom nucleus to which each is bound.*

The element carbon has the ability to form long chain-like molecules in which carbon atoms are covalently bonded to each other and to hydrogen atoms:

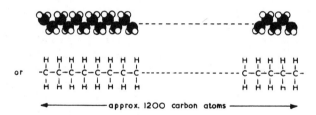

This is in fact a molecule of the material *poly-ethylene*, better known as polythene. By using chemical means to replace some of the hydrogen atoms by atoms, or groups of atoms, of other elements, a wide variety of plastics materials can be produced.

An important feature of these materials is that, unlike metals in which the outer-shell electrons can travel freely within the electron cloud so making metals conductors of electricity, the outer-shell electrons in these covalent substances are securely held to the atoms to which they belong. These electrons are not free to move away and so the materials are excellent insulators since they cannot conduct electricity. The initial use of polythene was as an electrical insulator in electronics equipment used in radar during the Second World War. In fact the commercial production of polythene began on 2nd September 1939 – the day on which the German armies invaded Poland.

Intermolecular forces

1.30   However large the molecules in covalent compounds they will always contain equal numbers of protons and electrons contributed by their constituent atoms. Therefore they will be electrically neutral and carry no resultant charges. So why do these molecules stick together to form a coherent solid mass?

Fig. 1.8 *The principle of dipole moments. Like charges repel (i) tending to cause realignment of molecules, so that unlike charges then attract (ii) producing bonds (van der Waals forces) between the molecules.*

1.31 The short answer is that electrons and protons are not necessarily equally distributed over the molecule, often due to interaction with electrons and protons of neighbouring molecules. Consequently, due to this irregular distribution of electrons and protons the molecule acquires what is sometimes called a *dipole moment*, that is it has a 'negative end' and a 'positive end' rather like the north and south poles of a magnet (Fig. 1.8). Since unlike charges attract (as do the unlike poles of magnets) then molecules will be attracted to each other. The greater the unevenness of distribution of the charges within molecules the greater the dipole moments and forces of attraction between them – and therefore the 'stronger' the material. Uneven distribution of electrons within a molecule may occur for a number of reasons and will give rise to intermolecular forces of different strengths. This fact is dramatically illustrated in the case of the two compounds water (H_2O) and methane (CH_4). Both substances are of similar molecular weight, i.e. 18 and 16 respectively, yet liquid methane boils at $-161\,°C$ (112 K) whilst water remains liquid until $100\,°C$ (373 K) – a temperature difference of $261\,°C$. This indicates that the dipole moments attracting water molecules to each other are very much greater than those attracting methane molecules to each other. Therefore much lower thermal vibrations are required to separate methane molecules than is the case with water molecules.

The difference in shapes of the two molecules is the reason for this. In the methane molecule (Fig. 1.7(iv)) the central carbon atom is 'shielded' by four *symmetrically spaced* hydrogen atoms and so the distribution of valence electrons is uniform around the outer fringes of the molecule. Its dipole

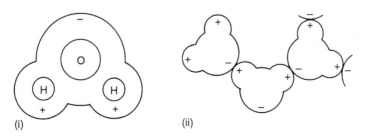

Fig. 1.9 *In the water molecule (i) the valence electrons are attracted to the larger oxygen nucleus giving it a resultant negative charge whilst the two hydrogen nuclei are left 'exposed' so that the water molecule has a strong dipole moment. Hence there is a strong force of attraction between the positively charged region of one water molecule and the negatively charged region of a neighbour (ii).*

moment is therefore very small. The water molecule on the other hand is not completely symmetrical and the distribution of valence electrons is uneven. Due to the shape of the molecule these electrons tend to be attracted towards the larger oxygen nucleus, the 'exposed' side of which acquires a resultant negative charge, leaving the two hydrogen nuclei, which are positively charged, also 'exposed'. The water molecule therefore has a strong dipole moment. But for this strong dipole moment which leads to strong attraction between water molecules, water would boil at a much lower temperature like other substances in which simple molecules contain few atoms of low atomic weights. If that were so could biological life as we know it on Earth ever have begun? However we are getting into 'deep waters' – sorry about that pun!

These forces of attraction between molecules are known collectively as *van der Waals forces*. In the case of small molecules the total van der Waals forces operating between molecules is small. Hence many such substances remain as liquids or even gases, but with large molecules like those of polyethylene the sum total of all the van der Waals forces operating between a molecule and its neighbours is great and the solid so formed is relatively strong.

Polymorphism

1.40   Many solid elements can exist in more than one different crystalline form and are said to be *polymorphic* (the term *allotropy* is also used to describe this phenomenon). Generally these different crystalline forms are stable over different temperature ranges so that transition from one form to another takes place as the transition temperature is passed. As we shall see later it is the polymorphism of iron which enables us to harden suitable steels.

1.41   Tin is also polymorphic existing as 'grey tin', ordinary 'white tin' and as 'brittle tin'. White tin, the form with which we are generally familiar, is stable above 13 °C whilst 'grey tin' is stable below 13 °C, but the change from one form to the other is very sluggish and white tin will not normally change to the powdery grey form unless the temperature falls well below 13 °C. At one time organ pipes were manufactured from tin and the story is told that during one particularly cold winter in St. Petersburg the organ pipes collapsed into a pile of grey dust as the organist played his first chord. This particular affliction to which tin was prone was known as 'tin pest' but was in fact a manifestation of a polymorphic change. On a more sombre note the final tragedy which befell Scott's Polar party in 1912 was due in no small measure to the loss of valuable fuel which had been stored in containers with *tin-soldered* joints. These had failed – due to 'tin pest' – during the intensely cold Antarctic winter and precious fuel oil seeped away.

Tin is one of a chemical family of elements which includes carbon, silicon and germanium. All of these are polymorphic elements and have one crystal structure in common – that of the carbon polymorph *diamond*. Moreover all elements of the group have unique properties. Silicon and germanium are used in the manufacture of transistors whilst the development of the silicon chip has revolutionized the modern technological world.

1.42   Polymorphism is exhibited in a most spectacular way by the element carbon which can exist as diamond and as graphite. It was shown in 1.23 that an atom of carbon can combine with four atoms of hydrogen to form a molecule of methane. That is, an atom of carbon is capable of forming four separate bonds – or *valences* as chemists call them – with other atoms. Under conditions of extremely high temperature and pressure (much greater than we can achieve industrially) carbon atoms will link up with each other to form *giant molecules* in which the carbon atoms form a rigid crystalline structure showing the same tetrahedral pattern as is exhibited in the distribution of hydrogen atoms in a methane molecule. This tetrahedral crystalline structure containing only carbon atoms is the substance *diamond*.

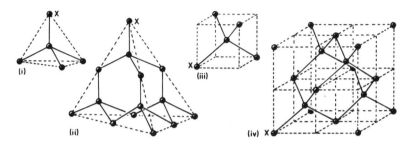

Fig. 1.10　*The crystal structure of diamond.*

Fig. 1.10(i) indicates the arrangement of the centres of carbon atoms in the basic unit of the diamond crystal. There is an atom at the geometrical centre of the tetrahedron and one at each of the four points of the tetrahedron. The structure is of course continuous (Fig. 1.10(ii)) such that each atom of carbon in the structure is surrounded by four other atoms covalently bonded to it and spaced equidistant from it. Since all carbon atoms in diamond are 'joined' by strong covalent bonds to four other carbon atoms it follows that diamond is a very strong, hard material. It is in fact the hardest substance known.

1.43 Under a different set of conditions of temperature and pressure carbon atoms will combine to form layer-like molecules (Fig. 1.11) in another polymorph, *graphite*. Here the layers are held together by relatively weak van der Waals forces generated by the spare electrons not used by the

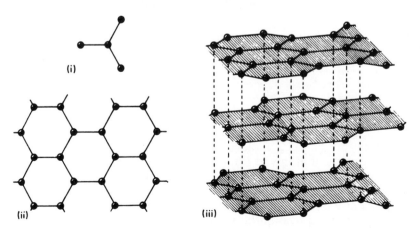

Fig. 1.11　*The crystal structure of graphite.*

primary bonding system. Consequently these layers will slide over each other quite easily so that graphite can be used as a lubricant. Coke, charcoal and soot, all familiar forms of carbon, contain tiny micro-crystals of graphite.

Since all of the carbon atoms in diamond are covalently bonded in such a manner that all valence bonds are used up no free electrons are present and so diamond does not conduct electricity. In graphite, however, spare electrons are present since not all of the valence electrons are used up in forming covalent bonds. Hence graphite is a relatively good conductor of electricity.

1.44   Discoveries in fundamental science become more rare as our knowledge of the Universe increases but since the last edition of this book was published a third allotrope of carbon was discovered. Imagine, if you will, a single graphite 'sheet' in which selected 'hexagons' have been replaced by *pentagons*. These pentagons will only 'fit' if the flat sheet is allowed to form a curved surface. This is the basis of the design of geodesic domes such as that used to house 'Science World' in Vancouver, BC. The architect responsible for this development was Richard Buckminster Fuller and this new form of carbon allotrope has been named *fullerene* in his honour. In fact there is a series of these fullerene molecules containing between 20 and 600 carbon atoms. Possibly the most common is the fullerene containing sixty carbon atoms, i.e. C_{60}. This is certainly the easiest of the fullerene molecules to visualize since its geometry is similar to that of a modern soccer ball (Fig. 1.12) for which reason it has become known affectionately as the 'Buckyball'.

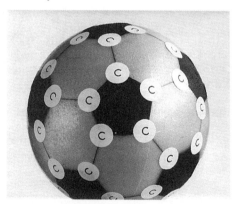

Fig. 1.12   *A model of the fullerene molecule C_{60} which I constructed using a child's soccer ball, a felt pen and some small labels. Each of the 12 pentagonal 'panels' (dark) is surrounded by 5 hexagonal panels (light). The white labels – each inscribed with a 'C' – indicate the positions of the nuclei of carbon atoms. Other fullerene molecules containing between 20 and 600 carbon atoms are spherical or ellipsoid in form.*

If the C_{60} fullerene is dissolved in benzene and allowed to crystallize out it forms magenta-coloured face-centred cubic (2.13) crystals but here the units are not atoms but large C_{60} molecules. The C_{60} molecule will react with other atoms and groups of atoms to form compounds but as yet the technology of these materials is in its infancy. Nevertheless one of my second cousins, a biochemist devoted to 'doing strange things to tomatoes' (I quote his father!) recently hinted darkly that some of the large fullerene molecules – roughly cigar-shaped – could be used as projectiles to carry other atoms or groups of atoms into larger molecules, possibly in genetic engineering.

In my far-off student days we were taught that three allotropes of carbon existed. The third one then was called 'amorphous carbon', a non-crystalline form – but later this was found to consist of tiny graphite crystals and we were left with two allotropes – diamond and graphite. The fullerene molecule was at that time undreamed of and so its discovery indicates that the romance of pure science is still not dead. As my old grandmother, her hair white with the snow of more than eighty winters, used to say, 'Sure an' the World is full of fairies yet – for those who have the eyes to see and the ears to hear them with!'

CHAPTER 2

The crystal structure of metals

2.10 All metals – and other elements, for that matter – can exist as either gases, liquids, or solids. The 'state' in which a metal exists depends upon the conditions of temperature and pressure which prevail at the time. Thus mercury will freeze to form a solid, rather like lead, if cooled to −39 °C; and will boil to form a gas or vapour if heated to 357 °C at atmospheric pressure. (Metals with low boiling-points – such as mercury, cadmium, and zinc – can be purified by distillation in the same way as can water.) At the other end of the scale tungsten melts at 3410 °C and boils at 5930 °C.

2.11 In any gas the particles (either atoms or molecules depending upon the constitution of the gas) are in a state of constant motion and the impacts which these particles make with the walls of the containing vessel constitute the pressure exerted by the gas. As the temperature increases the velocity and hence the number of impacts increases and therefore the pressure within the vessel increases. This is the basis of the kinetic theory of gases.

In a metallic gas the particles consist of single atoms[1] which are in a state of continuous motion. As the temperature falls condensation occurs at the boiling-point and in the resultant liquid metal the atoms are jumbled together willy-nilly, and since they are held together only by weak forces of attraction at this stage, the liquid lacks cohesion, and will flow. When the metal solidifies, the energy of each atom is reduced. This energy is given out as latent heat during the solidification process, which for a pure metal occurs at a fixed temperature (Fig. 2.1). During solidification, the atoms arrange themselves according to some regular pattern, or 'lattice structure'. Each atom becomes firmly bonded to its neighbours by stronger forces of attraction; so the solid metal acquires strength.

Since the atoms are now arranged in a regular pattern, they generally occupy less space. Thus most metals shrink during solidification, and the foundryman must allow for this, not only by making the wooden pattern a little larger than the required size of the casting, but also by providing adequate runners and risers, so that molten metal can feed into the body of a casting as it solidifies, and so prevent the formation of internal shrinkage cavities.

---

[1] Metallic gases are therefore said to be *monatomic*. Most non-metallic gases such as oxygen, hydrogen and nitrogen consist of molecules each of which contains two atoms. The *noble* – or inert – gases, helium, neon, etc. are monatomic because their atoms have completed outer electron shells.

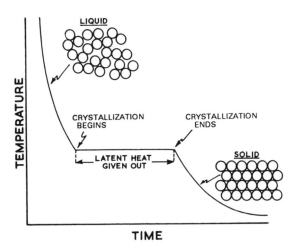

Fig. 2.1 *A pure metal solidifies at a fixed single temperature, and the atoms arrange themselves in some regular pattern.*

2.12 Most of the important metals crystallize into one of three different patterns as solidification takes place (Fig. 2.2A). It should be appreciated that the upper diagrams represent the simplest units in each case. Here the positions occupied by the centres of atoms are shown. In fact each outer 'face' of the figure is also part of the next adjacent unit, as shown in the lower diagrams (Fig. 2.2B). These lattice structures were derived by means of X-ray analysis.

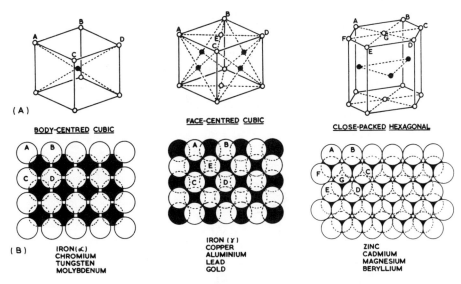

Fig. 2.2 *The principal types of lattice structure occurring in metals. (A) indicates the positions of the centres of atoms only, in the simplest unit of the structure; whilst (B) shows how these units occur in a continuous crystal structure viewed in 'plan'. (In each case the letters A, B, C, etc. indicate the appropriate atoms in both diagrams.) Although the atoms here are shown black or white, to indicate in which layer they are situated, they are, of course, all of the same type.*

2.13   Of these 'space lattices', the close-packed hexagonal arrangement represents the closest packing of atoms. It is the arrangement which would be produced if a second layer of snooker balls were allowed to fall into position on top of a set already packed in the triangle. The face-centred cubic arrangement is also a fairly close packing of atoms, but the body-centred cubic form is relatively 'open'; that is, the lattice contains relatively more free space.

Iron is a polymorphic element (1.40) which exists in two principal crystalline forms. The body-centred cubic form (α-iron) exists up to 910 °C, when it changes to the face-centred cubic form (γ-iron). On cooling again, the structure reverts to body-centred cubic. It is this fact which enables steel to be heat-treated in its own special way, as we shall see later in this book (12.21). Unfortunately, the sudden volume change which occurs as γ-iron changes to α-iron on being quenched gives rise to the formation of internal stresses, and sometimes distortion or even cracking of the component. However, we must not look a gift-horse in the mouth: but for this freak of nature in the form of the allotropic change in iron, we would not be able to harden steel. Without steel, as Man's most important metallic alloy, we might perhaps still be living in the Bronze age, for could many of our modern sophisticated materials and artefacts have been produced without the aid of a technology based on steel?

Dendritic solidification

2.20   When the temperature of a molten pure metal falls below its freezing-point, crystallization will begin. The nucleus of each crystal will be a single unit of the appropriate crystal lattice. For example, in the case of a metal with a body-centred cubic lattice, nine atoms will come together to form a single unit, and this will grow as further atoms join the lattice structure (Fig. 2.3). These atoms will join the 'seed crystal' so that it grows most quickly in those directions in which heat is flowing away most rapidly. Soon the tiny crystal will reach visible size, and form what is called a 'dendrite' (Fig. 2.4). Secondary and tertiary arms develop from the main 'backbone' of the dendrite – rather like the branches and twigs which develop from the trunk of a tree, except that the branches in a dendrite follow a regular geometrical pattern. The term 'dendrite' is, in fact, derived from the Greek *dendron* – a tree.

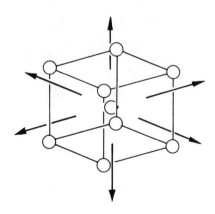

Fig. 2.3   *The nucleus of a metallic crystal (in this case a body-centred cubic structure).*

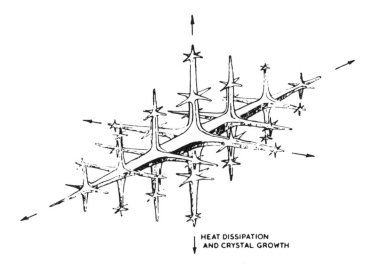

HEAT DISSIPATION
AND CRYSTAL GROWTH

Fig. 2.4   *The early stages in the growth of a metallic dendrite.*

2.21   The arms of the dendrite continue to grow until they make contact with the outer arms of other dendrites growing in a similar manner nearby. When the outward growth is thus restricted, the existing arms thicken until the spaces between them are filled, or, alternatively, until all the remaining liquid is used up. As mentioned earlier in this chapter, shrinkage usually accompanies solidification, and so liquid metal will be drawn in from elsewhere to fill the space formed as a dendrite grows. If this is not possible, then small shrinkage cavities are likely to form between the dendrite arms.

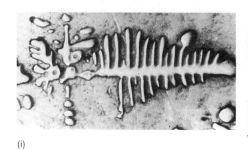

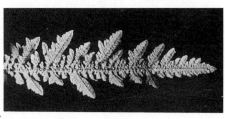

(i)   (ii)

Fig. 2.5   *Metallic dendrites. In (i) a molten solution of iron in copper was allowed to freeze. The iron, being of higher melting-point, crystallized first and was all used up before it could complete the formation of a crystal, so leaving a dendrite 'skeleton' of primary and secondary 'arms'. The molten copper then solidified filling in the spaces. To be more accurate the iron dendrite contains a small amount of dissolved copper whilst the solid copper matrix contains a small amount of dissolved iron (8.30).*

The zinc dendrite (ii) was grown by electrolysis from a water solution of zinc sulphate.

Note that the secondary 'arms' in the iron dendrite (i) are at right angles to the 'backbone' because the crystal structure in iron is basically cubic whilst in the zinc dendrite (ii) the secondary 'arms' are at angles of 60° to the backbone reflecting the close-packed hexagonal structure of zinc.

Fig. 2.6 *A section through a small aluminium ingot in the cast condition, etched in hydrofluoric acid (2 per cent). The crystals are all of the* same composition – *pure aluminium. Since each crystal has developed independently its lattice structure is 'tilted' at a random angle to the surface section, and so each crystal reflects light at a different intensity.*

2.22   Since each dendrite forms independently, it follows that outer arms of neighbouring dendrites are likely to make contact with each other at irregular angles, and this leads to the irregular overall shape of crystals, as indicated in Fig. 2.7. In a similar manner, the trees of a forest push and jostle

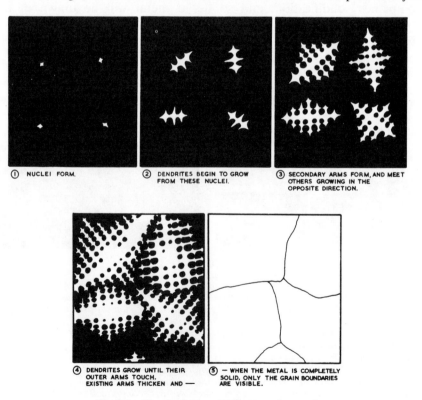

Fig. 2.7   *The dendritic solidification of a metal.*

each other as they reach towards the light, so that forest trees are rarely of regular geometrical shapes.

Impurities in cast metals

2.30   If the metal we have been considering is pure, then there will be no hint of the dendritic process of crystallization once solidification is complete, because all atoms in a pure metal are identical. If impurities were dissolved in the molten metal, however, these would tend to remain in solution until solidification was almost complete. They would therefore remain concentrated in that metal which solidified last; that is, between the dendrite arms. In this manner they would reveal the dendritic pattern (Fig. 2.8) when a suitably prepared specimen was viewed under the microscope. This concentration or *segregation* of impurities at crystal boundaries explains why a small amount of impurity can have such a devastating effect on mechanical properties, making the cast metal brittle and likely to fail along the crystal boundaries. In addition to this local segregation at all crystal boundaries, there is a general accumulation of impurities in the central 'pipe' of a cast ingot (Fig. 2.9). This is where metal solidifies last of all, and has become

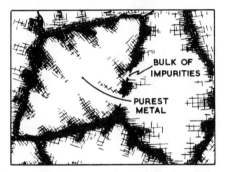

Fig. 2.8   *The local segregation of impurities. The heavily shaded region near the crystal boundaries contains the bulk of the impurities.*

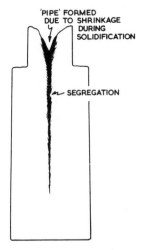

Fig. 2.9   *The segregation of impurities in the central 'pipe' of an ingot. In casting steel ingots, the central pipe is confined by using a fireclay collar on top of the mould (Fig. 3.1). The mould is tapered upwards, so that it can be lifted off the solid ingot.*

most charged with impurities, relatively pure metal having crystallized during the early stages of solidification.

2.31 Some impurities are insoluble in the molten metal and these generally float to the surface to form a slag. They include manganese sulphide in steels, residual from the combined deoxidation and desulphurization processes (11.30), and oxide dross in brass, bronzes and aluminium alloys. Some of this material may be carried down into the casting as a result of turbulence caused by pouring the charge and become entrapped by the growing dendrite arms. Such inclusions may occur anywhere in the crystal structure. Particles of manganese sulphide in steel are dove-grey in colour.

2.32 The formation of shrinkage cavities was mentioned above. They will form only between dendrite arms (Fig. 2.10) and along crystal boundaries during the final stages of solidification when, due to poor moulding design or adverse casting conditions, pockets of molten metal remain isolated. As these solidify and shrink small cavities remain into which molten metal is not available to feed.

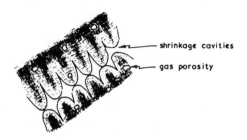

Fig. 2.10   *The distribution of shrinkage cavities and gas porosity.*

2.33 Other cavities may be found in cast metals and are generally due to the evolution of gas during the solidification process. This gas may be dissolved from the furnace atmosphere during the melting process. As solidification proceeds this gas tends to remain dissolved in the still molten part of the metal until a point is reached where the solubility limit is exceeded. Tiny bubbles of gas are then liberated and become trapped between the rapidly growing dendrites (Fig. 2.10).

The influence of cooling rates on crystal size

2.40 The rate at which a molten metal is cooling as it reaches the freezing-temperature affects the size of the crystals which form. A gradual fall in temperature results in the formation of few nuclei and so the resultant crystals grow unimpeded to a large size. A rapid fall in temperature, however, will lead to some degree of *undercooling* of the molten metal to a temperature *below* its actual freezing-point (Fig. 2.11). Due to this undercooling and the instability associated with it a sudden shower of nuclei is produced and because there are many nuclei the ultimate crystals will be tiny. As the foundryman says, 'Chilling causes fine-*grain*[2] castings'. Because

[2] In metallurgy the term 'grain' is often used to mean 'crystal' and does not imply any *directionality* of structure as does the term grain in timber technology.

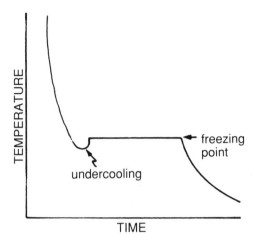

Fig. 2.11   *Undercooling of a molten metal due to a steep temperature gradient.*

of the difference in the rates of cooling, the resultant grain size of a die-casting is small as compared with that of a sand-casting. This is an advantage, since fine-grained castings are generally tougher and stronger than those with a coarse grain size.

2.41   When a large ingot solidifies, the rate of cooling varies from the outer skin to the core during the crystallization process. At the onset of crystallization, the cold ingot-mould chills the molten metal adjacent to it, so a layer of small crystals is formed. Due to the heat flow outwards, the mould warms up, and so chilling becomes less severe. This favours the growth inwards of elongated or columnar crystals. These grow inwards more quickly than fresh nuclei can form, and this results in their elongated shape. The residue of molten metal, at the centre, cools so slowly that very few nuclei form, and so the crystals in that region are relatively large. They are termed 'equi-axed crystals' – literally 'of equal axes' (Fig. 2.12).

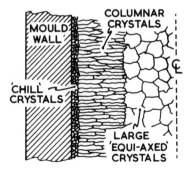

Fig. 2.12   *Zones of different crystal forms in an ingot.*

2.42   A cast material tends to be rather brittle, because of both its coarse grain and the segregation of impurities mentioned above. When cast ingots are rolled or forged to some other shape, the mechanical properties of the material are vastly improved during the process, for reasons discussed later

(6.50). Frequently, the engineer uses sand- or die-castings as integral parts of a machine. Often no other method is available for producing an intricate shape, though in many cases a sand-casting will be used because it involves the *cheapest* method of producing a given shape. The choice rests with the production engineer: he must argue the mechanical and physical properties required in a component with the Spectre of Production Costs – who is forever looking over his shoulder.

2.43   Rapid solidification processing (RSP) has been widely investigated in recent years, the aim being to produce even smaller crystals by increasing the rate of cooling as the freezing range of the metal or alloy is reached. In fact metallurgists have devised rapid solidification processes involving cooling rates in the region of 10^6 °C/second. Now cooling a molten metal so that its temperature falls at a rate equal to a range of a million degrees Celsius in one second taxes the ingenuity of the R and D staff (22.70) but under such conditions extremely tiny *micro-crystals* are produced and in some cases *amorphous* – that is, non-crystalline – *solid metals* are obtained. Thus in extreme cases the non-crystalline liquid structure is retained since the metal is given no opportunity to crystallize.

A most important feature of this development is that an amorphous solid metal contains no heavily segregated areas (2.30) since any impurity would be dispersed evenly throughout the liquid metal and will remain so in the amorphous solid. Segregation only occurs during crystallization. In RSP micro-crystalline metals any impurities will be widely dispersed because, as the crystals are tiny, there is a much increased area of grain-boundary per unit volume of metal over which these impurities are spread. This thinner spreading of impurities means that the metal will be more uniform in composition and materials which contain no regions of brittle segregate are stronger and tougher as well as being malleable and ductile.

Further it becomes possible to make alloys from metals which do not normally mix (dissolve in each other) as liquids, by quenching a mixture of them from the *vapour* state. RSP powders can also be produced in which the alloy is in a micro-crystalline or amorphous state. The powder is then consolidated under pressure and subsequently shaped by extrusion (7.23).

CHAPTER 3

Casting processes

3.10 Most metallic materials pass through a molten state at some stage during the shaping process. Much of this molten metal is cast into ingot form prior to being shaped by some mechanical working process. Some metals and alloys – and a number of non-metallic products such as concrete – can be shaped only by casting since they lack the necessary properties of either malleability or ductility (4.11) which make them amenable to working processes.

A few metallic substances are produced in powder form. The powder is then compressed and sintered to provide the required shape. This branch of technology is termed 'powder-metallurgy' (7.40).

Ingot-casting

3.20 Many alloys, both ferrous and non-ferrous, are cast in the form of ingots which are then rolled, forged, or extruded into strip, sheet, rod, tube, or other sections.

When produced as single ingots steel is generally cast into large iron moulds holding several tonnes of metal. These moulds generally stand on a flat metal base, and are tapered upwards very slightly so that the mould can be lifted clear of the solid ingot. A 'hot top' (Fig. 3.1) is often used in order that a reservoir of molten metal shall be preserved until solidification of the body of the ingot is complete. This reservoir feeds metal into the 'pipe' which forms as the main body of the ingot solidifies and contracts. Many non-ferrous metals are cast as slabs in cast-iron 'book-form' moulds, whilst some are cast as cylindrical ingots for subsequent extrusion.

3.21 Continuous casting This process is used to produce ingots of both ferrous and non-ferrous alloys (Fig. 3.2). Here the molten metal is cast into a short water-cooled mould, A, which has a retractable base, B. As solidification begins, the base is withdrawn downwards at a rate which will keep place with that of pouring.

With ordinary ingots, a portion of the top must always be cropped off and rejected, since it contains the pipe, and is a region rich in impurities. In continuous casting, however, there is little process scrap, since very long ingots are produced and consequently there is proportionally less rejected pipe. In fact pipe is virtually eliminated in modern steelmaking since the continuous-cast ingot is conveyed by a system of guide rolls direct to the rolling mill from which the emerging product (plate, sheet, rod, etc.) is cut to

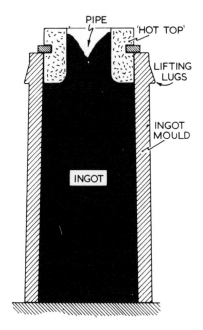

Fig. 3.1   *A typical mould for producing steel ingots. The 'hot top' restricts the formation of the 'pipe' which is caused by contraction of the metal as it solidifies.*

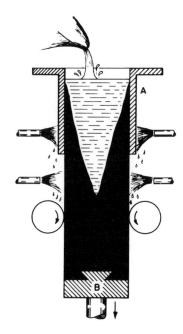

Fig. 3.2   *A method for continuous casting.*

the required lengths by a flying saw. The bulk of steel made in the developed world is produced by 'continuous' plant.

Sand-casting

3.30   The production of a desired shape by a sand-casting process first involves moulding foundry sand around a suitable pattern, in such a way that the pattern can be withdrawn to leave a cavity of the correct shape in the sand. To facilitate this, the sand mould is split into two or more parts, which can be separated so that the wooden pattern may be removed.

3.31   The production of a very simple sand mould is shown in Fig. 3.3. The pattern of the simple gear blank is first laid on a moulding-board, along with the 'drag' half of a moulding-box (i). Moulding sand is now riddled over the pattern, and rammed sufficiently for its particles to adhere to each other. When the drag has been filled, the sand is 'cut' level with the edge of the box (ii), and the assembly is turned over (iii).

A layer of parting sand (dry, clay-free material) is now sifted on to the sand surface, so that the upper half of the mould will not adhere to it when this is subsequently made. The 'cope' half of the moulding-box is now placed in position, along with the 'runner' and 'riser' pins, which are held steady by means of a small amount of moulding sand pressed around them (iv). The purpose of the runner in the finished mould is to admit the molten metal, whilst the riser provides a reservoir from which molten metal can feed back into the casting as it solidifies and shrinks.

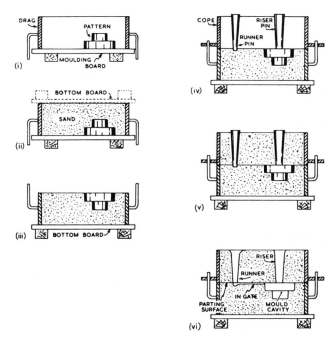

Fig. 3.3   *Moulding with a simple pattern.*

Moulding sand is now riddled into the cope, and rammed around the pattern, runner, and riser (v). The cope is then gently lifted off, the pattern is removed, and the cope is replaced in position (vi), so that the finished mould is ready to receive its charge of molten metal.

The foregoing description deals with the manufacture of a sand mould of the simplest type. In practice, a pattern may be of such a complex shape that the mould must be split into several sections, and consequently a multi-part box is used. *Cores* may be required to form holes or cavities in the casting.

3.32   Sand-casting is a very useful process, since very intricate shapes can be produced in a large range of metals and alloys. Moreover, relatively small numbers of castings can be made economically, since the necessary outlay on the simple equipment required is low. Wooden patterns are also cheap to produce, as compared with the metal die which is necessary in die-casting processes.

Die-casting

3.40   In die-casting, a permanent metal mould is used, and the charge of molten metal is either allowed to run in under the action of gravity (gravity die-casting), or is forced in under pressure (pressure die-casting). A number of different types of machine are employed in pressure die-casting, but possibly the most widely used is the cold-chamber machine (Fig. 3.4). Here a charge of molten metal is forced into the die by means of a plunger. As soon as the casting is solid, the moving platen is retracted, and as it comes against a fixed block, ejector pins are activated so that the casting is pushed out of the mould. Cycling time is rapid.

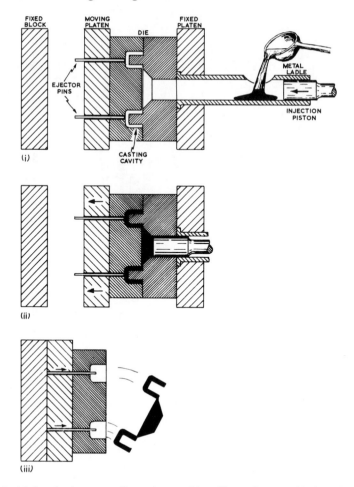

Fig. 3.4   *A 'cold-chamber' pressure die-casting machine. The molten metal is forced into the die by means of the piston. When the casting is solid, the die is opened, and the casting moves away with the moving platen, from which it is ejected by 'pins' passing through the platen.*

3.41   In gravity die-casting – or 'permanent-mould casting', as it is now generally called – the die is of metal, and may be of multi-part design if the complexity of shape of the casting demands it. Metal cores of complex shape must be split, in order to allow their removal from the finished casting; otherwise, sand cores may have to be used. The die cavity is filled under gravity, and the charge may be poured by hand, or it may be fed in automatically in modern high-speed plant.

3.42   The product of die-casting is metallurgically superior to that of sand-casting in that the internal structure is more uniform, and the grain much finer, because of the rapid cooling rates which prevail. Moreover, output rates are much higher when using a permanent metal mould than when using sand moulds. Greater dimensional accuracy and a better surface finish are also obtained by die-casting. However, some alloys which can be sand-cast cannot be die-cast, because of their high shrinkage coefficients. Such alloys would inevitably crack due to their contraction during solidification within the rigid, non-yielding metal mould. Die-casting is confined mainly to zinc- or aluminium-base alloys.

Centrifugal casting

3.50 This process is most commonly used in the manufacture of cast-iron pipes for water, sewage and gas mains (though for such purposes cement- or plastics-based materials are now more commonly used). A permanent cylindrical metal mould, *without* any central core, is spun at high speed, and molten metal is poured into it (Fig. 3.5). Centrifugal force flings the metal to the surface of the mould, thus producing a hollow cylinder of uniform wall thickness. The product has a uniformly fine-grained outer surface, and is considered superior to a similar shape which has been sand-cast. In addition to the types of pipe mentioned above, hollow cylinders from which cast-iron piston rings can be cut are also centrifugally cast.

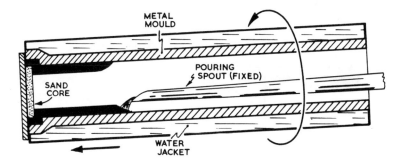

Fig. 3.5 *The principle of the centrifugal casting of pipes. The mould rotates and also moves to the left, so that molten metal is distributed along the mould surface.*

Shell-moulding

3.60 This process was developed in Germany by Johannes Croning during the Second World War. It was originally employed in the manufacture of hand-grenades, but has subsequently been used for the production of large numbers of castings requiring great dimensional accuracy.

3.61 It is fundamentally a sand-casting process in which the clay bond present in ordinary foundry sand is replaced by an artificial bonding material of the phenol-formaldehyde or urea-formaldehyde type (19.41.2). Find sands, free of clay, are mixed with about 5 per cent by weight of the plastic bonding-agent.

3.62 Each half of the shell-mould is made on a pattern plate, which must be of metal, because of the relatively high temperatures at which the bonding-agent 'sets'. The plate is first heated to about 250 °C, and is then coated with silicone oil, in order to facilitate the subsequent stripping of the shell from the pattern. The pattern plate is then placed on top of a dump-box containing the sand-resin mixture (Fig. 3.6(i)), and the box is inverted, so that the pattern becomes covered with the sand-resin mixture (ii).

The resin melts, and in approximately 30 seconds the pattern has become coated with a shell of resin-bonded sand. The shell soon becomes quite hard, because the phenolic-type resins used are of the thermosetting variety (19.41). The dump-box is turned back to its original position, so that surplus moulding material falls back into the bottom of the box (iii).

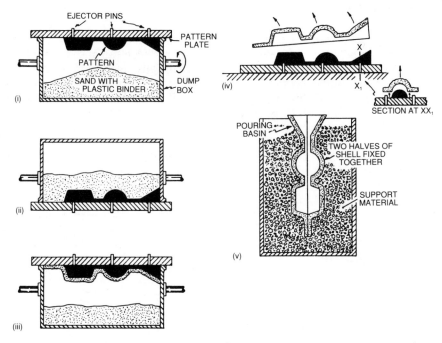

Fig. 3.6   *Stages in the production of a shell-mould.*

The pattern plate is then removed, and, with the shell still adhering to it, it is transferred to an oven, where the shell is hardened still further by curing for about 2 minutes at 315 °C. The shell is then stripped from the pattern plate by means of ejector pins built into the plate. Normally these pins bear on the surface of the shell at the edges of the mould cavity (iv), so as not to damage the surface of the mould cavity itself.

The two halves of the mould so produced are then joined together by adhesives, bolts, clamps, or – in the case of small moulds – by spring paper-clips. Sometimes the mould is supported with metal shot, coarse sand, or gravel (v), and is then ready to receive the molten charge.

The main advantage in using shell-moulding lies in the high dimensional accuracy of the product. Tolerances of the order of 5 mm/metre are claimed to be common practice, as compared with about 15 mm/metre in the average pressure die-casting. Moreover, the surface finish is far superior to that on an ordinary sand-casting, since loose sand and sand inclusions are absent. Since a shell-mould has a small heat capacity, as compared with a sand mould, the metal can be cast at a lower temperature. For this reason, oxidation losses and gas absorption during melting are reduced to a minimum. Shell-moulds will store well, so mould-making plant need not be sited near to the foundry. Labour costs too are lower than with green-sand moulding, and working conditions are much cleaner than with ordinary foundry sand. Most foundry metals can be successfully cast into shell-moulds.

3.63   The principal disadvantage of shell-moulding lies in the high cost of the metal patterns, mainly because these need to be of high dimensional accuracy. Obviously this disadvantage is largely offset by the consequent greater value of the resultant casting.

Investment-casting

3.70 Although investment-casting came into public eminence fairly recently, in connection with the production of turbine blades used in jet engines, it is in fact the most ancient casting process in use. It is thought that prehistoric Man had learned to fashion an image from beeswax, and then knead a clay mould around it. The mould was then hardened by firing, a process which also melted out the wax pattern, leaving a mould cavity without cores or parting lines.

3.71 The process was rediscovered in the sixteenth century by Benvenuto Cellini, who used it to produce many works of art in gold and silver. Cellini kept the process secret, so it was lost again until, during the latter part of the nineteenth century, it was rediscovered for a second time, and became known as the *cire perdue*, or 'lost wax', process. Since an expendable wax pattern is required for the production of each mould, it follows that a permanent mould must first be produced to manufacture the wax patterns – assuming of course that a large number of similar components is required. This master *mould* could be machined in steel, or produced by casting a low melting-point alloy around a master *pattern*.

3.72 To produce wax patterns, the two halves of the mould are clamped together, and the molten wax is injected at a pressure of about 3.5 N mm^{-2}. When the wax pattern has solidified, it is removed from the mould, and the wax 'gate' is suitably trimmed (Fig. 3.7(ii)), using a heated hand-tool, so that

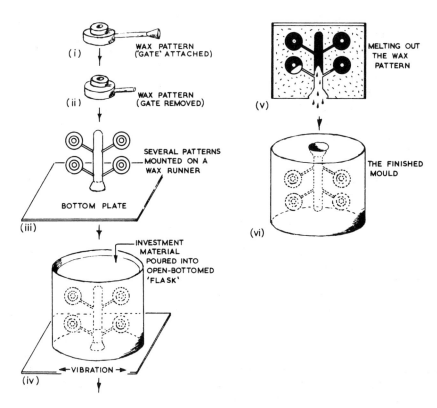

Fig. 3.7 *The investment-casting (or 'lost wax') process.*

it can be attached to a central 'runner' (iii). The assembled runner, with its 'tree', of patterns is then fixed to a flat bottom-plate by a blob of molten wax. A metal flask, lined with waxed paper and open at each end, is placed over the assembly. The gap between the end of the flask and the bottom-plate is sealed with wax, and the investment material is then poured into the flask. This stage of the process is conducted on a vibrating base, so that any entrapped air bubbles or excess moisture are brought to the surface whilst solidification is taking place (iv).

For castings made at low temperatures, an investment mixture composed of very fine silica sand and plaster of Paris is still sometimes used. A more refractory investment material consists of a mixture of fine 'sillimanite' sand and ethyl silicate. During moulding and subsequent firing of the mould, ethyl silicate decomposes to form silica, which knits the existing sand particles together, to give a strong, rigid mould.

The investment is allowed to dry in air for some eight hours. The baseplate is then detached, and the inverted flask is passed through an oven at about 150 °C, so that the wax melts and runs out, leaving a mould cavity in the investment material (v). When most of the wax has been removed, the mould is pre-heated prior to receiving its charge of molten metal. The pre-heating temperature varies with the metal being cast, but is usually between 700 and 1000 °C. The object of pre-heating is to remove the last traces of wax by volatilization, to complete the decomposition of the ethyl silicate bond to silica, and also to ensure that the cast metal will not be chilled, but will flow into every detail of the mould cavity.

3.73   Molten metal may be cast into the mould under gravity, but, if thin sections are to be formed, then it will be necessary to inject the molten metal under pressure.

The process is particularly useful in the manufacture of small components from metals and alloys which cannot be shaped by forging and machining operations. Amongst the best-known examples of this type of application are the blades of gas-turbine and jet engines.

3.74   Tolerances in the region of 3 to 5 mm/metre are obtained industrially by the investment-casting process, but it can also be used to produce complicated shapes which would be very difficult to obtain by other casting processes. A further advantage is the absence of a disfiguring parting line, which always appears on a casting made by any process involving the use of a two-part mould.

The main disadvantages of investment-casting are its high cost, and the fact that the size of components is normally limited to 2 kg or so.

3.75   In addition to jet-engine blades, other typical investment-cast products include special alloy parts used in chemical engineering; valves and fittings for oil-refining plant; machine parts used in the production of modern textiles; tool and die applications, such as milling-cutters, precision gauges and forming and swaging dies; and parts for various industrial and domestic equipment, such as cams, levers, spray nozzles, food-processing plant, parts for sewing machines and washing machines. The materials from which many of these components are made are extremely hard and strong over a wide range of temperatures making it impossible or, at best, expensive to shape them by orthodox mechanical methods. Investment casting provides a means of producing the required shapes whilst retaining the required high dimensional accuracy.

The full-mould process

3.80 This somewhat novel process resembles investment-casting in that a single-part flask is used, so that no parting lines – and hence no fins – appear on the finished casting. However, it is essentially a 'one-off' process, since the consumable pattern is carved from expanded polystyrene (20.45). This is a polymer derived from benzene and ethylene, and in its 'expanded' form it contains only 2 per cent actual solid polystyrene. Readers will be familiar with the substance, which is used in the manufacture of ceiling tiles, and also as a packaging material for fragile equipment. An expendable pattern, complete with runners and risers, is cut from expanded polystyrene (Fig. 3.8(i)), and is completely surrounded with silica sand containing a small proportion of thermosetting resin (ii). The molten metal is then poured *on to the pattern*, which melts and burns very quickly, leaving a cavity which is immediately occupied by the molten metal. No solid residue is formed, and the carbon dioxide and water vapour evolved in the combustion of the polystyrene do not dissolve in the molten metal, but escape through the permeable mould. The small amount of resin added to the sand is sufficient to form a bond between the sand grains preventing premature collapse of the mould. In some cases the resin is omitted and as the expanded polystyrene burns it produces a tacky bond between the sand grains for long enough for a skin of metal to form.

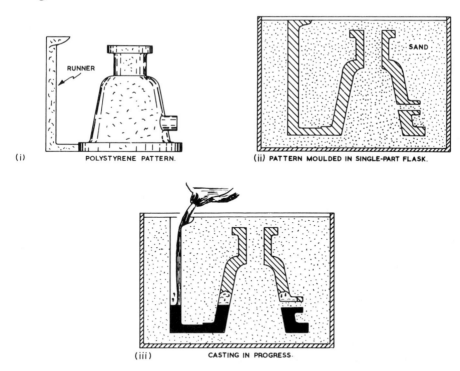

(i)   POLYSTYRENE PATTERN.

(ii) PATTERN MOULDED IN SINGLE-PART FLASK.

(iii)   CASTING IN PROGRESS.

Fig. 3.8 *The 'full-mould' process, which uses an expendable polystyrene pattern.*

A well-known example of full-mould casting is Geoffrey Clarke's cross which originally adorned the summit of the post-war Coventry Cathedral. In the engineering industries, the process is used in the manufacture of large press-tool die-holders, and similar components in the 'one-off' category.

3.81   The lost-foam process is a later modification of the full-mould process in which the expanded polystyrene pattern is invested with unbonded dry silica sand and the flask (container) then tapped or vibrated to consolidate the sand so that the bonding agent can be omitted. This modification of the original process has been adapted to mass production. The expanded polystyrene patterns are made by injecting polystyrene 'beads' into a heated split aluminium die which is then further heated so that the beads expand and fuse together forming an expendable pattern.

Semi-solid metal processing

3.90   This is a process whereby suitable alloys can be shaped to something near final form from a mushy, part solid/part liquid state (9.11). The semi-molten alloy (containing about 50 per cent solid) is stirred vigorously so that the dendrites tend to break up as they form and the viscosity of the slurry is greatly reduced. On standing the slurry will begin to stiffen again, becoming more viscous, but the viscosity can be reduced again by stirring. This reversible behaviour is similar to that of thixotropic (non-drip) paint.

3.91   Stirred slurries of this type can be die-cast in a normal cold-chamber machine (Fig. 3.4) provided relatively simple shapes are attempted. This type of process is used principally for manufacturing automobile components in aluminium alloys of suitably wide freezing range. It is an environment-friendly process since up to 35 per cent fuel saving is achieved over normal die-casting processes.

The choice of casting process

3.100   If dimensional accuracy is the main criterion in choosing a casting method then those processes described above could be considered in the following order:

1.   investment casting;
2.   shell moulding;
3.   die-casting;
4.   sand-casting;
5.   the full-mould process.

Relative costs per casting too will be roughly in that order.

   Also to be taken into account would be the cost of any subsequent machining (metal cutting) operations. With very hard alloys machining is either very expensive or impossible. In such cases investment-casting comes into its own. A further advantage of the process is that there is no intrusive parting line. In two-part moulds or dies pressure applied during the casting process may cause some separation of the mould parts giving rise to dimensional inaccuracy in the resulting casting. Cored holes which must be an accurate distance apart should never be separated by a parting line.

3.101   At the other end of the scale the full-mould process is less suitable for producing castings of high dimensional accuracy but is particularly suitable where a one-off feature is to be cast, e.g. a statue or other architectural object in aluminium. Simple one-off machine parts for subsequent machining can also be made economically by this process. Thus either investment-casting or shell-moulding may be chosen where a high dimensional accuracy is required in the resulting casting whilst the full-mould process is a useful method where single castings of no great accuracy are to be made.

3.102   'Long-run' production of castings usually follows either the well-established sand-moulding or die-casting route. In general die-casting will provide a better surface finish and greater dimensional accuracy as well as superior mechanical properties (due to a smaller crystal size) than are available with sand-castings and one or more of these features may determine the choice of die-casting. However, when none of these features are important then the manufacturer will obviously choose the cheaper process. This in turn will depend upon a forecast of the number of castings required. Generally speaking, due to the high cost of producing metal dies it is not economical to use die-casting unless a large number of castings is required. On the other hand the cost of making a wooden pattern (for sand-casting) is relatively low *but* the labour costs of an experienced sand-moulder are higher than those of a die-casting machine operator. Consequently, all other things being equal, the number of castings required will govern the choice of process (Fig. 3.9).

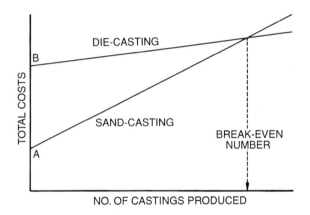

Fig. 3.9   *The initial costs of plant and equipment are higher in the case of die-casting (B) than for sand-casting (A). Labour costs however are continuous during operation of the process and are higher for sand-casting. The 'break-even value' – that is the number of castings which must be made before the use of die-casting is justified – may be of the order of several thousands. Once the break-even value has been reached the cost per die-casting becomes less than the cost per sand-casting.*

CHAPTER 4

Mechanical testing

4.10   To the layman the most important property of steel relates to what he thinks of as its 'strength'. Whilst this is of course an important property to be considered by the metallurgist he will also regard as very important the amount of *deformation* which the steel can withstand both in tension and in compression before it fractures.

4.11   We say that copper is a very *malleable* metal, meaning that it can be deformed a great deal by *compression* before it shows signs of cracking. Malleable metals can be rolled, forged, or extruded, since these are all processes where the metal is shaped under pressure. Malleability generally increases with temperature, and so processes involving pressure are invariably hot-working processes; that is, they are carried out on heated ingots or slabs of metal.

Copper is also very *ductile*; that is, it can be deformed considerably by *tension* before it fractures. Whilst all ductile metals are malleable, the converse is not always true. Some metals, although soft, are also weak in tension and therefore tend to tear apart whilst being stretched. The ductility of all metals decreases as the temperature rises, because they are weaker at high temperatures. Both malleability and ductility are reduced by the presence of impurities in the metal or alloy.

4.12   The *toughness* of a metal refers to its ability to withstand bending without fracture. Copper is a very tough material, since a piece of copper wire can be bent to and fro many times before it will fracture. Thus copper is very tough; whilst cast iron possesses the opposite property – *brittleness*.

These properties of malleability, ductility, and toughness are fundamental properties which are not easy to measure in simple numerical terms; all we can do is to arrange the metals in order of the property concerned. This is of very little use to the engineer. He is primarily concerned with the balancing of *forces* when he designs some structure or machine; he must therefore know the effects which the application of forces may have on a material, before he can use it.

4.13   Various mechanical tests have therefore been devised over the years, with the object of comparing the amount of deformation produced in a metal with the force which was employed to produce it. Thus, in a tensile test we measure the force required to stretch a specimen of the metal until it breaks; whilst in various hardness tests we produce a small dent in the surface of a test-piece by using a compressive force. The hardness number is then calculated as the force used divided by the surface area of the impression produced by it.

4.14 Complete information on the behaviour of metals and alloys when subjected to these derived tests has been compiled by such bodies as the British Standards Institution, who have drawn up their series of BS specifications. Now that we are 'in Europe' many European standards have been incorporated into our system and are designated as BSEN standards. In the USA such matters are dealt with by the American Society for Metals (ASM) and the American Society for Testing Materials (ASTM). It is on values specified by such bodies that the engineer bases his designs and accepts his materials.

4.15 In the type of test mentioned above, the test-piece is destroyed during the testing process. Such tests are therefore known as *destructive tests*, and can only be applied to individual test-pieces. These are taken from a batch of material which it is proposed to use for some specific purpose and they are therefore assumed to be representative of the batch. Tests of a different nature and purpose are used to examine manufactured components for internal flaws and faults; for example X-rays are used to seek internal cavities in castings. These tests are generally referred to as *non-destructive tests* (NDT), since the component, so to speak, 'lives to tell the tale'.

The tensile test

4.20 The tensile strength of a material is the stress required to cause fracture of a test-piece in tension. Many readers will be familiar with tensile-testing methods; nevertheless, a brief outline of these methods will be given here.

A test-piece of known cross-sectional area is gripped in the jaws of a testing-machine, and is subjected to a tensile force which is increased by suitable increments. For each increment of force, the amount by which the length of a known 'gauge length' on the test-piece increases is measured. This process continues until the test-piece fractures. Tensile-testing machines of more than fifty years ago were designed on the principle of a simple first-order lever in which a massive counterpoise weight was moved along a beam to provide the necessary increasing tensile force acting on the test-piece. The great disadvantage of such a system was that once a test-piece reached its *yield point* and began to stretch rapidly it was impossible to reduce the force quickly enough to enable the complete force-extension diagram to be plotted since fracture took place so rapidly once the maximum force had been reached. From a practical engineering consideration this did not matter very much since it was the maximum force the test-piece could withstand which was regarded as important. However in modern tensile-testing machines force is applied by hydraulic cylinders for machines of up to 1 MN in capacity whilst for small 'bench' machines of only 20 kN capacity the tensile force is applied by a spring beam. With such systems force relaxes automatically as soon as rapid extension of the test-piece begins. The complete force-extension diagram can then be plotted with accuracy until fracture occurs.

4.21 Examination of a typical force-extension diagram (Fig. 4.1) for an annealed carbon steel shows that at first the amount of extension is very

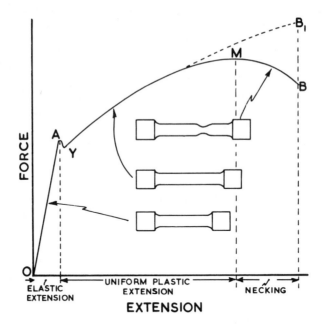

Fig. 4.1   *The force-extension diagram for an annealed low-carbon steel.*

small, compared with the increase in force. Such extension as there is is directly proportional to the force; that is, *OA* is a straight line. If the force is released at any point before *A* is reached, the test-piece will return to its original length. Thus the extension between *O* and *A* is *elastic*, and the material obeys Hooke's law:

$$\text{stress} \propto \text{strain}$$

or

$$\frac{\text{stress}}{\text{strain}} = \text{a constant } (E).$$

This constant, *E*, is known as Young's modulus of elasticity for the material.

4.22   If the test-piece is stressed past the point *A* (known as the *elastic limit* or *limit of proportionality*), the material suddenly 'gives'; that is, suffers a sudden extension for very little increase in force. This is called the *yield point* (*Y*), and, if the force is now removed, a small permanent extension will remain in the material. Any extension which occurs past the point *A* is of a *plastic* nature.

4.23   As the force is increased further, the material stretches rapidly – first uniformly along its entire length, and then locally to form a 'neck'. This 'necking' occurs just after the maximum force has been reached, at *M*, and since the cross-section decreases rapidly at the neck, the force at *B* required to break the specimen is *much less than* the maximum load at *M*.

This might be an appropriate moment to point out the difference between a force-extension diagram and a stress-strain diagram, since the terms are often loosely and imprecisely used. Fig. 4.1 clearly represents a force-extension diagram, since the total force is plotted against the total extension, and, as the force decreases past the point *M*, for the reasons mentioned above, this decrease is indicated by the diagram. If, however, we wished to plot *stress* (force applied *per unit area* of cross-section of the specimen), we would need to measure the minimum diameter of the specimen, as well as its length, *for each increment of force*. This would be particularly important for values of force after the point *M*, since in this part of the test the diameter is decreasing rapidly, due to the formation of the 'neck'. Just as a chain is only as strong as its weakest link, so the test-piece is only as strong as the force its minimum diameter will support.

4.24 Thus, if stress were calculated on this decreasing diameter, the resulting stress-strain diagram would follow a path as indicated by the broken line to B_1 (Fig. 4.1).

In practice, however, a *nominal* value of the *tensile strength* of a material is calculated, using the maximum force (at *M*) and the original cross-sectional area of the test-piece. Thus,

$$\text{tensile strength} = \frac{\text{maximum force used}}{\text{original area of cross-section}}$$

The term '*engineering* stress' is often used; it signifies the force at any stage of the loading cycle divided by the *original* area of cross-section of the material.

4.25 Tensile strength is a useful guide to the mechanical properties of a material. It is primarily an aid to quality control because it is a test which can be carried out under easily standardized conditions but it is not of paramount importance in engineering design. After all, the engineer is not particularly interested in the material once plastic flow begins – unless he happens to be a production engineer interested in deep-drawing, or some other forming process. In terms of structural or constructional engineering, the elastic limit, *A*, will be of far greater significance.

4.26 The force-extension diagram shown in Fig. 4.1 is typical of a low-carbon steel in the normalized condition. Unfortunately, force-extension diagrams for heat-treated steels, and for most other alloys, do not often show a well-defined yield point, and the 'elastic portion' of the graph merges gradually into the 'plastic section', as shown in the examples in Fig. 4.2.

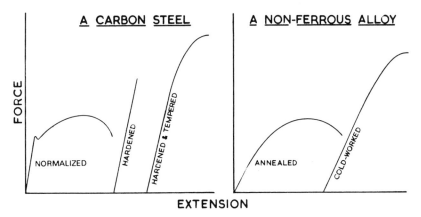

Fig. 4.2 *Typical force-extension diagrams for both carbon steels and non-ferrous materials.*

This makes it almost impossible to assess the yield stress of such an alloy, and, in cases like this, yield stress is replaced by a value known as *proof stress*. Thus the 0.1 per cent proof stress of an alloy (denoted by the symbol $R_{p0.1}$) is that stress which will produce a permanent extension of 0.1 per cent in the gauge length of the test-piece. This is very roughly equivalent to the permanent extension remaining in a normalized steel at its yield point.

The 0.1 per cent proof stress of a material is derived as shown in Fig. 4.3. The relevant part of the force-extension diagram is plotted as described earlier. A distance OA, equal to 0.1 per cent of the gauge length, is marked

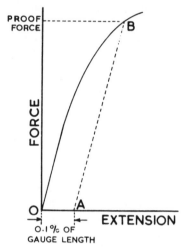

Fig. 4.3   *The determination of 0.1 per cent proof stress.*

along the horizontal axis. A line is then drawn from A, parallel to the straight-line portion of the force-extension diagram. The line from A intersects the diagram at B, and this indicates the proof force which would produce a permanent extension of 0.1 per cent in the gauge length of the specimen. From this value of force, the 0.1 per cent proof stress can be calculated:

$$0.1\% \text{ proof stress } = \frac{\text{proof force}}{\text{original cross-sectional area of test-piece}}$$

It will be obvious from the above that if the yield stress (or 0.1 per cent proof stress) is exceeded in a material then fairly rapid plastic deformation and failure are likely to occur.

4.27   In addition to determining the tensile strength and the 0.1 per cent proof stress (or, alternatively, the yield stress), the percentage elongation of the test-piece at fracture is also derived. This is a measure of the ductility of the material. The two halves of the broken test-piece are fitted together (Fig. 4.4), and the extended gauge length is measured:

$$\% \text{ elongation } = \frac{\text{increase in gauge length}}{\text{original gauge length}} \times 100$$

The two test-pieces in Fig. 4.4(i) are of similar material and of equal diameters. Consequently, the dimensions and shape of both 'necked'

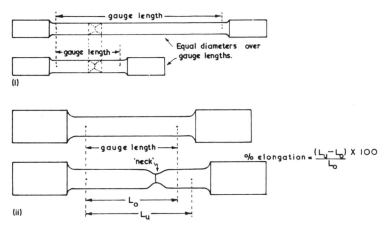

Fig. 4.4   *The determination of elongation (%).*

portions will also be similar, that is the *increase* in length will be the same in each test-piece. However, since different gauge lengths have been used it follows that the percentage elongation reported on gauge length would be different for each. Therefore in order that values of the percentage elongation shall be comparable, it is obvious that test-pieces should be geometrically similar; that is, there must be a standard relationship or ratio between cross-sectional area and gauge length. Test-pieces which are geometrically similar and fulfil these conditions are known as *proportional* test-pieces. They are generally circular in cross-section. BSI lays down that, for proportional test-pieces,

$$L_0 = 5.65\sqrt{S_0}$$

where L_0 is the gauge length and S_0 the original area of cross-section. This formula has been accepted by international agreement, and SI units are used. For test-pieces of circular cross-section, it gives a value

$$L_0 = 5d \quad \text{(approximately)}$$

where d is the diameter at the gauge length. Thus a test-piece 200 mm² in cross-sectional area will have a diameter of 15.96 mm (16 mm) and a gauge length of 80 mm.

4.28   Test-pieces must be as representative as is possible of the material under test. This applies to test-pieces in general and not only those used in a tensile test. Many materials are far from homogeneous. Thus the segregation of impurities and variations in grain size in castings will generally mean that tensile test-pieces should be taken from more than one position in a casting. Quite often test-pieces are made from 'runners' and 'risers' (3.31) of a casting and will generally give an adequate overall guide to quality.

In wrought materials impurities will be more evenly distributed and the grain size more uniform but there will inevitably be a directionality of properties caused by the formation of 'fibres' of impurity in the direction of working (6.50). The net result on the strength of a metal is similar to that of grain direction in wood. Thus wood is much stronger in the direction of the grain than it is across the grain. In a similar way wrought metals are stronger in the direction of rolling (or extrusion) than they are at right angles to the direction of rolling (6.50). Consequently tensile test pieces should, where

Fig. 4.5

possible, be made from material along the fibre direction (Fig. 4.5(i)) and also at right angles to the fibre direction (Fig. 4.5(ii)). For narrow strip material this is not possible, though test-pieces of rectangular cross-section are commonly used. General methods for tensile testing of metals are covered by BSEN 10 002:[1] *Methods of Tensile Testing Metals*, whilst the procedures for specific alloys are covered in the appropriate BS or BSEN specifications for those alloys.

The above paragraphs apply principally to the testing of metals. The tensile – and other – testing of plastics materials involves its own particular problems not the least of which is the influence of temperature. The mechanical testing of plastics materials is therefore considered separately (20.31) after an introduction to the physical properties of polymers has been outlined.

Hardness tests

4.30   A true definition of surface hardness is the capacity of that surface to resist abrasion as, for example, by the cutting action of emery paper. Thus early attempts to quantify hardness led to the adoption of Moh's scale which was used originally to assess the relative hardness of minerals. This consists of a list of materials arranged in order of hardness, with diamond, the hardest of all (with a hardness index of 10), at the head of the list and talc (with an index of 1) at the foot (Table 4.1). Any mineral in the list will scratch any one below it and in this way the hardness of any 'unknown' substance can be related to the scale by finding which substance on the scale will just scratch it (the substance beneath it will not) and a hardness index thus assigned to it.

Table 4.1   *Moh's scale of hardness.*

Material	Hardness index
diamond	10
sapphire	9
topaz	8
quartz	7
feldspar	6
apatite	5
fluorspar	4
calcite	3
gypsum	2
talc	1

Although this method of testing is useful in the classification of minerals rather than for the determination of hardness of metals, it nevertheless agrees with the classical meaning of surface hardness if we define hardness as the resistance of a surface to abrasion. In the Turner sclerometer,[2] devised

[1]   Formerly BS 18.
[2]   From Greek: 'skleros' meaning 'hard'.

in Birmingham by T. T. Turner in 1886, a diamond point, attached to a lever arm was loaded and drawn across the polished surface of the test-piece, the load being increased until a scratch was just visible. The 'hardness number' was quoted as the load in grams required to press the diamond point sufficiently to produce a 'normal' scratch. Obviously there was considerable room for error in judging what was a 'normal' scratch. For this reason modern methods of hardness testing really measure the material's resistance to penetration rather than to abrasion. They are therefore somewhat of a compromise on the true measuring of hardness but have the advantage of being easier to determine with accuracy.

4.31   The Brinell test, devised by a Swede, Dr J. A. Brinell, is probably the best known of the hardness tests. A hardened steel ball is forced into the surface of a test-piece by means of a suitable standard load. The diameter of the impression is then measured, using some form of calibrated microscope, and the Brinell hardness number (H) is found from:

$$H = \frac{\text{load } (P)}{\text{area of curved surface of the impression}}$$

If D is the diameter of the ball, and d that of the impression, it can be shown that:

$$\text{area of curved surface of the impression} = \frac{\pi}{2} D (D - \sqrt{D^2 - d^2})$$

(The reader may find the mathematical solution of this quite difficult.)

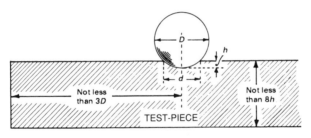

Fig. 4.6   *The relationship between ball diameter* (D), *depth of impression* (h) *and dimensions of the test-piece in the Brinell test.*

It follows that

$$H = \frac{P}{\frac{\pi}{2} D(D - \sqrt{D^2 - d^2})}$$

To make tedious calculations unnecessary, H is generally found by reference to tables which relate H to d – a different set of tables being used for each possible combination of P and D.

In carrying out a Brinell test, certain conditions must be fulfilled. First, the depth of impression must not be too great relative to the thickness of the test-piece, otherwise we may produce the situation shown in Fig. 4.7A. Here it is the table of the machine, rather than the test-piece, which is supporting the load. Hence it is recommended that the thickness of the test-piece shall be at least eight times the depth of the impression.

Materials of non-uniform cross-section also present a problem. Thus the surface skin of cold-rolled metal plate or strip will be much harder than the

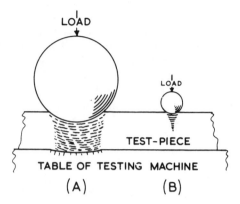

Fig. 4.7 *This illustrates the necessity of using the correct ball diameter in relation to the thickness of the test-piece.*

interior layers and the use of a small ball in the Brinell test would therefore suggest a higher hardness index than if a larger ball were used since in the latter case the ball would be supported by the softer metal of the interior. Similarly case-hardened steels (14.10) consist of a very hard skin supported by softer but tougher metal beneath and the only satisfactory way to deal with such a situation is to cut a slice through the component, polish and etch it (10.30), so that the extent of the case can be seen, and then make hardness measurements at appropriate points across the section using the smallest ball or, preferably, Vickers diamond pyramid test (4.33).

The *width* of the test-piece must also be adequate to support the load (Fig. 4.6) otherwise the edges of the impression may collapse due to lack of support and so give a falsely low reading (BS 240: *Methods for the Brinell Hardness Test*).

4.32 Balls of 10 mm, 5 mm, and 1 mm diameter are available; so one appropriate to the thickness of the test-piece should be chosen, bearing in mind that the larger the ball it is possible to use, the more accurate is the result likely to be. Having decided upon a suitable ball, we must now select a load which will produce an impression of reasonable proportions. If, for example, in testing a soft metal we use a load which is too great relative to the size of the ball, we shall get an impression similar to that indicated in Fig. 4.8(i). Here the ball has sunk to its full diameter, and the result is meaningless. On the other hand, the impression shown in Fig. 4.8(ii) would

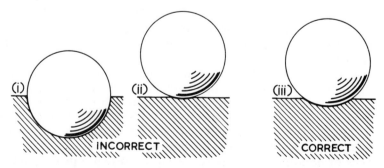

Fig. 4.8 *It is essential to use the correct P/D² ratio for the material being tested.*

be obtained if the load were too small relative to the ball diameter, and here the result would be uncertain. For different materials, then, the ratio P/D^2 has been standardized in order to obtain accurate and comparable results. P is still measured in 'kg force' and D in mm.

Material	P/D^2
steel	30
copper alloys	10
aluminium alloys	5
lead alloys and tin alloys	1

As an example, in testing a piece of steel, we can use a 10 mm ball in conjunction with a 3000 kgf load, a 5 mm ball with a 750 kgf load, or a 1 mm ball with a 30 kgf load. As mentioned above, the choice of ball diameter (D) will rest with the thickness of the test-piece; whilst the load to be used with it will be determined from the appropriate P/D^2 ratio.

4.33   The Vickers pyramid hardness test uses a square-based diamond pyramid (Fig. 4.9) as the indentor. One great advantage of this is that all impressions will be geometrically similar, and, within limits, the accuracy of the result will not vary with depth of the impression. Consequently, the operator does not have to choose a P/D^2 ratio as he does in the Brinell test, though he must still observe the relationship between depth of impression and thickness of specimen, for reasons similar to those indicated in the case of the Brinell test, and illustrated in Fig. 4.7. Here the thickness shall be at least 1.5 times the diagonal length of the indentation (BS 427: *Vickers Hardness Test*).

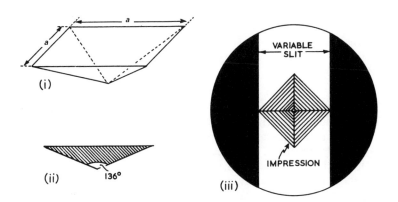

Fig. 4.9   *The Vickers pyramid hardness test. (i) The diamond indentor. (ii) The angle between opposite faces of the diamond is 136°. (iii) The appearance of the impression in the microscope eyepiece.*

A further advantage of the Vickers hardness test is that hardness values for very hard materials (above an index of 500) are likely to be more accurate than the corresponding Brinell numbers – a diamond does not deform under high pressure to the same extent as does a steel ball, and so the result will be less uncertain.

4.34   In this test, the diagonal length of the square impression is measured by means of a microscope which has a variable slit built into the eyepiece (Fig. 4.9(iii)). The width of the slit is adjusted so that its edges coincide with the corners of the impression, and the relative diagonal length of the impression is then obtained from a small instrument geared to the movement of the slit, and working on the principle of a revolution counter. The ocular reading thus obtained is converted to Vickers pyramid hardness number (VPN) by reference to tables. The size of the impression is related to hardness in the same way as is the Brinell number, i.e.

$$H = \frac{\text{load } (P)}{\text{surface area of indentation}}$$

which, in the case of the Vickers test,

$$H = \frac{2P. \sin \frac{136°}{2}}{d^2}$$

where P is the load (kgf) and d is the arithmetic mean of the diagonals (mm).
    In some quarters this test is referred to as the diamond pyramid hardness (DPH) test.

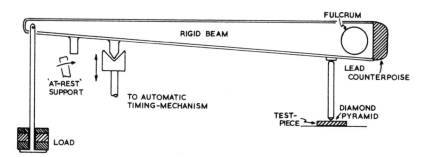

Fig. 4.10   *The loading system for the Vickers pyramid hardness machine. It is essentially a second-order lever system. The fifteen-second period of load application is timed by an oil dashpot system.*

4.35   Since the impression made by the diamond is generally much smaller than that produced by the Brinell indentor, a smoother surface finish is required on the test-piece. This is produced by rubbing with fine emery paper of about '400 grit'. At the same time, surface damage is negligible, making the Vickers test more suitable for testing finished components.
    The specified time of contact between the indentor and the test-piece in most hardness tests is 15 seconds. In the Vickers testing machine, this period of contact is timed automatically by a piston working in an oil dashpot.

4.36   The Rockwell test was devised in the USA. It is particularly useful for rapid routine testing of finished material, since the hardness number is indicated directly on a dial, and no subsequent measurement of the diameter of the impression is involved. Although the depth (h) (Fig. 4.11) of the impression is measured by the instrument, this is converted (on the dial) to hardness values in which the surface area of the impression is related to the load in the usual way. (BS 891: *Rockwell Hardness Test*; BS 4175: *Rockwell Superficial Hardness Test*.)

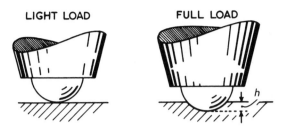

Fig. 4.11 *The Rockwell test.*

The test-piece, which needs no preparation save the removal of dirt and scale from the surface, is placed on the table of the instrument, and the indentor is brought into contact with the surface under 'light load'. This takes up the 'slack' in the system, and the scale is then adjusted to zero. 'Full load' is then applied, and when it is subsequently released (timing being automatic), the test-piece remains under 'light load' whilst the hardness index is read direct from the scale.

4.37 There are several different scales on the dial, the most important of which are:

1. scale B, which is used in conjunction with a ¹⁄₁₆″ diameter steel ball and a 100 kgf load;
2. scale C, which is used with a diamond cone of 120° angle and a 150 kgf load;
3. scale A, which is used in conjunction with the diamond cone and a 60 kgf load.

Of the scales available, possibly the most useful are scale C, which is used mainly for hardened steels and other very hard materials, and scale B, which is used for most other materials, including normalized steels and non-ferrous alloys.

The Rockwell machine is very rapid in action, and can be used by relatively unskilled operators. Since the size of the impression is also very small, it is particularly useful for the routine testing of stock or individual components on a production basis.

4.38 The Shore scleroscope (Greek *skleros* meaning 'hard') is a small portable instrument which can be used for testing the hardness of large components such as rolls, drop-forging dies, castings, and gears. Such components could not be placed on the table of one of the more orthodox machines mentioned above. The scleroscope embodies a small diamond-tipped 'tup', or hammer, of mass approximately 2.5 g, which is released so that it falls from a standard height of about 250 mm inside a graduated glass tube placed on the test surface. The height of rebound is taken as the hardness index.

Soft materials absorb more of the kinetic energy of the hammer, as they are more easily penetrated by the diamond point, and so the height of rebound is less. Conversely, a greater height of rebound is obtained from hard materials.

4.39 Wear resistance It was suggested at the beginning of this section that the well-established hardness tests do not represent the classical idea of surface hardness, that is the resistance of a surface to abrasion as is recorded in Moh's test for minerals or by the long-obsolete Turner sclerometer. Instead modern 'hardness' tests measure the resistance to penetration which, however, is closely connected to resistance to abrasion. In some instances it is desirable to obtain a more accurate assessment of the *wear resistance* of a material. Then specific tests can be devised which estimate the amount of wear which occurs when two surfaces rub together under standardized conditions. Thus a disc of the material under examination can be rotated against a standard abrasive disc for a given number of revolutions under a standard pressure. The loss in mass suffered by the test disc is a direct measure of the abrasion it has suffered.

Obviously the coefficient of friction (μ) between the rubbing surfaces is involved and hard materials have a relatively low μ. Interatomic bonding forces in metals are high and so are melting-points. Both factors resist the local removal of metal by friction. Since metals can also deform plastically under pressure contacting 'high spots' can disappear due to deformation. Moreover due to this deformation *work-hardening* (6.12) occurs and this further resists abrasion. Surface abrasion is important in the design of bearing surfaces in particular (18.60).

Impact tests

4.40   These tests are used to indicate the toughness of a material, and particularly its capacity for resisting mechanical shock. Brittleness, resulting from a variety of causes, is often *not* revealed during a tensile test. For example, nickel-chromium constructional steels suffer from a defect known as 'temper brittleness' (13.23). This is caused by faulty heat-treatment, yet a tensile test-piece derived from satisfactorily treated material and one produced from similar material but which had been incorrectly heat-treated might both show approximately the same tensile strengths and elongations. In an impact test, however, the unsatisfactory material would prove to be extremely brittle as compared with the correctly treated one, which would be tough.

4.41   **The Izod impact test** employs a standard notched test-piece (Fig. 4.12) which is clamped firmly in a vice. A heavy pendulum, mounted on ball-bearings, is allowed to strike the test-piece after being released from a fixed height (Fig. 4.13). The striking energy of approximately 163 J is partially absorbed in breaking the test-piece, and as the pendulum swings past, it carries with it a drag pointer which it leaves at its highest point of swing. This indicates the amount of mechanical energy used in fracturing the test-piece. To set up stress concentrations which will ensure that fracture does occur, the test-piece is notched. It is essential, however, that this notch always be standard, for which reason a standard gauge is supplied to test the dimensional accuracy of the notch, both in this and the other impact tests dealt with below (BS 131: *Notched-bar Tests*).

Examination of the fractured cross-section of the test-piece reveals further useful information. In the most ductile materials the fractured surface is likely to be of a 'fibrous' nature and will be rather dull and 'silky' in appearance (Fig. 4.14(i)) since plastic flow of the crystalline structure has occurred. With very brittle metals the fractured surface will be relatively bright, sparkling and 'crystalline' since crystals have not been plastically deformed. As fracture has followed the crystal boundaries each small crystal

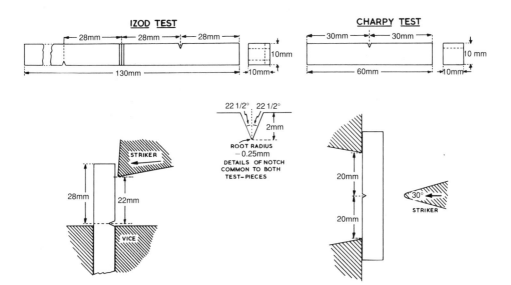

Fig. 4.12   *Details of standard test-pieces used in both the Izod and Charpy tests.*

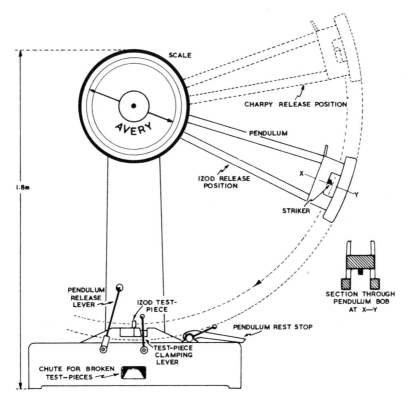

Fig. 4.13   *The Avery-Denison universal impact-testing machine. This machine can be used for either Charpy or Izod impact tests. For Izod tests, the pendulum is released from the lower position, to give a striking energy of 170 J; and for the Charpy test it is released from the upper position, to give a striking energy of 300 J. (The scale carries a set of graduations for each test.) The machine can also be used for impact-tension tests.*

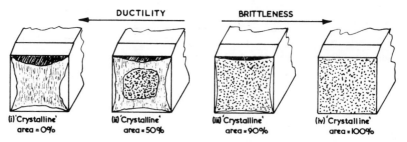

Fig. 4.14   *The nature of the fractured surface in the Izod test.*

reflects a bright point of light (Fig. 4.14(iv)). Many metals will fall somewhere between these extremes and the fractured surface will show a combination of ductile and 'crystalline' areas. For steels in particular it is possible to estimate quite accurately the percentage crystalline area of the fractured surface of the test-piece and to use this as a measure of the notch-ductility.

4.42   The Charpy impact test is of continental origin, and differs from the Izod test in that the test-piece is supported at each end (Fig. 4.12); whereas the Izod test uses a test-piece held cantilever fashion. Here the load on the pendulum can be varied so that the impact energy is either 150 J or 300 J.

4.43   Whilst the presence of minute cavities, slag particles, grain-boundary segregates and other flaws is likely to have a limited effect on the tensile strength of a metal and relatively none on its hardness, it will very seriously impair its impact properties. Defects such as these will act as stress raisers, particularly under conditions of shock loading, by introducing points at which a sudden high concentration of stress builds up. Thus there will tend to be a much greater 'scatter' of results in the impact test on a single cast material than there would be in the tensile test for the same material since a casting is likely to be much less homogeneous in its structure than is a wrought material. Nevertheless in wrought materials directionality of fibre (4.28) will have a much more significant effect on impact toughness than it does on either tensile strength or ductility (Table 6.2).

Creep

4.50   When stressed over a long period of time, some metals extend very gradually, and may ultimately fail at a stress *well below* the tensile strength of the material. This phenomenon of slow but continuous extension under a steady force is known as 'creep'. Such slow extension is more prevalent at high temperatures, and for this reason the effects of creep must be taken into account in the design of steam and chemical plant, gas and steam turbines, and furnace equipment.

Creep occurs generally in three stages (Fig. 4.15). At low stress and/or low temperature (Curve I) some *primary* creep may occur but this falls to a negligible amount in the *secondary* stage (the creep curve becomes almost horizontal). With increased stress and/or temperature (Curves II and III) the rate of secondary creep increases leading into *tertiary* creep and inevitable catastrophic failure.

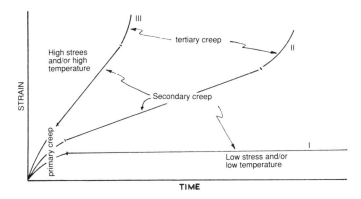

Fig. 4.15   *The variation of creep rate with stress and temperature.*

4.51   The *limiting creep stress* of a material at any given temperature is the maximum stress it can withstand without showing any measurable extension. Creep tests are carried out on test-pieces which are similar in form to ordinary tensile test-pieces. A test-piece is enclosed in a thermostatically controlled electric tube furnace which can be maintained accurately at a fixed temperature over the long period of time occupied by the test. The test-piece is statically stressed, and some form of sensitive extensometer is used to measure the extremely small extension at suitable time intervals. A set of creep curves, obtained for different static forces at the same temperature, is finally produced, and from these the limiting creep stress is derived.

4.52   Since we are dealing with such very small extensions, it is difficult to estimate the stress which just fails to produce any *measurable extension*. This makes it almost impossible to assess limiting creep stress with any accuracy. We are, in fact, restricted by the sensitivity of the extensometer. Moreover creep tests of this type can involve periods of time stretching into months. Hence it is often more practical to determine the stress which will produce a definite amount of creep in a fixed time – say two or three days; for example, a short-time creep test to determine the stress which will produce an extension of not more than 0.5 per cent of the gauge length during the first 24 hours and not more than 10^{-6} m per m per hour over the next 48 hours. Such a value is generally referred to as the 'time-yield stress' for those conditions. Other values relating to a given amount of *plastic* strain with both time and stress are also used.

Fatigue

4.60   Whenever failure of some structure or machine leads to a disaster which is worthwhile reporting, the press invariably hints darkly at 'metal fatigue'. Unfortunately, the technical knowledge of these gentlemen of the press is often suspect, as is evidenced by that oft-repeated account of the legendary cable 'through which a *current* of 132 000 *volts* was flowing'.

4.61   Although the public was first made aware of the phenomenon of fatigue following investigations of the Comet airliner disasters some years ago, the underlying principles of metal fatigue were appreciated more than a century ago by the British engineer Sir William Fairbairn, who carried out his classical experiments with wrought-iron girders. He found that a girder which would support a static load of 12 tonf for an indefinite period would nevertheless *fail* if a load of only 3 tonf were raised and lowered on it some three million times. Fatigue is associated with the effects which a fluctuating or an alternating force may have on a member – or, in everyday engineering terms, be subjected to the action of a 'live load'.

4.62   Following the work of Fairbairn, the German engineer Wöhler, with native inventiveness, produced the well-known fatigue-testing machine which still bears his name. This is a device (Fig. 4.16(i)) whereby alternations of stress can be produced in a test-piece very rapidly, and so reduce to a reasonable period the time required for a fatigue test. As the test-piece turns through 180°, the force W acting at a point on the specimen falls to zero, and then increases to W in the opposite direction. To find the fatigue limit, a number of similar specimens of the material are tested in this way, each at a different value of W, until failure occurs, or, alternatively, until about 20 million reversals have been endured. (It is, of course, not possible to subject the test-piece to the ideal infinite number of reversals.) From these results, an *S/N* curve is plotted; that is, stress (*S*) against the number of reversals (*N*) endured (Fig. 4.16(ii)). The curve becomes horizontal at a stress which will be endured for an infinite number of reversals. This stress is the *fatigue limit* or *endurance limit*. Some non-ferrous materials do not show a well-defined fatigue limit; that is, the *S/N* curve slopes gradually down to the horizontal axis.

In the Wöhler test it is convenient to make use of stresses of a torsional nature but in practice fatigue may result from stresses which are either tensile, compressive or torsional so long as loading is 'live' and either fluctuates or alternates.

4.63   A fatigue fracture has a characteristic type of surface, and consists of two parts (Fig. 4.16(iii)). One is smooth and burnished, and shows ripple-like

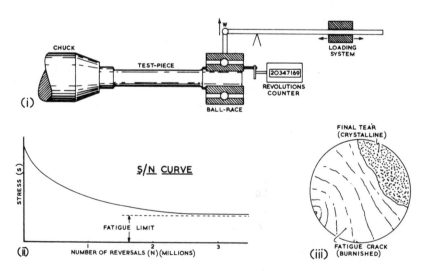

Fig. 4.16   *(i) The principle of a simple fatigue-testing machine. (ii) A typical S/N curve obtained from a series of tests. (iii) The appearance of the fractured surface of a shaft which has failed due to fatigue.*

marks radiating outwards from the centre of crack formation; whilst the other is coarse and crystalline, indicating the final fracture of the remainder of the cross-sectional area which could no longer withstand the load.

4.64   Fatigue failure will ultimately occur in any member which is stressed above its fatigue limit in such a way that the operating stress fluctuates or alternates. Such failure can be due simply to bad design and lack of understanding of fatigue, but is much more likely to be due to the presence of unforeseen high-frequency vibrations in a member which is stressed above the fatigue limit. This is possible since the fatigue limit is well below the tensile strength for all materials.

Some time ago, the author's opinion was sought on the possible cause of failure of a short length of copper tube which had held, cantilever fashion, a small pressure gauge, in the manner shown in Fig. 4.17. The broken tube was sent with the information that it had broken off at XX_1 under the action of no other force than the weight of the pressure gauge (P), which was very

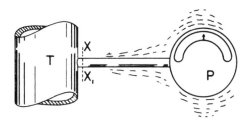

Fig. 4.17

small. The fracture appeared, on examination, to be of a typical fatigue nature. This tentative suggestion brought forth derisory comments from the engineer in charge, who claimed that the copper tube must be of poor quality, since failure had taken place after only a few days of service in a *static* condition. On ultimately visiting the site, the author's suspicions were soon confirmed – the pipe (T) was connected to a large air-compressor, the nerve-shattering vibrations from which were apparent at a distance of several hundred metres.

4.65   Any feature which increases stress concentrations may precipitate fatigue failure. Thus a fatigue crack may start from a keyway, a sharp fillet, a microstructural defect, or even a bad tool mark on the surface of a component which has otherwise been correctly designed with regard to the fatigue limit of the material from which it was made.

Factor of safety

4.66   By now it will be apparent to the reader that the working stress in an engineering component must be well below the ultimate tensile strength of the material and, in fact, below its elastic limit or proof stress (whichever value applies). Factors of safety have long been applied in engineering design to cope with this situation such that:

$$\text{factor of safety} = \frac{\text{ultimate tensile stress}}{\text{safe working stress}}$$

Wöhler (4.62) was instrumental in establishing factors of safety based on the above criteria. For 'dead loads' he suggested a factor of 4; for 'live loads' a factor of 6 or 8 and for alternating loads a safety factor between 12 and 16.

In modern practice a factor of safety is influenced by many features such as cost, uncertainty of working conditions and consequences of possible failure. Thus in structural steelwork a factor of 3 is commonly used whilst in steam boilers where creep, fluctuating stresses and the dire consequences of failure must be considered, a much higher factor of safety of 8 or 10 may be involved. Nevertheless in aircraft construction where it is important to keep weight to a minimum the factor of safety may be as low as 2 in some cases. In view of the catastrophic consequences of failure this means that rigorous quality control must be applied to materials and components used as well as thorough stress analysis to all members involved. In addition periodic non-destructive investigation to detect the onset of such faults as fatigue cracks are necessary.

Some materials are less uniform in composition and structure – and consequently less uniform in mechanical properties – than are others. Cast materials in particular are less reliable than are wrought materials because of the probable presence of segregation, gas porosity, shrinkage cavities and slag inclusions (2.30). For this reason it is necessary to use a higher factor of safety with cast metals than with much more homogeneous wrought metals.

4.67   It was mentioned earlier (4.25) that the elastic limit (or proof stress) is in reality a much more useful value than ultimate tensile stress in engineering design since, once the elastic limit has been exceeded, the metal has already begun to fail. Thus it has become more frequent practice to define factors of safety in terms of the elastic limit or proof stress. Obviously it becomes necessary to specify whether the factor of safety is based on *ultimate tensile stress* or on *proof stress*. Thus:

$$\text{proof factor of safety} = \frac{\text{proof stress}}{\text{safe working stress}}$$

Whichever value (proof stress or ultimate tensile stress) the factor of safety is based upon, a mixture of mathematical design criteria and *experience* of the behaviour of materials is generally used to enable a reliable estimate of safe working stress to be made.

Some other mechanical tests

4.70   Most of these tests are designed to evaluate some particular property of a material which is not revealed adequately during any of the preceding tests.

4.71   The Erichsen cupping test Materials used for deep-drawing are inevitably those of very high ductility. However, a simple measurement of ductility in terms of percentage elongation (obtained during the tensile test) does not always give a complete assessment of deep-drawing properties. A test which imitates the conditions present during a deep-drawing operation is often preferable. Such a test is the Erichsen cupping test, and it is commonly used to assess the deep-drawing quality of soft brass, aluminium, copper, or mild-steel sheet. In this test, a hardened steel ball is forced into the test-piece, which is clamped between a die-face and a blank-holder (Fig. 4.18). When the test-piece splits, the height of the cup which has been

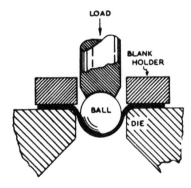

Fig. 4.18   *The principle of the Erichsen cupping machine.*

formed is measured, and this height (in mm) is taken as the Erichsen value. Unfortunately, the results from such a test can be variable, even with material of uniform quality. The depth of cup which can be drawn depends largely upon the pressure between the blank-holder and the die-face. Light pressure will allow metal to be drawn between the die and holder, and a deeper cup is formed. The continentals tend to use firm pressure between die and holder, whereas in Britain we 'cheat' by using lighter pressure. Nevertheless BSI recognizes the Erichsen test in BS 3855 where test-piece dimensions, testing apparatus, lubricants and precautions are dealt with.

4.72   Possibly the most useful aspect of the test is that it gives some idea of the grain size of the material, and hence its suitability for deep-drawing, as indicated in Fig. 4.19. Coarse grain, always associated with poor ductility in a drawing operation, will show up as a rough rumpled surface in the dome of the test-piece. This resembles the 'lumpy' outside skin of an orange – hence the description 'orange-peel effect'.

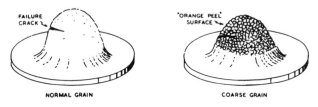

Fig. 4.19   *Erichsen test-pieces.*

4.73   **Bend tests** are often used as a means of judging the suitability of a metal for similar treatment during a production process. For example, copper, brass, or bronze strip used for the manufacture of electrical switch-gear contacts by simple bending processes may be tested by a bending operation which is somewhat more severe than that which will be experienced during production.

Simple bend tests are probably more closely related to the fundamental ideas of toughness (4.12) than are the widely used impact tests. The latter are more easily measured in numerical terms and in any case deal specifically with conditions of *shock* loading. Figs 4.20(i) and (ii) illustrate simple tests requiring little equipment other than a vice, whilst (iii) represents a more widely accepted test in which the wire is bent through 90° over a cylinder of

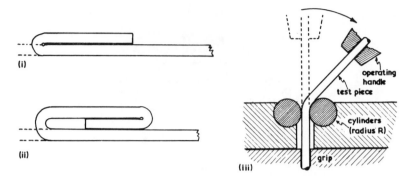

Fig. 4.20 *Simple bend tests. (i) The material is merely bent back upon itself. (ii) Here it is doubled over its own thickness, the second bend being the test bend. (iii)Here a specific radius* R *is used.*

specified radius *R*, then back through 90° in the opposite direction. This is continued until the test-piece breaks, the number of bending cycles being counted.

The surface affected by the bending process is examined for cracks, and, if necessary, for coarse grain ('orange peel').

4.74   Compression tests are used mainly in connection with cast iron and concrete since these are materials more likely to be used under the action of compressive forces than in tension. A cylindrical block, the length of which is twice its diameter, is used as a test-piece. This is compressed (using a tensile-testing machine running in 'reverse') until it fails.

Malleable metals do not show a well-defined point of failure (Fig. 4.21(ii)) but with brittle materials (iii), the *ultimate compressive stress* can be measured accurately, since the material fails suddenly, usually by multiple shear at angles of 45° to the direction of compression.

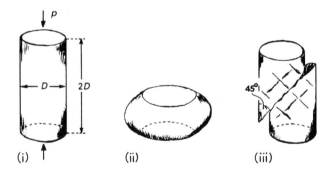

Fig. 4.21 *The behaviour of brittle and ductile materials during a compression test.*

4.75   Torsion tests are often applied to wire – steel wire in particular. The test consists of twisting a piece of wire in the same direction round its own axis until it breaks or until a specified number of twists has been endured. The simple machine used (Fig. 4.22) consists essentially of two grips which remain in the same axis and do not apply any bending moment to the test-piece. One

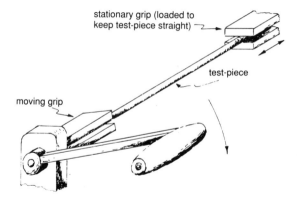

Fig. 4.22   *A very simple 'twist' testing machine.*

grip remains stationary whilst the other is rotated, the distance between the grips being adjustable so that test-pieces of different lengths can be tested. The machine is so constructed that a suitable small tensile force can be applied to the test-piece to keep it straight and also allow for change in length during the test.

The test-piece is held in the machine so that its longitudinal axis coincides with the axis of the grips and so that it remains straight during the test. This is achieved by applying a constant tensile force just sufficient to straighten it but not exceeding 2 per cent of the nominal tensile strength of the specimen. One grip is rotated at a constant speed which shall be sufficiently slow to prevent any rise in temperature of the test-piece. This continues until the test-piece breaks or until the specified number of turns is registered. The number of complete turns of the rotating grip is counted.

CHAPTER 5

Non-destructive testing

5.10 Those tests described in the previous chapter are destructive tests carried out on samples which – it is hoped – are representative of a batch of manufactured material. In many cases such tests will be adequate particularly where wrought products like rolled strip and drawn rod are involved, since properties are generally uniform throughout a large batch of material derived from a single cast ingot. Components which are produced individually, however, such as castings and welded joints, may vary in quality, for even in these days of rigorous on-line inspection faulty components *can* arise since the production methods for casting and weldments fall under the influences of many variable factors such as working temperature, surrounding atmosphere – and the skill of the operator. If the quality of such components is very important – as for example castings used in aircraft construction – it may be necessary to test each component individually using some type of non-destructive examination. Such investigation seeks to detect faults and flaws either at the surface or below it and a number of suitable methods are available in each case. Since tests of this type give an overall assessment of the quality of the product the term 'non-destructive testing' (NDT) is often replaced by 'non-destructive evaluation' (NDE).

The detection of surface cracks and flaws

5.20 Surface cracks may arise in a material in a number of ways. Some cracks show up during inspection using a simple hand magnifier, whilst others are far harder to detect. Thus steel tools, which have been water quenched (12.25), may develop hair-line cracks which are not apparent during ordinary visual inspection. Castings and weldments may crack due to contraction during the period of solidification and cooling.

5.21 **Penetrant methods** In these methods, the surface to be examined is first cleaned adequately to remove grease, and then dried. The penetrant liquid is then sprayed or swabbed on to the surface, which should be warmed to about 90 °C. Small components may be immersed in the heated penetrant. After sufficient time has elapsed for the penetrant to fill any cracks, the excess is carefully washed from the surface with warm water (the surface tension of water is too high to allow it to enter the fine cracks). Alternatively other solvents may be used to remove excess penetrant, whilst some oily penetrants are treated with an emulsifier which enables them to

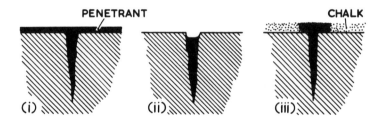

Fig. 5.1 *The penetrant method of crack detection. (i) The cleaned surface is coated with the penetrant, which seeps into any cracks. (ii) Excess penetrant is cleaned from the surface. (iii) The surface is coated with chalk. The penetrant is expelled from the crack by the contraction of the cooling metal, and stains the chalk.*

be washed off with water. The test surface is then carefully dried and coated with a 'developer' such as powdered chalk. This developer can be blown on as a powder or sprayed on as a solution or suspension. In some cases the component is dipped into a suspension or solution of the developer which is allowed to dry on the surface. The component is then set aside for some time. As the coated surface cools it contracts and penetrant tends to be squeezed out of any cracks, so the chalk layer will become stained, thus revealing the presence of the cracks. Most penetrants of this type contain a scarlet dye which renders the stain immediately noticeable.

5.22   In some cases, the penetrant contains a compound which becomes fluorescent under ultra-violet light. The use of chalk is then unnecessary. When the prepared surface is illuminated by ultra-violet light – or 'black light', as it is sometimes called – in a darkened cubicle, the cracks containing the penetrant are revealed as bright lines on a dark background.

Penetrant methods in general are particularly useful for the examination of non-ferrous metals and austenitic (non-magnetic) steels. Aluminium-alloy castings are frequently examined in this way.

5.23   Magnetic dust methods These methods can be applied only to magnetic materials, but nevertheless provide a quick and efficient method of detecting cracks. A further advantage over the penetrant method is that flaws immediately *below* the surface are also detected; so the magnetic dust method is particularly suitable for examining machined or polished surfaces, where the mouth of a crack may well have become 'burred' over.

The magnetic method involves making the component part of a 'magnetic circuit' (Fig. 5.2(ii)). This magnetic field can be induced in the component either by permanent magnets or by electromagnetic means. Alternatively, using probes, a heavy current at low e.m.f. is passed through the component so that a magnetic field will be induced in it (Fig. 5.2(iii)).

No matter which method is used to produce the magnetic field, the object is to 'saturate' the component with lines of force. These lines of force pass easily through a magnetic material, which is, so to speak, a 'good conductor' of lines of force. In air, they repel each other to the extent that they spread as indicated in Fig. 5.2(i). Consequently, on meeting a gap or other discontinuity at or just below the surface, the lines of force spread outwards (Fig. 5.2(ii) and (iii)). If some iron dust is now sprinkled on the surface of the component, it will stick to the surface where the lines of force break out, thus revealing the site of the fault. Alternatively, the magnetized component

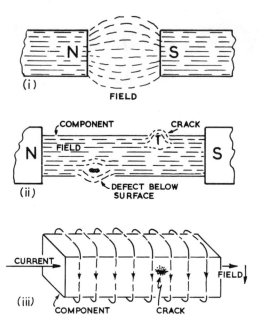

Fig. 5.2   *Magnetic methods of crack detection. (i) The principle of the process – lines of force tend to 'spread' in an air gap. (ii) A magnetic 'circuit'. (iii) Using a high current to induce a field.*

can be placed in paraffin containing a suspension of tiny iron particles. The particles will be attracted to the surface of the component at any points where, due to the presence of a fault, lines of force cut through. The sensitivity of this method can be improved by coating the surface to be treated with a carrier liquid containing magnetic particles which have been treated with a fluorescent compound. The surface is then examined under 'black light' (5.22) in a darkened cubicle.

5.24   Acid pickling methods Surface defects in steel castings can generally be revealed by pickling the casting in a 10 per cent solution of sulphuric acid at about 50 °C for up to two hours. Alternatively a 20–30 per cent solution of hydrochloric acid at ambient temperature for up to twelve hours can be used, whilst stainless steels require mixtures of concentrated nitric and hydrofluoric acids. The latter mixture is *extremely corrosive* and tests involving this and other acids should be carried out only by a competent technician who will be aware of the very dangerous nature of these chemicals.

After removal from the acid bath the casting is washed thoroughly and, if necessary, any residual acid neutralized by immersing it in a hot suspension of slaked lime, after which the casting is washed again and dried quickly. This simple procedure provides a cheap but effective way of dissolving oxide layers and revealing most surface defects.

The detection of internal defects

5.30   Castings are liable to contain unwanted internal cavities, in the form of gas blow-holes and shrinkage porosity (2.33). Wrought materials may contain slag inclusions and other flaws, whilst welded joints can suffer from any of these defects. Unfortunately, metals are opaque to light, so these internal faults are hidden from us. Other forms of electromagnetic radiation, however, will penetrate metals, and so enable us to 'see' into the interior of the material. Of these forms of radiation, X-rays and γ-rays ('gamma' rays) are the most widely used in the detection of internal faults.

5.31   The railway employee whose function is to tap the wheels of railway rolling-stock was for generations the subject of music-hall humour. In fact he employs what we may term 'sonic testing' – if a wheel is cracked, it no longer emits the correct 'ring' when tapped. Methods such as this have also been used industrially to check the internal soundness of components, but much more sophisticated methods of testing, using 'ultrasonic' vibrations, are now in use.

5.32   X-ray methods   X-rays are produced when high-velocity electrons strike a metal target, so the main requirements for generating X-rays are (1) a source of electrons, (2) a means of accelerating these electrons to a high velocity and (3) a suitable target. In an X-ray tube (Fig. 5.3) the electrons are given off from a filament by passing through it a current (stream of electrons) sufficient to heat it and so thermally activate the electrons that they 'escape' from the filament. The filament also acts as the cathode in a high-tension circuit and since the electrons are negatively charged they are accelerated *away* from the negatively charged cathode and *towards* the positively charged 'anti-cathode' which serves as the target. This is made of tungsten which will resist melting at the high temperature produced by the impacting electrons, much of the kinetic energy of which is absorbed to generate heat as well as X-rays. The interior of the tube is evacuated since the presence of relatively large molecules of nitrogen and oxygen (air) would obstruct the passage of small electrons. The greater the potential difference between anti-cathode and cathode the greater the velocity (and hence kinetic energy) of the electrons and the shorter the wavelength of the X-rays produced. These short-wavelength X-rays are able to penetrate metals more easily and are referred to as 'hard' X-rays.

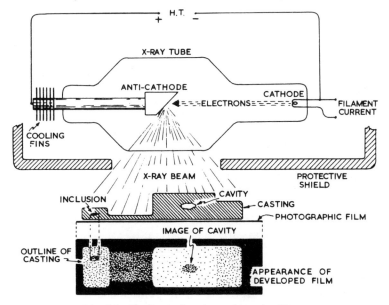

Fig. 5.3   *Radiography of a casting using X-rays.*

5.33   X-rays travel in straight lines as do light rays but whilst metals are opaque to light they are transparent to X-rays provided that the latter are 'hard' enough and that the metal is not so thick in cross-section that the X-rays are completely absorbed. X-rays affect a photographic film in a manner similar to that of light so the most efficient method of detecting faults in a body of metal is to take an X-ray photograph of its interior. The reader may be familiar with a similar application of X-rays in medicine. There is however an important difference in the type of X-rays used since those employed for the radiography of metals need to be much 'harder' so that they will penetrate the metal successfully. These hard X-rays would cause very serious damage to our body tissues if exposed to them and for this reason such X-ray equipment must be well shielded to prevent stray radiations from reaching the operator. For this purpose a shielding wall of 'barium cement' is often used. Barium and its compounds are very effective in absorbing X-rays. The 'barium meal' used in diagnostic medicine serves a similar purpose. A method of shielding is indicated in Fig. 5.3. Here the X-ray tube is emitting radiation which passes through the casting, and forms an image on a photographic film placed behind the casting. X-rays will penetrate that region of the casting containing the cavity much more easily, and so produce a *denser* image on the film. The barium meal of diagnostic medicine has the reverse effect in that it absorbs X-rays more effectively and so will cause a *less dense* image to be formed on the film. (Readers who have an interest in photography will know that those areas on a monochrome negative which receive the most light are densest and darkest when the negative has been developed.)

A fluorescent screen may be substituted for the photographic film, so that the resultant radiograph may be viewed instantaneously. This type of fluoroscopy is much cheaper and quicker, but is less sensitive than photography, and its use is limited to the less-dense metals.

5.34   **γ-ray methods** γ-rays can also be used in the radiography of metals. They are 'harder' than X-rays, and are therefore able to penetrate a greater thickness of metal or, alternatively, a more dense metal. Hence they are particularly useful in the radiography of steel, which absorbs radiation more effectively than do the light alloys. As might be expected, exposure to γ-radiation is extremely dangerous. It is in fact the lethal radiation emitted from the notorious strontium-90, amongst other radioactive isotopes, present in the fall-out products of some nuclear explosions.

5.35   Originally radium was used as a source of γ-rays. Since it is very scarce and consequently expensive artificially activated isotopes are now employed. These are prepared by irradiating a suitable element in an atomic pile. One of the most useful activated isotopes is cobalt-60.

Manipulation of the isotope as a source of γ-rays in radiography is in many respects simpler than when using X-rays though security arrangements need to be even more stringent. Not only are γ-rays harder but, unlike X-rays, cannot be 'switched off'. A radioactive isotope emits continuously for a period varying from a few seconds to thousands of years, which is why nuclear accidents are to be avoided.

γ-rays can be used to radiograph considerable thicknesses of steel and have the further advantage over X-rays that the equipment is more mobile and less cumbersome to transport. Thus γ-radiography is useful in positions difficult to access or where portable equipment needs to be used as in the examination of steelwork in motorway bridges.

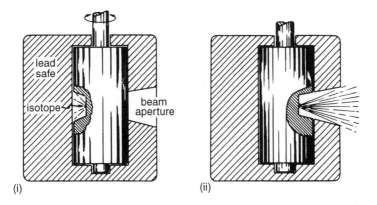

Fig. 5.4 *A lead safe suitable for storing and transporting radioactive isotopes: (i) an isotope in the shrouded position; (ii) the isotope in transmitting position.*

5.36   Ultrasonic testing It is fun to yodel in the mountains, and listen to a succession of echoes from mountain-sides both far and near. Ultrasonic testing is somewhat similar, except that the sound waves generated by a yodel are replaced by very high-frequency vibrations, which are beyond the acoustic range which can be received by our ears. In ultrasonic testing frequencies between 500 000 and 10 000 000 Hz are commonly used whereas our ears can detect sounds at frequencies only between 30 and 16 000 Hz.

Figure 5.5 represents the principles of ultrasonic testing. A probe containing a quartz crystal, which can both transmit and receive high-frequency vibrations, is passed over the surface of the material to be tested. The probe is connected to an amplifier, which converts and amplifies the signal, before it is recorded on the cathode-ray tube.

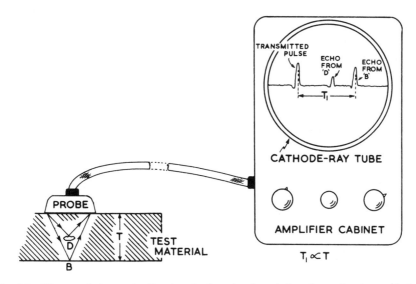

Fig. 5.5   *The use of ultrasonic vibrations in detecting flaws below the surface in metal plate.*

5.37   Under normal conditions, the vibrations will pass from the probe, unimpeded through the metal, and be reflected from the bottom inside surface at B back to the probe, which also acts as a receiver. Both the transmitted pulse and its echo are recorded on the cathode-ray tube, and the distance, T_1, between the 'blips' is proportional to the thickness, T, of the test material. If any discontinuity is encountered, such as the blowhole, D, then the pulse is interrupted, and reflected back as indicated. Since this echo returns to the receiver in a shorter time, an intermediate 'blip' appears on the cathode-ray tube. Its position relative to the other 'blips' indicates the distance of the fault below the surface. Different types of probe are available for materials of different thickness. This method is particularly suitable for detecting faults in sheet, plate and strip materials more than 6 mm thick, whilst modified equipment is used for detecting faults in welds.

CHAPTER 6

Mechanical deformation and recrystallization of metals

6.10 When the shape of a piece of metal is changed by the application of forces deformation takes place in two stages as is shown during a tensile test (4.22). At first crystals within the metal are distorted in an elastic manner and this distortion increases proportionally with the increase in stress. If the stress is removed during this stage, the metal returns to its original shape, illustrating the elastic nature of the deformation. On the other hand, if the stress is increased further, a point is reached (the *yield point*) where the forces which bind together the atoms in the lattice structure are overcome to the extent that layers or planes of atoms begin to slide over each other. This process of 'slip', as metallurgists call it, is *not* reversible; so, if the stress is now removed, permanent deformation remains in the metal (Fig. 6.1). This type of deformation is termed 'plastic'.

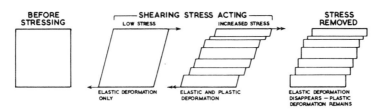

Fig. 6.1 *Deformation of a metal by both elastic and plastic means.*

6.11 Visual evidence of the nature of slip is fairly easily obtained. If a piece of pure copper or aluminium is polished and etched, and then squeezed in a vice with the polished face uppermost, so that it is not directly damaged, examination under the microscope shows a large number of parallel, hair-like lines on the polished and etched surface. These lines are in fact shadows cast by minute ridges formed when blocks or layers of atoms within each crystal slide over each other (Fig. 6.2). These hair-like lines are known as *slip bands*.

6.12 Slip of this type can occur along a suitable plane, until it is prevented by some fault or obstacle within the crystal. A further increase in the stress will then produce slip on another plane or planes, and this process goes on until all available slip planes in the piece of metal are used up. The metal is then said to be work-hardened, and any further increase in stress will lead to fracture. Microscopic examination will show that individual crystals have become elongated and distorted in the direction in which the metal was

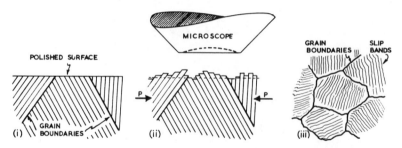

Fig. 6.2   *The formation of slip bands: (i) indicates the surface of the specimen before straining, and (ii) the surface after straining. The relative slipping along the crystal planes produces ridges, which cast shadows on the surface of the specimen when this is viewed through the microscope (iii).*

deformed. In this condition, the metal is hard and strong; but it has lost its ductility, and, if further shaping is required, it must be softened by annealing.

6.13   We know precisely the distance apart of the atoms (or more properly, ions) within a metallic crystal – this can be measured with great accuracy using X-ray methods. We also know the magnitude of the forces of attraction acting between these positively charged ions and the negatively charged electrons within the metallic bond (1.22). It is therefore possible to calculate the *theoretical* strength of a metal, that is, the sum total of all the net forces of attraction acting between ions and electrons across a given plane in a metal.

Under controlled laboratory conditions a single crystal of a metal can be grown and if this is subjected to a tensile test its *true* strength is found to be only about one-thousandth of the theoretical when the latter is calculated as outlined briefly above.

It is now realized that instead of slip taking place simultaneously by one block of atoms sliding wholesale over another block across a complete crystal plane, it occurs *step-by-step* by the movement of faults or dis-continuities within the cystallographic planes. These faults where, in effect, half-planes of atoms are missing within a crystal (Fig. 6.3) are known as *dislocations*. When the crystal is stressed to its yield point these dislocations will move step-by-step along the crystallographic plane until they are halted or 'pegged' by some obstruction such as the atom of some alloy metal which is larger (or smaller) than those of the parent metal. Alternatively the dislocation will move along until it is stopped by a crystal boundary or another dislocation moving across its path.

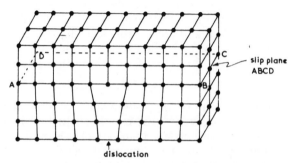

Fig. 6.3   *An 'edge' dislocation. This is the most simple form of dislocation and can actually be seen in some materials when they are examined under very high magnifications in electron microscopy.*

If the surface of the metal cuts through the slip plane a minute 'step' will be formed there (Fig. 6.4). If several hundreds of these dislocations follow each other along the same slip plane then the step at the surface may be large enough to be visible as a 'slip band' (Fig. 6.2(iii)) using an ordinary optical microscope.

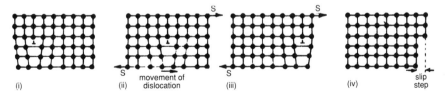

(i) (ii) movement of dislocation (iii) (iv) slip step

Fig. 6.4 *The movement of an edge dislocation under the action of a stress, S. In illustrations, dislocations are usually denoted thus: ⊥.*

The force necessary to move dislocations *step-by-step* through a metallic crystal is much less than the force which would be necessary to overcome the total forces acting between ions and electrons along the complete slip plane if this slip occurred *simultaneously in a single jerk*. Thus the true strength of a metal is only a small fraction of that which was calculated on the assumption that slip took place by a single sudden movement.

6.14 Professor N. Mott's well-known analogy with the means available for smoothing the wrinkles from a carpet helps to explain this process of slip in metallic crystals. Imagine a heavy wall-to-wall carpet which, when laid, proves to be wrinkled near to one wall (Fig. 6.5(i)). Attempts to remove the wrinkle by pulling on the opposite edge (Fig. 6.5(ii)) will be fruitless and lead only to broken finger nails and the utterance of unseemly expletives, since an attempt is being made to overcome the *sum* of the forces of friction between the *whole* of the carpet and the floor and so to produce slip over the whole surface at the same instant. Instead it requires very little force to ease the wrinkle along with one's toe (Fig. 6.5(iii)) until it has been moved progressively to the opposite edge of the carpet (Fig. 6.5(iv)). In this way only the small force of friction between the carpet and the floor *adjacent to the wrinkle* has to be overcome during any instant when that small part of the carpet is moving.

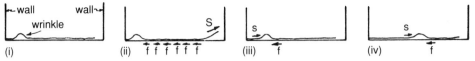

(i) (ii) (iii) (iv)

Fig. 6.5 *Prof. Mott's 'carpet analogy' in which a wrinkle in the carpet is likened to a 'dislocation' in a metallic crystal.*

6.15 Similarly the step-by-step movement of a dislocation through the crystal structure of a metal requires a much smaller force than would the movement of a large block of ions by simultaneous slip. The real strength of a metal is therefore much less than that calculated from considerations of a perfect crystal. A metal becomes *work-hardened* when all dislocations present have become jammed against each other or against various other obstacles within the crystals or indeed against the crystal boundaries themselves. Many dislocations will move out to the surface of the metal and be lost as slip bands.

6.16   Deformation by twinning   Whilst most metals deform plastically by the process of slip described above, a few deform by what is known as 'twinning'. Whereas during slip all ions in a block have moved the same distance when slip is complete, in deformation by twinning (Fig. 6.6) ions within each successive plane in a block will have moved different distances. When twinning is complete the lattice direction will have altered so that one half of the twinned crystal is a mirror image of the other half, the twinning plane corresponding to the position of the mirror.

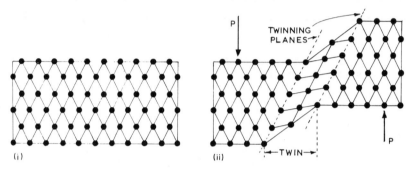

Fig. 6.6   *Deformation by twinning (i) before stressing, and (ii) after stressing. In the deformed crystal one half is a mirror image of the other about the twin plane.*

The stress required to cause deformation by twinning is generally higher than that needed to produce deformation by slip. Twinning is more commonly found in close-packed hexagonal (CPH) metals such as zinc. When a bar of tin is bent quickly the sound of the layers of ions moving relative to each other is emitted as an audible 'squeak'. This was known by the Ancients as 'The Cry of Tin'.

These twins formed during mechanical deformation are not to be confused with 'annealing twins' occurring in copper, brasses, bronzes and austenitic stainless steels. In these instances the twin crystals grow from a *plane* during recrystallization of a cold-worked metal instead of from a point source.

Annealing and recrystallization

6.20   A cold-worked metal, that is, one which has been deformed without the application of heat, is in a state of considerable internal stress due to the presence of elastic strains internally balanced within the distorted crystal structure. During annealing these stresses are removed as the thermal activation supplied allows ions to migrate into positions approaching nearer to equilibrium and the original ducility of the metal returns. The changes which accompany an annealing process occur in three stages.

6.21   The relief of stress   As the temperature of the cold-worked metal is gradually raised, some of the internal stresses disappear, as atoms and probably whole dislocations move through small distances into positions nearer equilibrium. At this stage, there is no alteration in the generally distorted appearance of the structure, and, indeed, the strength and hardness produced by cold-working remain high. Nevertheless, some hard-drawn materials, such as 70–30 brass, are given a low-temperature anneal in order to relieve internal stresses, as this reduces their tendency towards 'season cracking' during service, that is, the opening up of cracks along grain boundaries, due to the combined effects of internal stresses and surface corrosion.

6.22   Recrystallization   With further increase in temperature, a point is reached where new crystals begin to grow from nuclei which form within the structure of the existing distorted crystals. These nuclei are generally produced where internal stress is greatest, that is where dislocations are jammed together at grain boundaries and on slip planes. As the new crystals grow they take up atoms from the old distorted crystals which they gradually replace. Unlike the old crystals, which had become elongated in one direction by the cold-working process, these new crystals are small and equi-axed.

This phenomenon, known as *recrystallization*, is used to obtain a tough, fine-grained structure in most non-ferrous metals. The minimum temperature at which it will occur is called the 'recrystallization temperature' for that metal, though it is not possible to determine this temperature precisely, because it varies with the amount of cold-work to which the metal had been subjected before the annealing process. The more heavily the metal is cold-

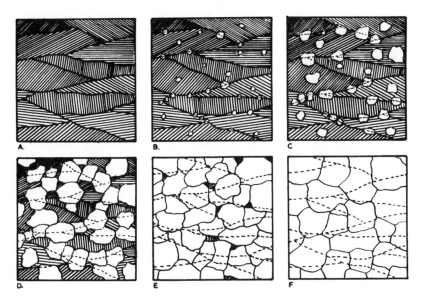

Fig. 6.7   *Stages in the recrystallization of a metal during an annealing process. (A) represents the metal in the cold-rolled state. At (B) recrystallization has just begun, with the formation of new crystal nuclei. These grow by absorbing the old crystals, until at (F) recrystallization is complete.*

Table 6.1   *Recrystallization temperatures of some metals.*

Metal	Recrystallization temperature (°C)
tungsten	1200
nickel	600
pure iron	450
copper	190
aluminium	150
zinc	20
lead ⎫	below room-temperature; hence
tin ⎭	they cannot be 'cold-worked'

worked, the greater the internal stress, and the lower the temperature at which recrystallization will begin. Alloying, or the presence of impurities, raises the recrystallization temperature of a metal.

6.23   Grain growth   If the annealing temperature is well above that for recrystallization of the metal, the new crystals will increase in size by absorbing each other cannibal-fashion, until the resultant structure becomes relatively coarse-grained. This is undesirable, since a coarse-grained material is generally less ductile than a fine-grained material of similar composition. Moreover, if the material is destined for deep-drawing, coarse grain tends to disfigure a stretched surface by giving it a rough, rumpled appearance known as 'orange peel' (4.72). Both the time and temperature of annealing must be controlled, in order to limit grain growth; though, as indicated by Fig. 6.8, temperature has a much greater influence than does time.

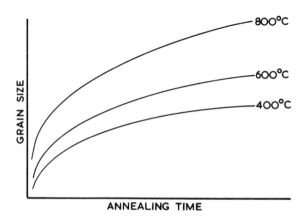

Fig. 6.8   *The effects of time and temperature during annealing on the grain size of a previously cold-worked metal.*

6.24   The amount of cold-work the material receives prior to being annealed also affects the grain size produced. In heavily cold-worked metal, the amount of locked-up stress is great and so, when thermal activation is supplied during annealing, a large number of new crystal nuclei will form instantaneously as the recrystallization temperature is reached and so the resultant crystals will be small since there will be less space in which individual crystals can grow.

On the other hand, very light cold-work (the 'critical' amount) gives rise to few nuclei, because the metal is not highly stressed. Consequently the grain size will be large, and the ductility poor.

6.25   Alloy additions are often made to limit the grain growth of metals and alloys during heat-treatment processes. Thus, up to 5 per cent nickel is added to case-hardening steels in order to reduce grain growth during the carburising process. When we say that nickel toughens a steel, we must remember that it does so largely by limiting its grain growth during heat treatment.

Cold-working processes

6.30   Most metals and alloys are produced in wrought form by hot-working processes, because they are generally softer and more malleable when hot, and consequently require much less energy to shape them. Hot-working processes make use of *compressive* forces to shape the work-piece. The reason for this is that metals become weak in tension at high temperatures, so their ductility decreases. Any attempt to pull a metal through a die at high temperature would fail, because the metal would be so weak as to tear. Consequently, those shaping processes in which tension is employed are cold-working processes. Since most metals work-harden quickly during cold-working operations, frequent inter-stage annealing is necessary. This increases the cost of the process; and, as far as possible, operations involving the use of hot-working by compression are used, rather than cold-working processes, which make frequent annealing stages necessary. Wire used to be made by drawing down a previously rolled bar. This involved many drawing operations, interspersed with annealing stages. Now, a cast billet is hot-extruded as thick wire in a single operation; this thick wire is then drawn down in a minimum of cold-working stages, through dies.

6.31   Other reasons for using cold-working processes are:

- To obtain the necessary combination of strength, hardness and toughness for service. Mild steel and most non-ferrous materials can be hardened only by cold-work.
- To produce a smooth, clean surface finish in the final operation. Hot-working generally leaves an oxidized or scaly surface, which necessitates 'pickling' the product in an acid solution.
- To attain greater dimensional accuracy than is possible in hot-working processes.
- To improve the machinability of the material by making the surface harder and more brittle (7.57).

6.32   The main disadvantage of cold-working is that the material quickly work-hardens, making frequent inter-stage annealing processes necessary. A typical sequence of operations for the manufacture of a cartridge-case by deep-drawing 70–30 brass sheet is shown in Fig. 6.9. Here the inter-stage annealing processes will be carried out in an oxygen-free atmosphere ('bright annealing'), in order to make acid-pickling unnecessary.

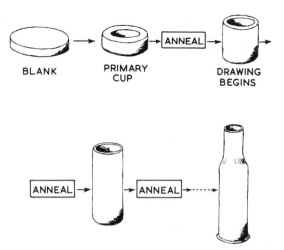

Fig. 6.9 *Typical stages in the deep-drawing of a cartridge-case. In practice, many more intermediate operations may be necessary, only the principle being indicated here.*

6.33   Other cold-working processes include:

- the drawing of wire and tubes through dies;
- the cold-rolling of metal plate, sheet, and strip;
- spinning and flow-turning, as in the manufacture of aluminium kitchenware;
- stretch forming, particularly in the aircraft industry;
- cold-heading, as in the production of nails and bolts;
- coining and embossing.

Hot-working processes

6.40   A hot-working process is one which is carried out at a temperature well above the recrystallization temperature of the metal or alloy. At such a temperature, recrystallization will take place simultaneously with deformation, and so keep pace with the actual working process (Fig. 6.10). For this reason, the metal will not work-harden, and can be quickly and continuously reduced to its required shape, with the minimum of expended energy. Not only is the metal naturally more malleable at a high temperature, but it remains soft, because it is recrystallizing continuously during the working process.

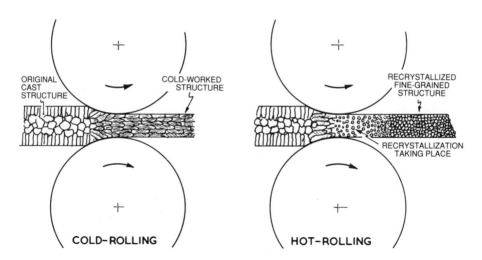

Fig. 6.10   *During cold-rolling a metal becomes work-hardened, but during a hot-rolling process recrystallization can take place.*

6.41   Thus hot-working leads to a big saving both in energy used and in production time. It also results in the formation of a uniformly fine grain in the recrystallized material, replacing the original coarse cast structure. For this reason, the product is stronger, tougher, and more ductile than was the original cast material.

6.42   The main disadvantage of hot-working is that the surface condition is generally poor, due to oxidation and scaling at the high working temperature. Moreover, accuracy of dimensions is generally more difficult to attain because the form tools need to be of simpler design for working at high temperatures. Consequently, hot-working processes are usually followed by some surface-cleaning process, such as water-quenching (to detach scale), shot-blasting and/or acid pickling prior to at least one cold-working operation which will improve the surface quality and accuracy of dimensions.

6.43 The principal hot-working processes are:

- hot-rolling, for the manufacture of plate, sheet, strip, and shaped sections such as rolled-steel joists;
- forging and drop-forging, for the production of relatively simple shapes, but with mechanical properties superior to those of castings;
- extrusion, for the production of many solid and hollow sections (tubes) in both ferrous and non-ferrous materials.

6.44 Many metal-working processes involve preliminary hot-working, followed by cold-forming or finishing. Mild-steel sheet destined for the production of a motor-car body is first hot-rolled down from the ingot to quite thin sheet. This is cooled, and then pickled in acid, after which it is lightly cold-rolled to give a dense smooth surface. Finally, it is cold-formed to shape by means of presswork. This final shaping also hardens and strengthens the material, making it sufficiently rigid for service. Similarly, a drop-forged connecting rod will be finally 'sized' by 'tapping' it in a cold die, in order to improve both dimensions and finish.

Grain flow and fibre

6.50 During any hot-working process, the metal is moulded in a plastic manner, with a result similar to that produced by a baker kneading his dough (though a machine now does this for him). Segregated impurities in the original casting are mixed in more uniformly, so that brittle films no longer coincide with grain boundaries. Obviously these impurities do not disappear completely, but form 'flow lines' or 'fibres' in the direction in which the material has been deformed. Since new crystals grow independently of these fibres, the latter weakens the structure to a much smaller extent than did the original intercrystalline films. Consequently, the material becomes stronger and tougher, particularly along the direction of the fibres. At right angles to the fibres, the material is weaker (Table 6.2), since it tends to pull apart at the interface between the metal and each fibre.

Table 6.2   *The effects of hot-working on mechanical properties.*

Condition	Tensile strength (N/mm²)	% elongation	Izod impact (J)
as cast	540	16	3
hot-rolled (in the direction of rolling)	765	24	100
hot-rolled (at right angles to the direction of rolling)	750	15	27

6.51 Stages in the manufacture of a simple bolt are shown in Fig. 6.11. The bolt has been produced from steel rod in which a fibrous structure (A) is the result of the original hot-rolling process. The rod will be either hot- or cold-headed, and this operation will give rise to plastic flow in the metal, as indicated by the altered direction of the fibre in the head (B). The thread will then be rolled on, producing a further alteration in the direction of the fibre in that region (C). The mechanical properties of a bolt manufactured in

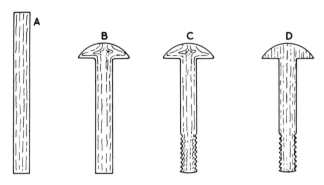

Fig. 6.11 *The direction of 'fibre' in a cold-headed or forged bolt as compared with that in a bolt machined from a solid bar (D).*

this way will be superior to those of one machined from a solid bar (D). The latter would be much weaker, and it is highly likely that, in tension, the head may shear off, due to weakness along the flow lines – even supposing the thread had not already stripped for the same reason. This bolt (D) would be extremely anisotropic[1] in its mechanical properties – in this instance strong parallel to the fibre direction but weak at right angles to the fibre direction. For the mass production of bolts, the forging/thread-rolling method would be less costly in any case; so the choice of process here is very simple.

6.52 It is thus bad engineering practice to use any shaping process which exposes a cut fibrous structure such that shearing forces can act *along* exposed fibres, and so cause failure. Similarly, in a drop-forging it is essential that fibres are formed in those directions in which they will give rise to maximum strength and toughness.

The macro-examination of fibre direction

6.60 It is often necessary for the drop-forger to examine a specimen forging to ensure that grain flow is taking place in the desired direction. This will be indicated by fibre direction. Similarly, the engineer may wish to find out whether or not a component has been manufactured by forging – assuming that any such evidence as 'flash lines' and the like has been obliterated by subsequent light machining.

6.61 Whereas *micro*-examination involves the use of a microscope to view the prepared surface (10.50), *macro*-examination implies that no such equipment is necessary, other than, possibly, a hand magnifier. Hence the macro-examination of flow lines or fibre is a fairly simple matter. A section is first cut so that it will reveal fibre on a suitable face – in the case of the bolt mentioned above, a section cut symmetrically along its axis is used.

6.62 The section is then filed perfectly flat, or, better still, ground flat with a linishing machine. It is then rubbed on successively finer grades of emery paper, the 'grits' most generally used being '120', '180', '220', and '280'. Finer grades than these are not necessary, though the method of grinding

---

[1] Anisotropic – a substance having different physical properties in different directions.

requires a little care, to ensure that deep scratches are eliminated from the surface. This is achieved by grinding the surface such that, on passing from the linisher to the coarsest paper, the specimen is turned through 90°. The new set of grinding marks will then be at right angles to the previous ones, and in this way it is easy to see when the old grinding marks have been removed. The same procedure is adopted in passing from one paper to the next. Small specimens are ground by rubbing the specimen on paper supported on a piece of plate glass. Large specimens are more difficult to manipulate, and it is often more convenient to grind such specimens by rubbing a small area at a time, using a wooden block faced with emery paper.

6.63 When a reasonably smooth surface with no deep scratches has been produced, the specimen is wiped free of swarf and dust. It is then immersed in a 50 per cent solution of hydrochloric acid, and gently heated. Considerable effervescence takes place as hydrogen gas is liberated, and the surface layers of steel are dissolved. Those areas containing the most impurity tend to dissolve more quickly, and this helps to reveal the fibrous structure. The specimen is examined from time to time, being lifted from the solution with laboratory tongs. This deep-etching process may take up to twenty minutes or so, depending upon the quality of the steel.

6.64 When the flow lines are suitably revealed (Fig. 6.12) the specimen is washed in running water for a few seconds, and then dried as quickly as possible, by immersion in 'white' methylated spirit, followed by holding in a current of warm air from a hair-dryer.

Fig. 6.12 *The fibrous structure of a hot-forged component (6.50). The 'flow-lines' indicate the direction in which the metal moved during the forging operation. Etched in boiling 50 per cent hydrochloric acid for 15 minutes.*

6.65   If a section has been really deeply etched, it is possible to take an ink print of the surface. A blob of printer's ink is rolled on a flat glass plate, so that a thin but even film is formed on the surface of the rubber roller or squeegee used for this purpose. The roller is then carefully passed over the etched surface, after which a piece of paper is pressed on to the inked surface. Provided the paper does not slip in the process, a tolerable print of the flow lines will result.

Furnace atmospheres used during the heat-treatment of metals

6.70   Most common engineering metals become badly scaled during heat-treatment unless preventative precautions are taken. Scaling – that is, oxidation – occurs when a metal is heated to a temperature which allows it to react with any oxygen in the surrounding atmosphere. In many cases this can be prevented by burning the fuel gas with a restricted air supply so that *no free oxygen* remains in the gas which circulates within the furnace chamber (assuming that a fairly simple furnace is being used in which the burning gas is in direct contact with the charge).

The principal fuel gases are based on hydrocarbons, i.e. methane (natural gas) and propane. When these burn they produce mixtures of nitrogen (from the air used for combustion); carbon dioxide (CO_2); carbon monoxide (CO) and water vapour (H_2O), e.g.

$$\text{methane} \quad \boxed{\text{air}}$$
$$CH_4 + \boxed{2O_2 + 8N_2} = CO_2 + 2H_2O + 8N_2$$

The resultant gas mixture from the above reaction contains *no free oxygen*. Nevertheless at high temperatures some metals, including steel, will react with the carbon dioxide it contains. This carbon dioxide is acting here as an oxidizing agent by releasing some of its oxygen to oxidize the metal surface:

$$Fe + CO_2 = FeO + CO$$

Moreover, the carbon content of steel surfaces can also be decreased by a similar reaction:

$$C + CO_2 = 2CO$$

6.71   Inert atmospheres   Possibly the ideal furnace atmosphere for the prevention of surface deterioration would be the gas argon.[2] Unfortunately argon is too expensive to be used in the large volumes required, so nitrogen is the most widely used, reasonably inert gas. It can be produced by the decomposition of liquid ammonia. From the resultant nitrogen/hydrogen mixture, hydrogen is burned to form water vapour which is removed by condensation, leaving almost pure nitrogen but allowing about 1 per cent hydrogen to remain, thus ensuring that no excess oxygen has been admitted to the resulting mixture. This gas is still comparatively expensive to produce and is used mainly where *bright annealing* is required. Even nitrogen is reactive to some metals at high temperatures, notably titanium which must

---

[2]   The most plentiful of the inert gases in the atmosphere, argon is used to surround the tungsten filament of an electric lamp – such a heated filament would oxidize or burn out very quickly if surrounded by air.

be heat-treated in complete vacuum thus increasing its production costs even further.

6.72   Exothermic atmospheres are produced by *exothermic* combustion, that is, without the addition of external heat. Hydrocarbon gases are incompletely burned, the air supply being restricted so that no free oxygen remains in the atmosphere produced. Since the product contains large amounts of water vapour and carbon dioxide these are removed to give an atmosphere containing mainly carbon monoxide, hydrogen and nitrogen.

6.73   Endothermic atmospheres are provided by *endothermic* combustion, that is, external heat must be supplied. The initial gas, usually propane, is mixed with a controlled amount of air and passed through a heated gas generator. The so-called *carrier gas* (containing mainly carbon monoxide, hydrogen and nitrogen) produced is useful as an atmosphere in which many steels can be heat-treated without damage or change in carbon content of the surface. Moreover if extra propane is added to this carrier gas an atmosphere is produced which will *carburise* (14.23) the steel.

6.74   Whatever atmosphere is used to protect components from oxidation it is obvious that they must not be withdrawn from this atmosphere whilst still hot. Consequently furnaces must be either of the batch type (in which the charge cools down whilst still surrounded by a protective atmosphere), or of the continuous type (where the charge is carried on a moving hearth, through a cooling tunnel – both of which contain the protective atmosphere – before being discharged, cold, from the exit). Steel components to be hardened are tipped from the end of the moving hearth into a quench bath sited within the protective atmosphere.

CHAPTER 7

The mechanical shaping of metals

7.10 It has been suggested that the Romans conquered their known world with the aid of the bronze sword. Be that as it may it is certain that long before that time Man had found that hacking his enemy apart with sword and spear was a more efficient way of killing him than by the earlier method of clubbing him to death. The smith who shaped the sword and spear – and later the ploughshare[1] – was a very important member of society. Using goat skin bellows to raise the temperature of his hearth he was able to forge metals to a variety of shapes. Almost as soon as he discovered bronze he found that a tolerable cutting edge could be developed by cold-forging to harden the alloy.

7.11 Modern shaping processes can be divided into hot-working operations and cold-working operations. The former tend to be used wherever possible, since less power is required, and working can be carried out more rapidly. Cold-working processes, on the other hand, are used in the final stages of shaping some materials, so that a high-quality surface finish can be obtained, or suitable strength and hardness developed in the material.

Hot-working processes

7.20 A hot-working process is one which is carried out at a temperature above that of recrystallization for the material. Consequently, deformation and recrystallization take place at the same time, so the material remains malleable during the working process. Intermediate annealing processes are therefore not required, so working takes place very rapidly.

7.21 **Forging** The simplest and most ancient metal-working process is that of hand-forging, mentioned above. With the aid of simple tools called 'swages', the smith can produce relatively complex shapes, using either a hand- or a power-assisted hammer.

7.21.1 *Drop-forging* If large numbers of identically shaped components are required, then it is convenient to mass-produce them by drop-forging. A shaped two-part die is used, one half being attached to the hammer, whilst the other half is carried by a massive anvil. For complicated shapes, a series of dies may be required.

　　The hammer, working between two vertical guides, is lifted either

---

[1]　Early ones were of wood.

mechanically or by steam pressure, and is then allowed to fall, or is driven down (Fig. 7.1) on to the metal to be forged. This consists of a hot bar of metal, held on the anvil by means of tongs. As the hammer comes into contact with the metal, it forges it between the two halves of the die.

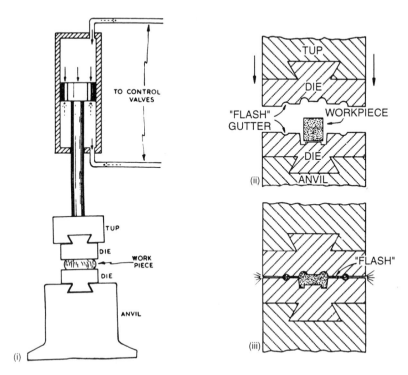

Fig. 7.1   *(i) Drop-forging using a double-acting steam hammer. (ii) A typical die arrangement. (iii) To ensure that the die is completely filled the work-piece must contain excess metal. This excess is forced out into the 'flash gutter' and is trimmed off.*

7.21.2   *Hot-pressing.* This is a development of drop-forging which is generally used in the manufacture of simpler shapes. The drop hammer is replaced by a hydraulically driven ram, so that, instead of receiving a rapid succession of blows, the metal is gradually squeezed by the static pressure of the ram. The downwards thrust is sometimes as great as 500 MN.

The main advantage of hot-pressing over drop-forging is that mechanical deformation takes place more uniformly throughout the work-piece, and is not confined to the surface layers, as it is in drop-forging. This is important when forging large components like marine propellor-shafts, which may otherwise suffer from having a non-uniform internal structure.

7.22   Hot-rolling Until the Renaissance in Europe, hand-forging was virtually the only method available for shaping metals. Rolling seems to have originated in France in about 1550, whilst in 1680 a sheet mill was in use in Staffordshire. In 1783, Henry Cort adapted the rolling mill for the production of wrought-iron bar. The introduction of mass-produced steel by Bessemer, in 1856, led to the development of bigger and faster rolling mills. The first reversing mill was developed at Crewe, in 1866, and this became standard equipment for the initial stages in 'breaking-down' large ingots to strip, sheet, rod, and sections.

Traditionally a steel-rolling shop consists of a powerful 'two-high' reversing mill (Fig. 7.2) to reduce the section of the incoming white-hot ingots, followed by trains of rolls which are either plain or grooved according to the product being manufactured. Now that the bulk of steel produced in the Developed World originates from continuous-cast ingots (3.21) the two-high *reversing* mill is no longer suitable. Instead the continuous ingot leaves the casting unit via a series of guide rolls which convey it to a train of reduction rolls followed by a set of finishing rolls. The finished strip passes into a coiling machine and is cut by a flying saw as required.

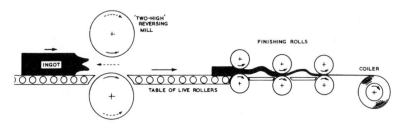

Fig. 7.2 *The rolling of steel strip. The ingot is first 'broken down' by the two-high reversing mill (the piped top is usually cropped after several passes through this mill). The work-piece is then delivered to the train of rolls which roll it down to strip.*

Hot-rolling is also applied to most non-ferrous alloys in the initial shaping stages but finishing is more likely to be a cold-working operation.

7.23   Extrusion The extrusion process is used for shaping a variety of both ferrous and non-ferrous alloys. The most important feature of the process is that, in a single operation from a cast billet, quite complex sections of reasonably accurate dimensions can be obtained. The billet is heated to the required temperature (350–500 °C for aluminium alloys, 700–800 °C for brasses, and 1100–1250 °C for steels), and placed in the container of the extrusion press (Fig. 7.3). The ram is then driven forward hydraulically, with sufficient force to extrude the metal through a hard alloy-steel die. The *solid* metal section issues from the die in a manner similar to the flow of toothpaste from its tube.

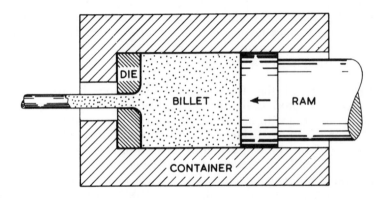

Fig. 7.3   *The extrusion process.*

A wide variety of sections can be extruded, including round rod, hexagonal brass rod (for parting off as nuts), brass curtain rail, small-diameter rod (for drawing down to wire), stress-bearing sections in aluminium alloys (mainly for aircraft construction), and tubes in carbon and stainless steels, as well as in aluminium alloys and copper alloys. Here some form of mandrel (7.32) is used to support the bore of the tube.

Cold-working processes

7.30   The surface of a hot-worked component tends to be scaled, or at least heavily oxidized; so it needs to be sand-blasted or 'pickled' in an acid solution if its surface condition is to be acceptable. Even so, a much better surface quality is obtained if the work-piece is cold-worked *after* being pickled. Consequently some degree of cold-work is applied to most wrought metallic materials as a final stage in manufacture. However, cold-working is also a means of obtaining the required mechanical properties in a material. By varying the amount of cold-work in the final operation, the degree of hardness and strength can be adjusted.

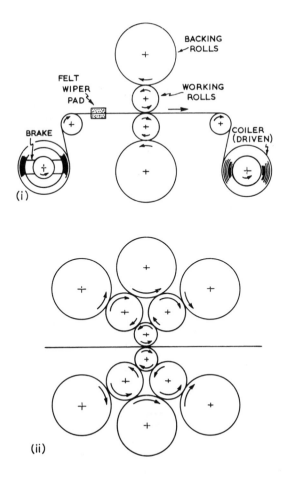

Fig. 7.4   *Roll arrangements in mills used for the production of thin foil. (i) A four-high mill. (ii) A 'cluster' mill.*

Moreover, some operations can be carried out only on cold metals and alloys. Those processes which involve *drawing* – or pulling – the metal must generally be carried out on cold material, since ductility is usually less at high temperatures. This is because tensile strength is reduced, so the material tears apart very easily when heated. So, whilst malleability is increased by rise in temperature, ductility is generally reduced. Hence there are more cold-working operations than there are hot-working processes, because of the large number of different final shapes which are produced in metallic materials of varying ductility. Finally, cold-working allows much greater accuracy of dimensions to be obtained in the finished material.

7.31   Cold-rolling Cold-rolling is used during the finishing stages in the production of both strip and section, and also in the manufacture of foils. The types of mill used in the manufacture of the latter are shown in Fig. 7.4. To roll very thin material, small-diameter rolls are necessary; and, if the material is of great width, this means that the working rolls must be supported by backing rolls, otherwise they will bend to such an extent that reduction in thickness of very thin material becomes impossible. For rolling thicker material, ordinary two-high mills are generally used. The production of mirror-finished metal foil necessitates the use of rolls with a highly polished surface; only by working in perfectly clean surroundings with highly polished rolls can really high-grade foil be obtained.

7.32   Drawing Drawing is exclusively a cold-working process, because it relies on high ductility of the material being drawn. Rod, wire, and hollow sections (tubes) are produced by drawing them through dies. In the manufacture of wire (Fig. 7.5), the material is pulled through the die by winding it on to a rotating drum or 'block'; whilst in the production of tube, the bore is maintained by the use of a mandrel. Rods and tubes are drawn at a long draw-bench, on which a power-driven 'dog' pulls the material through the die (Fig. 7.6). In each case, the material is lubricated with oil or soap before it enters the die aperture.

Drawing dies are made from high-carbon steel; from tungsten-molybdenum steels; from tungsten carbide; and, for very fine-gauge copper wire, from diamond.

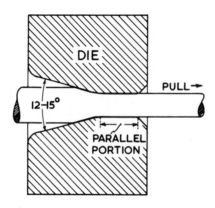

Fig. 7.5   *A wire-drawing die.*

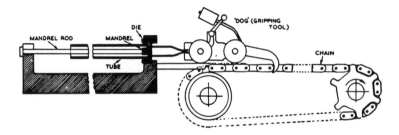

Fig. 7.6 *A draw-bench for the drawing of tubes. Rod could also be drawn at such a bench, the mandrel then being omitted.*

7.33   Cold-pressing and deep-drawing These processes are so closely allied to each other that it is often difficult to define each separately; however, a process is generally termed deep-drawing if some thinning of the walls of the component occurs under the application of tensile forces. Thus the operations range from making a pressing in a single-stage process to cupping followed by a number of drawing stages. In each case, the components are produced from sheet stock, and range from pressed mild-steel motor-car bodies to deep-drawn brass cartridge-cases (6.32), cupro-nickel bullet-envelopes, and aluminium milk-churns.

Only very ductile materials are suitable for deep-drawing. The best known of these are 70–30 brass (16.50), pure copper, cupro-nickel (16.80), pure aluminium and some of its alloys (17.50), and some of the high-nickel alloys. Mild steel of excellent deep-drawing quality is now being produced in

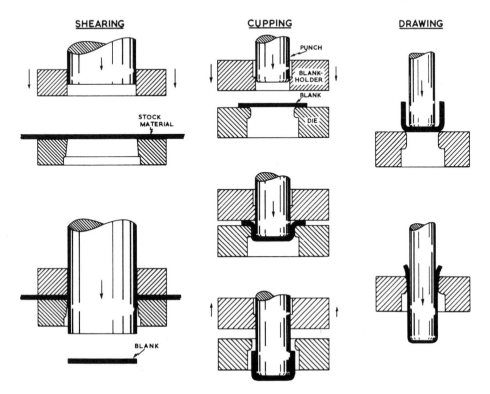

Fig. 7.7   *Stages in the deep-drawing of a component.*

increasing quantities by the 'oxygen processes' (11.20) and is used in a large number of motor-car parts as well as domestic equipment. It has in fact replaced much of the very expensive materials such as 70–30 brass for deep-drawn articles.

Typical stages in a pressing and deep-drawing process are shown in Fig. 7.7. Wall-thinning may or may not take place in such a process. If wall-thinning is necessary, then a material of high ductility must be used. Although Fig. 7.7 shows the processes of shearing and cupping being carried out on different machines, usually a combination tool is used, so that both processes take place on one machine.

Simple cold-pressing is widely used, and alloys which are not sufficiently ductile to allow deep-drawing are generally suitable for shaping by simple presswork. Much of the bodywork of a motor car is produced in this way, from mild steel.

7.34   Spinning This is one of the oldest methods of shaping sheet metal, and is a relatively simple process, in which a circular blank of metal is attached to the spinning chuck of a lathe. As the blank rotates, it is forced into shape by means of hand-operated tools of blunt steel or hardwood, supported against a fulcrum pin (Fig. 7.8). The purpose of the hand-tool is to press the metal blank into contact with a former of the desired shape. The former, which is also fixed to the rotating chuck, corresponds to the internal shape of the finished component, and may be made from a hardwood such as maple, or, in some cases, from metal. Formers may be solid; if the component is of re-entrant shape (as in Fig. 7.8), then the former must be segmented, to enable it to be withdrawn from the finished product. Adequate lubrication is necessary during spinning. For small-scale work, beeswax or tallow are often used, whilst for larger work, soap is the usual choice.

Large reflectors, components used in chemical plant, stainless-steel dairy-utensils, aluminium teapots and hot-water bottles, ornaments in copper and brass, and other domestic hollow-ware are frequently made by spinning.

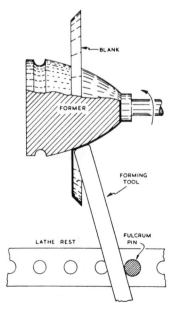

Fig. 7.8   *Spinning.*

A mechanized process somewhat similar to spinning, but known as 'flow-turning', is used for the manufacture of such articles as stainless-steel or aluminium milk-churns. In this process, thick-gauge material is made to flow plastically, by pressure-rolling it in the same direction as the roller is travelling, so that a component is produced in which the wall thickness is much less than that of the original blank (Fig. 7.9). Aluminium cooking utensils in which the base is required to be thicker than the side walls are made in this way.

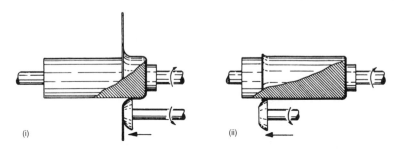

Fig. 7.9   *The principles of flow-turning.*

7.35   Stretch-forming Stretch-forming was introduced in the aircraft industry just before the Second World War, and soon became important to the production of metal-skinned aircraft. The process is also used in the coach-building trade, where it is one of the principal methods of forming sheet metal and sections.

In any forming process, permanent deformation can only be achieved in the work-piece if it is *stressed beyond the elastic limit*. In stretch-forming, this is accomplished by applying a tensile load to the work-piece such that the elastic limit is exceeded, and plastic deformation takes place. The operation is carried out over a form-tool or stretch-block, so that the component assumes the required shape.

In the 'rising-table' machine (Fig. 7.10), the work-piece is gripped between jaws, and the stretch-block is mounted on a rising table which is actuated by a hydraulic ram. Stretching forces of up to 4 MN are obtained with this type of machine. Long components, for example aircraft fuselage panels up to 7 m long and 2 m wide, are stretch-formed in a similar machine in which the stretch-block remains stationary, and the jaws move tangentially to the ends of the stretch-block.

Stretch-blocks are generally of wood or compressed resin-bonded

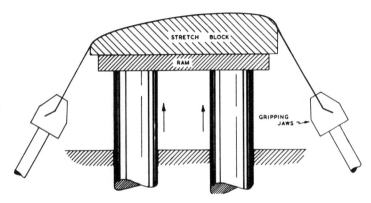

Fig. 7.10   *The principles of stretch-forming.*

plywoods, though other materials, such as cast synthetic resins, zinc-base alloys, or reinforced concrete, are also used. Lubrication of the stretch-block is, of course, necessary.

Although the process is applied mainly to the heat-treatable light alloys, stainless steel and titanium are also stretch-formed on a commercial scale.

7.36   Coining and embossing Coining is a cold-forging process in which deformation takes place entirely by compression. It is confined mainly to the manufacture of coins, medals, keys, and small metal plaques. Frequently, pressures in excess of 1500 N mm^{-2} are necessary to produce sharp impressions, and this limits the size of work which is possible.

The coining operation is carried out in a closed die (Fig. 7.11). Since no provision is made for the extrusion of excess metal, the size of the blanks must be accurately controlled to prevent possible damage to the dies, due to the development of excessive pressures.

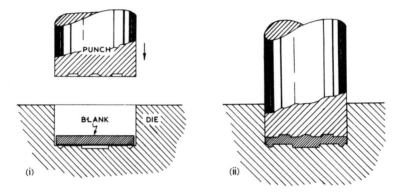

Fig. 7.11   *Coining.*

Embossing differs from coining in that virtually no change in thickness of the work-piece takes place during pressing. Consequently, the force necessary to emboss metal is much less than in coining, since little, if any, lateral flow occurs. The material used for embossing is generally thinner than that used for coining, and the process is effected by using male and female dies (Fig. 7.12). Typical embossed products include badges and military buttons.

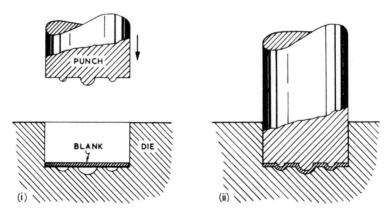

Fig. 7.12   *Embossing.*

7.37   Impact-extrusion Extrusion as a hot-working process was described earlier in this chapter (7.23). A number of cold-working processes also fall under the general description of extrusion. Probably the best-known of these is the method by which disposable collapsible tubes are manufactured. Such tubes were produced in lead, for containing artists' colours, as long ago as 1841. A few years later, similar tubes were produced in tin, but it was not until 1920 that the impact-extrusion of aluminium was established on a commercial scale.

Heavily built mechanical presses are used in the impact-extrusion of these collapsible tubes. The principles of die and punch arrangement are illustrated in Fig. 7.13. A small unheated blank of metal is fed into the die cavity. As the ram descends, it drives the punch very rapidly into the die cavity, where it transmits a very high pressure to the metal, which then immediately fills the cavity. Since there is no other method of exit, the metal is forced upwards through the gap between punch and die, so that it travels along the surface of the punch, forming a tube-shaped shell. The threaded nozzle of the collapsible tube may be formed during the impacting operation, but it is more usual to thread the nozzle in a separate process.

The impact-extrusion of tin and lead is carried out on cold metal, but aluminium blanks may be heated to 250°C for forming. Zinc, alloyed with 0.6 per cent cadmium and used for the extrusion of dry-battery shells, is first heated to 160°C, so that the alloy becomes malleable.

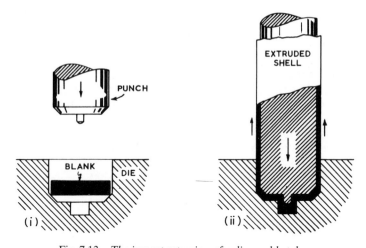

Fig. 7.13   *The impact extrusion of a disposable tube.*

Disposable tubes in lead, tin, and aluminium are used as containers for a wide range of substances, such as shaving cream, toothpaste, medicines, shoe-polish, adhesives, and condensed milk, though in recent years thermoplastics materials have replaced aluminium in many such containers. Impact extension is also used for the production of many other articles, principally in aluminium. These include canisters and capsules for food, medical products, and photographic films; and shielding cans for radio components.

7.38 The Hooker process (Fig. 7.14) closely resembles hot-extrusion, mentioned earlier, in so far as the flow of metal in relation to the die and punch is concerned. The Hooker process, however, is a cold-working operation of the impact type, and its products include small brass cartridge-

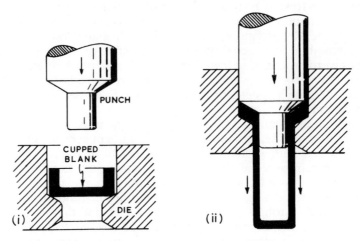

Fig. 7.14   *Impact extrusion by the Hooker process.*

cases, copper tubes for radiators and heat-exchangers, and other short tubular products. Flat 'slugs' are sometimes used in the Hooker process, but cupped blanks are usually considered to be more satisfactory. The blank is placed in the die; and, as the punch descends, metal is forced down between the body of the punch and the die, producing a tubular extrusion as shown.

Powder-metallurgy

7.40 Powder-metallurgy processes were originally developed to replace melting and casting for those metals – the so-called 'refractory' metals – which have very high melting-points. For example, tungsten melts at 3410 °C, and this is beyond the softening temperature of all ordinary furnace-lining materials. Hence tungsten is produced from its ore as a fine powder. This powder is then 'compacted' in a die of suitable shape at a pressure of approximately 1500 N mm^{-2}. Under such high pressure, the particles of tungsten become joined together by 'cold-welding' at the points of contact between particles.

7.41 The compacts are then heated to a temperature above that of recrystallization – about 1600 °C in the case of tungsten. This treatment causes recrystallization to occur, particularly in the highly deformed regions where cold-welding has taken place, and in this way the crystal structure becomes regular and continuous as grain-growth takes place across the original boundaries where cold welds have formed between particles. This heating process is known as 'sintering'.

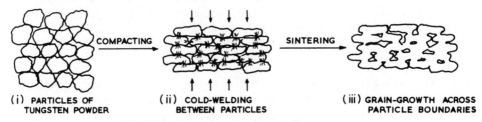

(i) PARTICLES OF TUNGSTEN POWDER   (ii) COLD-WELDING BETWEEN PARTICLES   (iii) GRAIN-GROWTH ACROSS PARTICLE BOUNDARIES

Fig. 7.15   *Stages in a powder-metallurgy process.*

At the end of this process, the slab of tungsten has a strong, continuous structure, though it will contain a large number of tiny cavities. It is then rolled, and drawn down to wire, which is used in electric lamp filaments. Most of the cavities are welded up by the working processes.

7.42　Although powder-metallurgy was originally used to deal with metals of very high melting-point, its use has been extended for other reasons, such as:

- to produce metals and alloys of *controlled* porosity, e.g. stainless-steel filters to deal with corrosive liquids, and also oil-less bronze bearings;
- to produce 'alloys' of metals which do not mix in the molten state, e.g. copper and iron for use as a cheap bearing material.
- for the production of small components such as the G-frame of a micrometer screw-gauge where the negligible amount of process scrap makes the method competitive.

7.43　One well-known use of powder-metallurgy is in the manufacture of cemented carbides for use as tool materials. Here, tungsten powder is heated with carbon powder at about 1500 °C, to form tungsten carbide. This is ground in a ball mill, to produce particles of very small size (about 20 μm), and the resultant tungsten carbide powder is mixed with cobalt powder, so that the particles of tungsten carbide become coated with powdered cobalt. The mixture is then compacted in hardened steel dies at pressures of about 300 N mm⁻², to cause cold-welding between the particles of cobalt. The compacts are then sintered at about 1500 °C, to cause recrystallization and grain growth in the cobalt, so that the result is a hard, continuous structure consisting of particles of very hard tungsten carbide in a matrix of hard, tough cobalt.

7.44　The principles involved in the manufacture of sintered-bronze bearings are slightly different. Here the main reason for using a powder-metallurgy process is to obtain a bearing with a controlled amount of porosity, so that it can be made to absorb lubricating oil. Powders of copper and tin are mixed in the correct proportions (about 90 per cent copper, 10 per cent tin), and are then compacted in a die of suitable shape (Fig. 7.16).

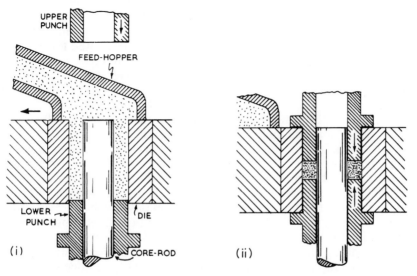

Fig. 7.16　*The compacting process used in the manufacture of an 'oil-less' bronze bearing-bush.*

The compacts are then sintered at 800 °C for a few minutes. This, of course, is *above* the melting-point of tin, so the process differs from true powder-metallurgy processes, in which no fusion occurs. As the tin melts, it percolates between the copper particles, and alloys with them to produce a continuous mass. The compact, however, still retains a large amount of its initial porosity; and, when it is quenched into lubricating oil, the latter is drawn into the pores of the bearing as it cools down. Sometimes the bearings are then placed under vacuum, whilst still in the oil bath. This causes any remaining air to be drawn out of the bearing, being replaced by oil when the pressure is allowed to return to that of the atmosphere.

7.45   The final structure resembles that of a metallic sponge which, when saturated with lubricating oil, produces a self-oiling bearing. In some cases, the amount of oil in the bearing lasts for the lifetime of the machine. Such bearings are used in the motor-car industry, but are also particularly useful in many domestic machines such as vacuum cleaners, refrigerators, electric clocks, and washing machines, in all of which long service with a minimum of attention is desirable.

Machining and machinability of metals

7.50   Metal-cutting processes are expensive so it is usual to shape a component to as near its final form as possible by casting, forging or powder-metallurgy processes. The need for accuracy of dimensions will nevertheless often demand some amount of machining. The ease with which a metal can be cut depends upon the design of tools, method of lubrication and so on but also upon the microstructure of the work-piece itself and it is this aspect which concerns us briefly here. After all, complete books are available dealing with the various techniques of metal-cutting processes!

7.51   Machining is a cold-working process in which the cutting edge of a tool forms shavings or chips of the material being cut. During cold-working processes heat is invariably generated as mechanical energy is being absorbed. In a machining process heat arises from this source as well as from energy dissipated by ordinary friction between tool and work-piece. Not only is energy wasted in this way but the heat generated tends to soften the cutting tool so that its 'edge' is lost. This involves expensive tool maintenance or the use of expensive high-speed steels (13.32) or ceramic tool materials (23.20).

7.52   As Fig. 7.17(i) indicates, ductile materials will generally machine badly. Not only does the material tend to tear under the pressure of the tool, giving a poor surface finish, but since the displaced metal 'flows' around the cutting edge heat generated by friction between the two will be great and the tool life short. Brittle materials on the other hand generally machine well, producing a fine powdery swarf (Fig. 7.17(ii)) in contrast to the curly slivers from a ductile metal. Unfortunately metals which are brittle overall are unsuitable for most engineering purposes but it is possible to improve machinability by introducing a degree of *local* brittleness in the material without impairing unduly its overall toughness. This local brittleness then causes minute chip cracks to form just in advance of the edge of the cutting tool.

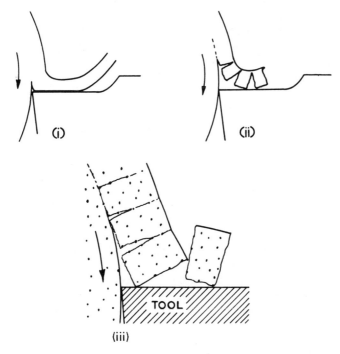

Fig. 7.17 *The influence of microstructure upon machinability. In (i) a soft ductile metal flows around the cutting tool and heat builds up at the cutting edge due to friction between the metal and the tool. (ii) Indicates a fragmented swarf from a brittle material with consequent lower friction losses. (iii) A typical free-cutting alloy containing isolated particles which initiate chip cracks in advance of the cutting edge.*

Such local brittleness can be imparted in a number of ways:

7.53   By the presence of a separate constituent in the microstructure. Small isolated particles in the structure behave as tiny holes and have the effect of setting up local stress concentrations just ahead of the advancing cutting edge. These stress raisers cause minute fractures to travel between the particles and the tool edge, thus reducing friction between tool and work-piece (Fig. 7.17(iii)). As well as a reduction in tool wear there will be a reduction in power consumption. Moreover since a fine powdery swarf is generated there will be a smaller remelting loss.

7.54   Some alloys contain microconstituents which put them in the 'free-cutting' category. The graphite flakes in grey cast iron (15.51) and the particles of hard compound in tin bronzes (16.60) are examples. Generally however small alloy additions are made to improve machinability in this way. Thus up to 2.0 per cent lead, which is insoluble in both liquid and solid copper alloys, is added to brasses and bronzes (16.51; 16.65) to improve machinability. Being insoluble this lead is present as tiny particles in the microstructure. 'Ledloy' steels contain no more than 0.2 per cent lead yet this improves machinability by up to 25 per cent whilst causing very little loss in mechanical properties due to the presence of the tiny lead globules.

7.55   In the cheaper free-cutting steels the presence of a high sulphur content is utilized by adding an excess of manganese so that isolated globules of manganese sulphide, MnS, are formed in preference to the brittle intercrystalline network of ferrous sulphide, FeS, which would otherwise be present. These small isolated globules of MnS, scattered throughout the structure, act as stress raisers during a cutting process. So, whilst a good-quality steel contains a maximum of only 0.06 per cent sulphur, a free-cutting steel may contain up to 0.3 per cent sulphur along with up to 1.5 per cent manganese to ensure its existence as MnS globules rather than as a brittle intercrystalline network of FeS.

7.56 By suitable heat-treatment of the material prior to the machining process. Low carbon-steels machine more easily when normalized (11.51). This process produces small patches of pearlite which break up the continuity of the structure in much the same way as MnS globules in a free-cutting steel. High-carbon steels can be spheroidized (11.52.2) to improve machinability. Again this has the effect of providing isolated globules which precipitate chip formation.

7.57 By cold-working the material before the machining operation. Materials supplied in rod or bar form for machining operations are generally cold-drawn or 'bright-drawn'. Free-cutting steels containing manganese sulphide are usually supplied as 'bright-drawn bar'. This treatment improves machinability by reducing the ductility of the surface layers, and introducing local brittleness there.

Metallurgical furnaces

7.60 These vary in size from small bench units used for the heat-treatment of dies and tools to the iron blast furnace some 60 m high (11.11). Designs also differ widely depending upon factors like fuel economy, operational temperature and chemical purity required in the charge. Either gas, oil, coke or electricity may be used as fuel. Metallurgical furnaces can be conveniently classified into three groups according to the degree of contact between the fuel and the charge:

7.61   Furnaces in which the charge is in direct contact with the burning fuel and its products of combustion Such furnaces include the blast furnace and foundry cupola. Due to intimate contact between the charge and burning fuel and the fact that operation is continuous, the fuel efficiency is very high *but* impurities in the fuel may contaminate the charge.

7.62   Furnaces in which the charge is out of contact with the fuel but is in contact with the products of combustion. This group includes a large variety of gas- and oil-fired installations used in the melting, smelting and heat-treatment of metals. Heat efficiency is still reasonably high but the charge may still be contaminated by gaseous impurities originating in the burning fuel.

7.63   Furnaces in which the charge is totally isolated from the fuel and from its combustion products Here we include crucible furnaces used for melting metals, the molten surface of which is often protected by a layer of molten flux; and gas- or oil-fired muffle furnaces where the solid charge is heated in a totally enclosed chamber. Fuel efficiency is obviously lower than in the previous groups but the purity of the charge can be maintained and there is complete control over the furnace atmosphere (6.70) so that bright annealing can be employed.

7.64   Electricity has the obvious advantage as a metallurgical fuel in that it can be used to generate heat either by electrical resistance methods; by high- or low-frequency alternating current; or by direct-arc methods and at the same time produce no chemical contamination of the charge; or of the natural environment! Unfortunately electricity is still rather more expensive than most other fuels.

CHAPTER 8

Alloys

8.10 I still have my grandmother's copper kettle. It was a wedding present from her young brother in 1877. At that time copper was used for the manufacture of kettles, pots and pans because it was ductile and reasonably corrosion-resistant. Now aluminium, unknown commercially when my grandmother was married, has replaced copper for the manufacture of such kitchenware. In the meantime the production of increasingly high-purity dead-mild steel – which is as near as we get to pure iron commercially – has confirmed its use in the manufacture of the bodywork of motor cars, domestic refrigerators, washing machines and a multitude of other items of everyday equipment.

In all of these products strength is developed in otherwise soft metals by cold-work. When greater strength or hardness is required it can only be achieved by alloying, that is by adding an element or elements which, by modifying the crystal structure in some way, will oppose the process of slip when stress is applied.

An alloy is a mixture of two or more metals, made with the object of improving the properties of one of these metals, or, in some cases, producing new properties not possessed by either of the metals in the pure state. For example, pure copper has a very low electrical resistance, and is therefore used as a conductor of electricity; but, with 40 per cent nickel, an alloy, 'Constantan', with a relatively high electrical resistance is produced. Again, pure iron is a ductile though rather weak material; yet the addition of less than 0.5 per cent carbon will result in the exceedingly strong alloy we call steel. In this chapter we shall examine the internal structures of different types of alloy, and show to what extent the structures of these alloys influence their mechanical properties.

8.11 Oil and water do not 'mix'. If we shake a bottle containing a quantity of each liquid droplets of each may form but they will soon separate into two distinct layers – oil floating on top of water. Alcohol and water however will 'mix' – or 'dissolve' – in each other completely in all proportions. Beer is a solution of about 5 per cent alcohol in water whilst whiskey is a solution of some 40 per cent alcohol in water. In each case various flavourings are also present but the result is a single homogeneous *solution* in the glass.

Molten metals behave in much the same way. Thus molten lead and molten zinc are, like oil and water, *in*soluble in each other and would form two separate layers in a crucible, molten zinc floating on top of the more dense molten lead. Clearly such metals as do not dissolve in each other as liquids are unlikely to form a useful alloy since in such an alloy the atoms of each metal must mingle intimately together. This can only occur when they form a homogeneous liquid solution. However it is to the manner in which this liquid solution subsequently solidifies that we must turn our attention.

Eutectics

8.20 Sometimes on solidification, the two metals cease to remain dissolved in one another, but separate instead, each to form its own individual crystals. In a similar way, salt will dissolve in water; though when the solution reaches its freezing-point, separate crystals of pure ice and pure salt are formed. Another fact we notice is that the freezing-temperature of the salt solution is much lower than that of pure water (which is why the Local Authority scatters damp salt on our roads after a night of frost). This phenomenon is known as 'depression of freezing-point', and it is observed in the case of many metallic alloys. Thus, the addition of increasing amounts of the metal cadmium to the metal bismuth will cause its freezing-point to be depressed proportionally; whilst, conversely, the addition of increasing amounts of bismuth to cadmium will have a similar effect on its melting-point, as shown in Fig. 8.1.

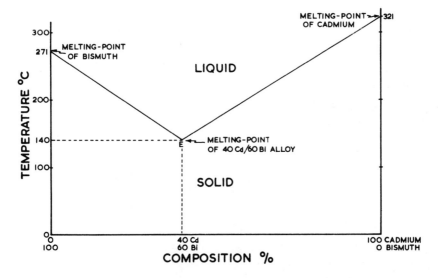

Fig. 8.1 *The freezing-points (melting-points) of both bismuth and cadmium are depressed by adding each to the other. A minimum freezing-point – or 'eutectic point' – is produced.*

8.21 It will be noticed that the two lines meet at the point *E*, corresponding to an alloy containing 60 per cent bismuth and 40 per cent cadmium. This alloy melts and freezes at a temperature of 140 °C, and is the lowest melting-point alloy which can be made by mixing the metals cadmium and bismuth. The point *E* is called the *eutectic*[1] *point*, and the composition 60 per cent bismuth/40 per cent cadmium is the *eutectic mixture*. If a molten alloy of this composition is allowed to cool, it will remain completely liquid until the temperature falls to 140 °C, when it will solidify by forming alternating thin layers of pure cadmium and pure

[1] Pronounced 'you-tek-tick', this word is derived from the Greek *eutektos*, meaning 'capable of being melted easily'.

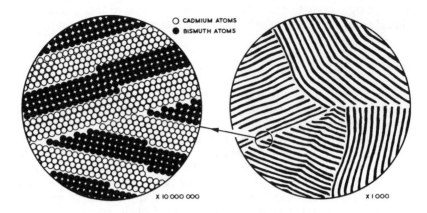

○ CADMIUM ATOMS
● BISMUTH ATOMS

X 10 000 000   X 1 000

Fig. 8.2   *If we had a microscope which gave a magnification of about ten million times, the arrangement of atoms of cadmium and bismuth would look something like that in the left-hand part of the diagram, except that the bands in the eutectic would each be many thousands of atoms in width.*

Fig. 8.3   *The dendritic structure of a cast alloy. This is a photomicrograph of cast 70–30 brass at a magnification of ×39. The dendrites would not be visible were it not for the coring of the solid solution (8.30).*

bismuth (Fig. 8.2) until solidification is complete. The metallic layers in this eutectic structure are extremely thin, and a microscope with a magnification of at least 100 is generally necessary to be able to see the structure.

Since the structure is laminated, something like plywood, the mechanical properties of eutectics are often quite good. For example, when the material forming one type of layer is hard and strong, whilst the other is soft and ductile, the alloy will be characterized by strength and toughness, since the strong though somewhat brittle layers are cushioned between soft but tough layers. Thus the eutectic in aluminium-silicon alloys (17.61) consists of layers of hard, brittle silicon sandwiched between layers of soft, tough aluminium, and the tensile strength of these cast aluminium-silicon alloys is much higher than that of pure aluminium.

Solid solutions

8.30   Sometimes two metals which are completely soluble in each other in the liquid state remain dissolved in each other during and after solidification, forming what metallurgists call a 'solid solution'. This is generally the case when the two metals concerned are similar in properties, and have atoms which are approximately equal in size. During solidification, the crystals which form are built from atoms of both metals. Inevitably, one of the metals will have a melting-point higher than that of the other, and it is reasonable to expect that this metal will tend to solidify at a faster rate than the one of lower melting-point. Consequently, the core of a resultant dendrite contains rather more of the metal with the higher melting-point; whilst the outer fringes of the crystal will contain correspondingly more of the metal with the lower melting-point (Fig. 8.4). This effect, known as 'coring', is prevalent in all solid solutions in the cast condition.

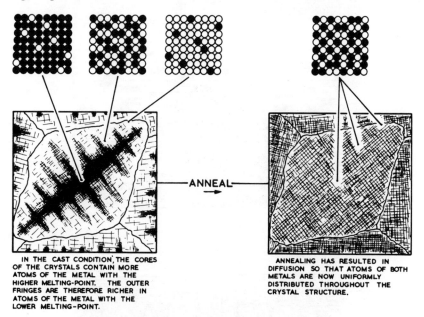

ANNEAL→

IN THE CAST CONDITION, THE CORES OF THE CRYSTALS CONTAIN MORE ATOMS OF THE METAL WITH THE HIGHER MELTING-POINT. THE OUTER FRINGES ARE THEREFORE RICHER IN ATOMS OF THE METAL WITH THE LOWER MELTING-POINT.

ANNEALING HAS RESULTED IN DIFFUSION SO THAT ATOMS OF BOTH METALS ARE NOW UNIFORMLY DISTRIBUTED THROUGHOUT THE CRYSTAL STRUCTURE.

● ATOMS OF THE METAL WITH THE HIGHER MELTING-POINT
○ ATOMS OF THE METAL WITH THE LOWER MELTING-POINT

Fig. 8.4   *The variations in composition in a cored solid solution. The coring can be dispersed by annealing.*

8.31 Many students appear to find the basic idea of solid solution one which is difficult to understand; so an analogy, in which bricks replace atoms as the building units, will be drawn here.

Suppose that, in building a high wall, a team of bricklayers is given a mixture of red and blue bricks with which to work. Further, let us suppose that, as building proceeds, each successive load of bricks which arrives at the site contains a slightly higher proportion of red bricks than the previous load. We will also assume that the 'brickies', being paid on a piece-work basis, lay whatever brick comes to hand first. Thus there will be no pattern in the laying of individual bricks; though, as the wall rises, there will be more red bricks and less blue ones in successive courses. By standing close to the wall, one would observe a small section, possibly like that shown in Fig. 8.5 (i), and this would give no indication of the overall distribution of red and blue bricks in the wall. On standing further away from the wall, the lack of any pattern in the laying of individual bricks would still be obvious, though the general relationship between the numbers of red and blue bricks at the top and bottom of the wall would become apparent (Fig. 8.5 (ii)). Provided that the red and blue bricks were of roughly equal size and strength, the wall would be perfectly sound, though it might look rather odd. Let us assume that we now view the wall from a distance of about half a kilometre. Individual bricks will now no longer be visible, though a gradual change in

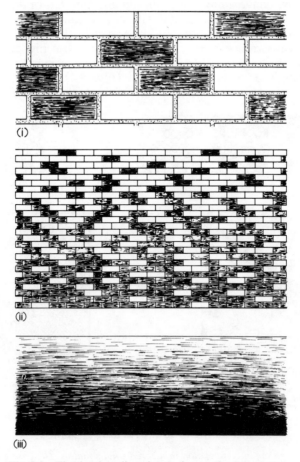

Fig. 8.5 *The 'brick-wall analogy' of a cored solid solution.*

colour from blue at the bottom, through various shades of purple, to red at the top will indicate the relative numbers of each type of brick at various levels in the wall (Fig. 8.5 (iii)).

8.32 In many metallic solid solutions the distribution of the two different types of atom follows a pattern similar to that of the bricks in the unlikely wall just described, that is, whilst the atoms in general conform to some overall geometrical arrangement – as do the bricks in the wall – there is usually no rule governing the order or frequency in which atoms of each type will be present within the overall pattern. The brick-wall analogy also illustrates the folly of viewing a microstructure at a high magnification, without first examining it with a low-power lens in the microscope. When using a high magnification, one will have only a restricted field of view of a very small part of the structure, so that no overall pattern is apparent; whereas the use of a low-power lens may reveal the complete dendritic structure, and show beyond doubt the nature of the material.

8.33 In the above discussion, it was assumed that the type of solid solution formed was one in which the atoms of two metals concerned were of roughly equal size; and this type of structure (Fig. 8.6 (i)) is termed a *substitutional* solid solution, since atoms of one metal have, so to speak, been substituted for atoms of the other. Many pairs of metals, including copper/nickel, silver/gold, chromium/iron, and many others, form solid solutions of this type in all proportions. A still greater number of pairs of metals will dissolve in each other in this way but in limited proportions. Notable examples include copper/tin, copper/zinc, copper/aluminium, aluminium/magnesium, and a host of other useful alloys.

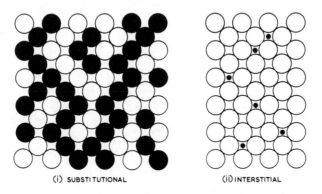

(i) SUBSTITUTIONAL   (ii) INTERSTITIAL

Fig. 8.6 *The two main types of solid solution: (i) a substitutional solid solution; (ii) an interstitial solid solution.*

8.34 There is, however, another type of solid solution, which is formed when the atoms of one element are so much smaller than those of the other that they are able to fit into the *interstices* (or spaces) between the larger atoms. Accordingly, this is known as an *interstitial solid solution* (Fig. 8.6 (ii)). Carbon dissolves in face-centred cubic iron in this way. Since the relatively small carbon atoms fit into the spaces between the much larger iron atoms, this explains why a piece of *solid* steel can be carburized (14.20) by being heated in the presence of carbon at a temperature high enough to make the steel face-centred cubic in structure. The carbon atoms 'infiltrate', so to speak, through the face-centred cubic ranks of iron atoms.

8.35   When a solid solution of either type is heated to a sufficiently high temperature so that atoms become thermally activated a process of *diffusion* takes place and atoms exchange positions so that coring gradually disappears and the structure becomes more uniform in composition throughout. This is achieved by atoms moving from one part of the crystal to another. It is easy to see that this can happen in an interstitial solution where tiny solute[2] atoms can move through the gaps between the large solvent atoms, but in a substitutional solid solution such movement would be impossible if crystal structures were as perfect as that shown in Fig. 8.6 (i). In the real world however crystal structures are never perfect and gaps exist where atoms or groups of atoms are missing. Examination of a piece of cast metal under the microscope will reveal many tiny cavities even in high-quality material. Each of these cavities represents some thousands of absent atoms so it is reasonable to assume that points where a single atom is missing are numerous indeed. Such *'vacant sites'* as they are called will allow the movement of individual atoms through the crystal structure of a substitutional solid solution.

It is assumed that the process of diffusion takes place by a series of successive 'moves' as suggested in Fig. 8.7. Since the solute atom will be either larger or smaller than the solvent atom it is likely to occur alongside a vacant site because in this way stress in the structure will be kept to a minimum. By following the series of six 'moves' shown in Fig. 8.7 the solute

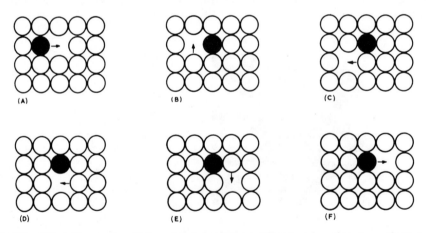

Fig. 8.7   *This is the way in which metallurgists believe diffusion takes place in a substitutional solid solution. A series of five 'moves' is necessary in order that the 'black' atom (and its accompanying 'vacant site') can go forward by one space.*

atom (black) advances to the right by one lattice space. The series of moves will of course be repeated every time the solute atom and its accompanying vacant site move forward by one lattice space.

The driving force behind this process of diffusion is really the lattice strains (and the stresses they produce) caused by irregularities in the lattice due to the presence of atoms of different sizes. During diffusion atoms move into positions of lower strain and consequently lower stress.

---

[2]   A solution consists of a *solute* dissolved in a *solvent*.

8.36 Solid solutions are the most important of metallic alloy structures since they provide the best combination of strength, ductility and toughness. The increase in strength is a result of distortions arising in the crystal structure when atoms of different sizes accommodate themselves into the same lattice pattern (Fig. 8.8). Slip (6.10) then becomes more difficult as dislocations must follow a less direct path and so a greater force must be used to produce it – which is another way of saying that the yield stress has increased. Since slip is nevertheless still ultimately able to take place, much of the ductility of the original pure metal is retained. Most of our useful metallic alloys are basically solid solution in structure.

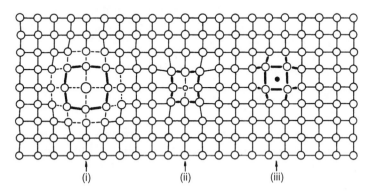

Fig. 8.8 *Crystal lattice distortions caused by the presence of solute atoms: (i) a large substitutional atom; (ii) a small substitutional atom; (iii) an interstitial atom. In each case the distortion produced will oppose the passage of a dislocation through the system.*

Intermetallic compounds

8.40 When heated, many metals combine with oxygen to form compounds which we call oxides, whilst some metals are attacked by sulphur gases in furnace atmospheres, to form sulphides. This is a general pattern in which metals (said to be electropositive elements) combine with non-metals (said to be electronegative elements) to form compounds which usually bear no physical resemblance to the elements from which they are formed. Thus, the very reactive metal sodium (which is silvery-white in appearance) combines with the very reactive gas chlorine (which is greenish in colour) to form sodium chloride – alias table-salt. Metallic oxides, sulphides and chlorides are of course typical ionic compounds (1.21) but sometimes two metals, when melted together, will combine to form a definite chemical compound called an *intermetallic compound*. This often happens when the two metals are very unlike in their physical and chemical properties, and when one metal is strongly electropositive and the other weakly electropositive.

8.41 When a solid solution is formed, it usually bears at least some resemblance to the parent metals, as far as colour and other physical properties are concerned. Thus, the colour of a low-tin bronze (16.60) is a blend of the colours of its parent metals, copper and tin, as one might expect; but if the amount of tin is increased, so that the alloy contains 66 per cent copper and 34 per cent tin, a hard and extremely brittle substance is produced, bearing no resemblance whatever to either copper or tin. What is more, this intermetallic compound – for such it is – is of a pale blue colour. This is due mainly to the fact that an intermediate compound generally crystallizes in a different pattern to that of either of the parent metals. This particular intermetallic compound has the chemical formula $Cu_{31}Sn_8$, which, like many such *intermetallics* does not obey the ordinary *valence laws* of chemistry. Moreover since an intermetallic compound is always of fixed composition, in common with all chemical compounds, there is never any coring in crystals of such a substance in the cast state.

8.42 Because of the excessive brittleness of most of these intermetallic compounds, they are used to only a limited extent as constituents of engineering alloys, and then only in the form of small, isolated particles in the microstructure. Since many intermetallic compounds are very hard, they also have very low coefficients of friction. Consequently, a typical use of these compounds is as a constituent in bearing metals (18.60). If present in an alloy in large amounts, an intermetallic compound will often form brittle intercrystalline networks. The strength and toughness of such an alloy would be negligible, and it would be of no use to the engineer.

Summary

8.50 In a study of metallurgy the term *phase* refers to any chemically stable, single homogeneous constituent in an alloy. Thus in a solid alloy a phase may be a solid solution, an intermetallic compound or, of course, a pure metal. A homogeneous liquid solution from which an alloy is solidifying also constitutes a phase. Any of the solid phases form the basic units of which metallic alloys are composed. It may be helpful therefore to summarize their properties:

- *Solid solutions* are formed when one metal is very similar to another, both physically and chemically, and is able to replace it, atom for atom, in its crystal structure. Alternatively, if the atoms of the second element are very small, they may be able to fit into the spaces between the larger atoms of the other metal.

  Solid solutions are stronger than pure metals, because the presence of atoms of the second metal causes some distortion of the crystal structure, thus making slip more difficult. At the same time, solid solutions retain much of the toughness and ductility of the original pure metal.
- *Intermetallic compounds* are formed by chemical combination, and the resultant substance generally bears little resemblance to its parent metals. Most intermetallic compounds are hard and brittle, and are of limited use only in engineering alloys.
- *Eutectics* are formed when two metals, soluble in each other in the liquid state, become insoluble in each other in the solid state. Then alternate layers or bands of each metal form, until the alloy is completely solid. This occurs at a fixed temperature, which is below the melting-points of either of the two pure metals.

  When two metals are only partially soluble in each other in the solid state, a eutectic may form consisting of alternate layers of two solid solutions. In some cases, a eutectic may consist of alternate layers of a solid solution and an intermetallic compound.

8.51 Engineering alloys often contain more than two elements and many contain six or even more, each contributing its own special effect. The structure of such an alloy is often much more complex. For example ternary or quaternary eutectics may be formed in which layers of three or four different phases are present. In useful alloys however, it is more likely that a uniform solid solution will be present, one metal acting as the solvent of all of the other additions. Thus the stainless steel 347S17 (Table 13.9) is composed almost entirely of a solid solution in which iron has dissolved 18 per cent chromium; 10 per cent nickel; 1 per cent niobium and 0.8 per cent manganese – a residual 0.04 per cent carbon existing as a few scattered undissolved carbide particles.

CHAPTER 9

Equilibrium diagrams

9.10   Among some engineering students the prospect of having to study the topic of equilibrium diagrams seems to be received with some dismay. Nevertheless, the topic need cause the reader no undue alarm, since for most purposes we can regard the equilibrium diagram as being no more than a graphical method of illustrating the relationship between the composition, temperature, and structure, or state, of any alloy in a series.

Much useful information can be obtained from these diagrams, if a simple understanding of their meaning has been acquired. For example, an elementary knowledge of the appropriate equilibrium diagram enables us to decide upon a suitable heat-treatment process to produce the required properties in a carbon steel. Similarly, a glance at the equilibrium diagram of a non-ferrous alloy system will often give us a pretty good indication of the structure – and hence the mechanical properties – a particular composition is likely to have. In attempting to assess the properties of an unfamiliar alloy, the modern metallurgist invariably begins by consulting the thermal equilibrium diagram for the series. There is no reason why the engineering technician should not be in a position to do precisely the same.

9.11   How are these equilibrium diagrams devised? Purely by a great deal of laborious routine laboratory work accompanied by some experience of the behaviour of metals in forming alloys, but, it is only possible to predict the general 'shape' of an equilibrium diagram with any certainty in the case of a limited number of alloy systems. Altogether there are some seventy different metallic elements and if these are taken in *pairs* to form *binary* alloys quite a large number of binary alloy systems is involved. Of course it would be extremely difficult to make alloys from some pairs of metals – for example high-melting point tungsten with very reactive caesium. Nevertheless a high proportion of the metallic elements have been successfully alloyed with each other and with some of the non-metallic elements like carbon, silicon and boron.

From Roman times until the middle of the twentieth century[1] much domestic and industrial pipework carrying water was of lead but, when its potentially toxic nature was realized, lead was replaced by copper and PVC. Nevertheless he who installs or repairs our pipework is still known as a 'plumber', though a plumber was originally one who worked in lead – the word being derived from the Latin *plumbum*, lead. In the days of lead

---

[1]   Omitting of course that period of almost a thousand years between the fall of Rome and the Renaissance.

piping the plumber joined lengths of pipe, or repaired fractures in them, using 'plumber's solder'. This contains roughly two parts by weight of lead to one of tin. On cooling, it begins to solidify at about 265 °C, but is not completely solid until its temperature has fallen to 183 °C (Fig. 9.1 (i)). It thus passes through a pasty, part solid/part liquid range of some 80 °C enabling the plumber to 'wipe' a joint using, traditionally, a moleskin 'glove' whilst the solder is in this 'mushy' state between 265 and 183 °C.

In the case of tinman's solder, used to join pieces of suitable metal, rather different properties are required. The solder must of course 'wet' (alloy with) the surfaces to be joined; but it will be an advantage if its melting-point is low, and, more important still, if it freezes quickly over a small range of temperature, so that the joint is less likely to be broken by rough handling in its mushy stage. A solder with these properties contains 62 per cent tin and 38 per cent lead. It freezes, as does a pure metal, at a single temperature – 183 °C in this case (Fig. 9.1 (ii)). Since the cost of tin is more than ten times that of lead, tinman's solder often contains less than the ideal 62 per cent. It will then freeze over a range of temperature which will vary with its composition. Thus, 'coarse' tinman's solder contains 50 per cent tin and 50 per cent lead. It begins to solidify at 220 °C, and is completely solid at 183 °C.

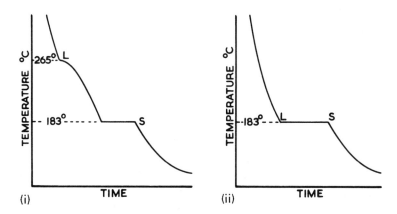

Fig. 9.1   *Cooling curves for plumber's solder and for tinman's solder.*

9.12   The freezing-range of any tin-lead solder can be determined by melting a small amount of it in a clay crucible, and then taking temperature readings of the cooling alloy every 15 seconds (Fig. 9.2). A thermocouple is probably the best temperature-measuring instrument to use for this; though a '360 °C' thermometer will suffice, provided it is protected by a fireclay sheath. Failure to use the latter will probably lead to the fracture of the thermometer as the solder freezes on to it, contracting in the process.

A temperature/time cooling curve can now be plotted in order to determine accurately the temperature at which freezing of the alloy begins (*L*) and finishes (*S*) (Fig. 9.1).

This procedure is repeated for a number of tin-lead alloys of different compositions. Representative values of *L* and *S* for some tin-lead alloys are shown in the following table.

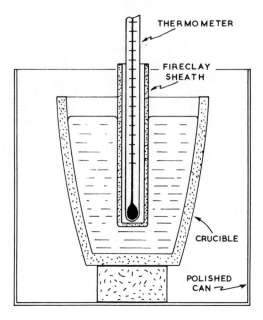

Fig. 9.2 *Simple apparatus for determining the freezing-range of a low-temperature alloy. The polished can prevents the alloy from cooling too quickly, and also protects the crucible from draughts.*

Composition Lead %	Tin %	Temperature at which solidification begins (L) °C	Temperature at which solidification ends (S) °C
67	33	265	183
50	50	220	183
38	62	183	183
20	80	200	183

The information obtained from the above table can now be plotted on a single diagram, as shown in Fig. 9.3, in order to relate freezing-range to composition of alloy.

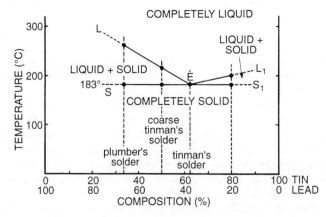

Fig. 9.3 *Part of the lead-tin equilibrium diagram. The limited information obtained from the cooling curves mentioned enables us to construct only so much of the diagram.*

The line LEL_1 (called the *liquidus*) joins all points (L) at which solidification of the various alloys begins; whilst the line SES_1 (called the *solidus*) joins all points (S) at which solidification of the alloys has finished. What we have plotted is only a *part* of the lead-tin thermal equilibrium diagram. The complete diagram contains other lines – or 'phase boundaries', as they are called. To determine these lines, other, more complex methods have to be used; but we are not concerned here with advanced metallurgical laboratory practice.

9.13   Even this restricted portion of the lead-tin equilibrium diagram provides us with some useful information. We can, for example, use it to determine the freezing-range of any alloy containing between 67 per cent lead/33 per cent tin and 20 per cent lead/80 per cent tin. Thus, reading from the diagram (Fig. 9.3), a solder containing 60 per cent lead/40 per cent tin would solidify between 245 and 183 °C. Similarly given the composition of an alloy and its temperature at any instant we can determine the state in which it exists. For example, an alloy containing 55 per cent lead/45 per cent tin at a temperature of 200 °C, will be in a pasty, part solid/part liquid state; whilst the same alloy at 250 °C will be completely molten. Conversely, when cooled below 183 °C, it will be completely solid.

Types of equilibrium diagram

9.20   Generally speaking a useful alloy will only be formed when two metals are completely soluble in each other in the liquid state, that is they form a single homogeneous liquid in the crucible. Thus, attempts to produce a lead-zinc alloy will fail because lead and zinc are only slightly soluble in each other as liquids. When melting a mixture of, say, equal parts of lead and zinc we find a layer of zinc-rich liquid solution floats on top of a layer of lead-rich liquid solution – rather like oil on water. On casting this mixture the two separate layers would form within the mould and solidify as such. Consequently complete mutual solubility in the liquid state is a prerequisite to producing alloys by the traditional melting-casting route.

9.21   There are a number of different types of thermal equilibrium diagrams, but we need deal with only three of them, namely those in which the characteristics of the diagram are governed by the extent to which one metal forms a solid solution with the other. The possibilities are that:

1.   the two metals are completely soluble in each other in all proportions in the solid state,
2.   the two metals are completely insoluble in each other in the solid state,
3.   the two metals are partially soluble in each other in the solid state.

Strictly speaking, equilibrium diagrams indicate only microstructures which will be produced when alloys cool under equilibrium conditions, and in most cases that means extremely slowly. Under industrial conditions, alloys often solidify and cool far too rapidly for equilibrium to be reached, and, as a result, the final structure deviates considerably from that shown by the diagram. The coring of solid solutions mentioned in the previous chapter (8.30) is a case in point, and will be discussed further in the section which follows.

An alloy system in which the two metals are soluble in each other in all proportions in both liquid and solid states

9.30   An example of this type of system is afforded by the nickel-copper alloy series. Atoms of nickel and copper are of approximately the same size, and, since both metals crystallize in similar face-centred cubic patterns (2.12), it is not surprising that they should form mixed crystals of a substitutional solid-solution type when a liquid solution of the two metals solidifies. The resulting equilibrium diagram (Fig. 9.4) will have been derived from a series of cooling curves, as described earlier in this chapter, except that a pyrometer capable of withstanding high temperatures would be required for taking the temperature measurements.

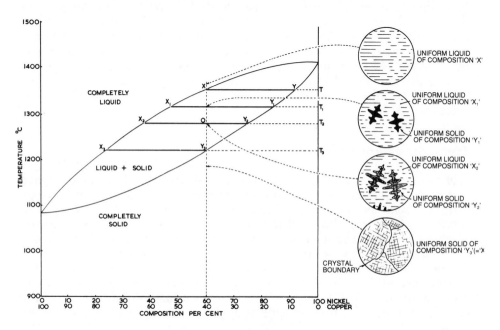

Fig. 9.4   *The nickel-copper thermal equilibrium diagram. The solidification of an alloy under conditions of equilibrium (slow cooling) is illustrated.*

9.31   This equilibrium diagram consists of only two lines:

- the upper, or *liquidus*, above which any point represents in composition and temperature an alloy in the completely molten state; and
- the lower, or *solidus*, below which any point represents in composition and temperature an alloy in the completely solid state.

Any point between the two lines will represent in composition and temperature an alloy in the pasty or part solid/part liquid state. From the diagram, we can read not only the compositions of the solid part and liquid part respectively, but also determine the relative proportions of the solid and liquid material.

9.32   Let us consider what happens when an alloy (*X*), containing 60 per cent nickel and 40 per cent copper, cools and solidifies *extremely slowly*, so that its structure is able to reach equilibrium at every stage of the process.

Solidification will begin when the temperature falls to T (the vertical line representing the composition X and the horizontal line representing the temperature T intersect on the liquidus line). Now, it is a feature of equilibrium diagrams that, when a horizontal line representing some temperature cuts two adjacent phase-boundaries in this way, the compositions indicated by those two intersections can exist in equilibrium together. In this case, liquid solution of composition X can exist in equilibrium with solid solution of composition Y at the temperature T. Consequently, when solidification begins, crystal nuclei of composition Y begin to form.

9.33   Since the solid Y contains approximately 92 per cent nickel/8 per cent copper (as read from the diagram), it follows that the liquid which remains will be less rich in nickel (but correspondingly richer in copper) than the original 60 per cent nickel/40 per cent copper composition. In fact, as the temperature falls slowly, solidification continues; and *the composition of the liquid changes along the liquidus line, whilst the composition of the solid changes – by means of diffusion* (8.35) *– along the solidus line*. Thus, by the time the temperature has fallen to T_1, the liquid solution has changed in composition to X_1, whilst the solid solution has changed in composition to Y_1. At a lower temperature – say T_2 – solidification will have progressed still further, and the composition of the liquid will have changed to X_2, whilst the composition of the corresponding solid will have changed to Y_2.

9.34   Clearly, since the alloy is gradually solidifying so that the *proportions* of solid and liquid are continually changing, the *compositions* of solid and liquid must also change, because the overall composition of the alloy as a whole remains at 60 per cent nickel/40 per cent copper throughout the process. The relative weights of solid and liquid – as well as their compositions – can be obtained from the diagram, assuming that the alloy is cooling slowly enough for equilibrium to be attained by means of diffusion. Thus, at temperature T_2,

weight of liquid (composition X_2) $\cdot OX_2$
$$= \text{weight of solid (composition } Y_2) \cdot OY_2$$

This is commonly referred to as the *lever rule*. Engineers will appreciate that this is an apt title, since, in this particular case, it is as though moments had been taken about the point O (the overall composition of the alloy). We will now substitute actual values (read from the equilibrium diagram, Fig. 9.4) in the above expression. Then,

weight of liquid (38% nickel/62% copper) $\cdot (60-38)$
$$= \text{weight of solid (74\% nickel/26\% copper)} \cdot (74-60)$$

or

$$\frac{\text{weight of liquid (38\% nickel/62\% copper)}}{\text{weight of solid (74\% nickel/26\% copper)}} = \frac{(74-60)}{(60-38)} = \frac{14}{22}$$

Thus, assuming that the alloy is cooling slowly, and is therefore in equilibrium, we can obtain the above information about it at the temperature T_2 (1280 °C).

The solidification process will finish as the temperature falls to T_3. Here the last trace of liquid (X_3) will have been absorbed into the solid solution, which, due to diffusion, will now be of uniform composition Y_3.

9.35   Composition Y_3 is of course the same as X – the composition of the original liquid. Obviously, it cannot be otherwise if the solid Y_3 has become uniform throughout due to diffusion. 'Why go through all this complicated procedure to demonstrate an obvious point?', the reader may ask. Unfortunately – whether in engineering or in other branches of applied science – it is not often possible to make a straightforward application of a simple scientific principle. Influences of other variable factors usually have to be taken into account, such as the effects of friction in a machine, or of the pressure of wind in a civil engineering project.

9.36   In the above description of the solidification of the 60 per cent nickel/40 per cent copper alloy, we have assumed that diffusion has taken place completely, resulting in the formation of a *uniform* solid solution. Under industrial conditions of relatively rapid cooling this is rarely possible; there just isn't time for the atoms to 'jiggle' around as described in the previous chapter (8.35). Consequently, the composition of the solid solution always lags behind that indicated by the equilibrium diagram for some particular temperature, and this leads to some residual coring in the final solid. If the

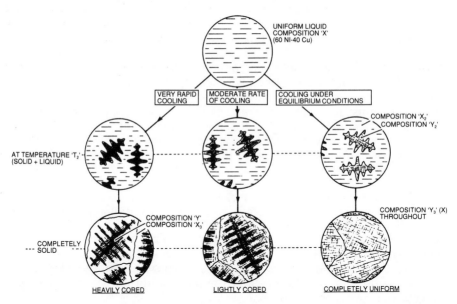

Fig. 9.5 *Illustrating the effects of cooling-rate on the extent of coring in the 60 per cent nickel/40 per cent copper alloy dealt with above.*

rate of solidification has been very rapid, the cores of the crystals may be of a composition almost as rich in nickel as Y; whilst the outer fringes of the crystals may be of a composition somewhere in the region of X_3. Slower rates of solidification will lead to progressively lesser degrees of coring, as, under these circumstances, the effects of diffusion make themselves felt. Alternatively, if this 60 per cent nickel/40 per cent copper alloy were annealed for some hours at a temperature just below T_3, that is, just below the solidus temperature, any coring would be dissipated by diffusion.

9.37   In this section we have been dealing with different modes of solidification of a 60 per cent nickel/40 per cent copper alloy. Since nickel and copper are soluble in each other in all proportions in the solid state, any other alloy composition of these two metals will solidify in a similar manner.

An alloy system in which the two metals are soluble in each other in all proportions in the liquid state, but completely insoluble in the solid state

9.40   In this case, the two metals form a single homogeneous liquid when they are melted together, but on solidification they separate again and form individual crystals of the two *pure* metals. Cadmium and bismuth form alloys of this type. Both metals have low melting-points, but, whilst cadmium is a malleable metal used to some extent for electroplating, bismuth is so brittle as to be useless for engineering purposes. It should be noted that the name 'bismuth' is often used to describe a compound of the actual metal which is sometimes used in medicine.

9.41   Again, the equilibrium diagram (Fig. 9.6) consists of only two boundaries: the liquidus *BEC*, and the solidus *AED* – or, more properly, *BAEDC*. As in the previous case, any point above *BEC* represents in composition and temperature an alloy in the completely molten state; whilst any point below *AED* represents an alloy in the completely solid state. Between *BEC* and *AED*, any point will represent in composition and temperature an alloy in the part liquid/part solid state.

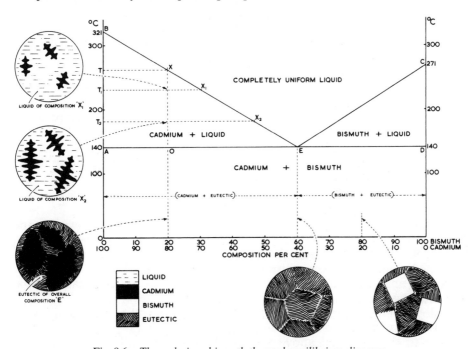

Fig. 9.6   *The cadmium-bismuth thermal equilibrium diagram.*

9.42   Let us consider a molten alloy of composition *X*, containing 80 per cent cadmium and 20 per cent bismuth. This will begin to solidify when the temperature falls to *T* (Fig. 9.6). In this case, the appropriate 'temperature horizontal' through *T* cuts that part of the solidus *BA* which represents a composition of 100 per cent cadmium. Consequently, nuclei of pure cadmium begin to crystallize. As a result, the remaining molten alloy is left less rich in cadmium and correspondingly richer in bismuth; so, as the temperature falls and cadmium continues to solidify, the liquid composition follows the liquidus line from X to X_1. This process continues, and, by the time the temperature has fallen to T_2, the remaining liquid will be of composition X_2.

The crystallization of pure cadmium continues in this manner until the temperature has fallen to 140 °C (the final solidus temperature), when the remaining liquid will be of composition E (40 per cent cadmium/60 per cent bismuth). Applying the lever rule at this stage:

$$\text{weight of pure cadmium} \cdot AO = \text{weight of liquid (composition } E) \cdot OE$$

or $\quad \dfrac{\text{weight of pure cadmium}}{\text{weight of liquid (composition } E)} = \dfrac{OE}{AO} = \dfrac{(60-20)}{(20-0)} = \dfrac{40}{20} = 2$

Thus, there will be twice as much solid cadmium by weight as there is liquid at this stage.

9.43   The two metals are now roughly in a state of equilibrium in the remaining liquid, which is represented in composition and temperature by the point E (the eutectic point). Until this instant, the liquid has been adjusting its composition by rejecting what are usually called 'primary' crystals of cadmium. However, due to the momentum of solidification, a little too much cadmium solidifies, and this causes the composition of the liquid to swing back across point E, by depositing for the first time a thin layer of bismuth. Since this now upsets equilibrium in the other direction, a layer of cadmium is deposited, and so the liquid composition continues to swing to-and-fro across the eutectic point, by depositing alternate layers of each metal until the liquid is all used up. This see-saw type of solidification results in the laminated structure of eutectics (8.21), and takes places whilst the temperature remains constant – in this case at the eutectic temperature of 140 °C.

9.44   It was shown above that, just before solidification of the eutectic began,

$$\text{weight of pure cadmium} = 2 \cdot (\text{weight of liquid composition } E)$$

Since the liquid of composition E has changed to eutectic, it follows that the final structure will contain two parts by weight of primary cadmium to one part by weight of eutectic. By the same reasoning, an alloy containing 70 per cent cadmium/30 per cent bismuth will contain equal parts of primary cadmium and eutectic; whilst an alloy containing 40 per cent cadmium/60 per cent bismuth will consist entirely of eutectic.

Alloys containing more than 60 per cent bismuth will begin to solidify by depositing primary crystals of bismuth, and the general procedure will be similar to that outlined above for the cadmium-rich alloy. Whatever the *overall* composition of the alloy, the eutectic it contains will always be of the same composition; that is, 40 per cent cadmium/60 per cent bismuth. If the overall composition contains more than 40 per cent cadmium, then some primary cadmium must deposit first; whilst if the overall composition has less than 40 per cent cadmium, then some primary bismuth will deposit first.

9.45 It must be admitted here that *complete insolubility* in the solid state probably does not exist in metallic alloys – there is always some small amount of one metal dissolved in the other. However the solubility of solid cadmium and bismuth in each other is so small that their equilibrium system is used here to illustrate this case which serves as a useful introduction to the more general case which follows (9.50). Some older textbooks in fact use the lead-antimony system as an example of complete insolubility in the solid state whereas quite high mutual solubility of several per cent is now known to exist between these two metals.

An alloy system in which the two metals are soluble in each other in all proportions in the liquid state, but only partially soluble in each other in the solid state

9.50 This is a situation intermediate between the two previous cases, that is complete solid solubility in all proportions on the one hand and total insolubility on the other. As might be expected very many alloy systems fall into this category. In the early paragraphs of this chapter, part of the lead-tin thermal equilibrium system was used to give a general idea of a method by which these thermal equilibrium diagrams can be produced. We shall now explore the lead-tin system more fully, by reference to the complete equilibrium diagram (Fig. 9.7).

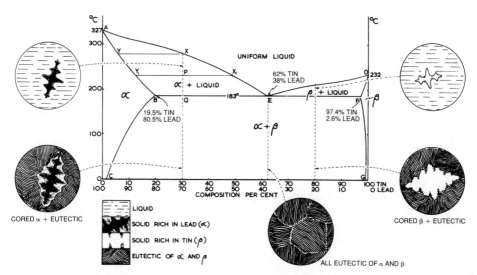

Fig. 9.7   *The lead-tin thermal equilibrium diagram, showing the effects of rapid cooling on representative microstructures.*

This diagram indicates that at 183 °C lead will dissolve a maximum of 19.5 per cent tin in the solid state, giving a solid solution designated α;[2] whilst, at the same temperature, tin will dissolve a maximum of 2.6 per cent lead, forming a solid solution β. Any alloy whose composition falls between *B* and *F* will show a structure consisting of primary crystals of either α or β, and also some eutectic of α and β. The overall composition of the eutectic

---

[2]   Metallurgists use letters of the Greek alphabet to indicate different solid phases occurring in an alloy system.

part of the structure will be given by E. In fact, an alloy containing precisely 62 per cent tin and 38 per cent lead will have a structure which is entirely eutectic, consisting of alternate layers of α and β.

9.51   Let us consider what happens when an alloy containing 70 per cent lead and 30 per cent tin solidifies, and cools *slowly* to room temperature. Solidification will commence at X (at about 270 °C), when nuclei of α (composition Y) begin to form. By the time the temperature has fallen to, say, 220 °C, the α will have changed in composition to Y_1, due to diffusion (8.35), whilst the remaining liquid will be of composition X_1. By applying the lever rule we have that, at 220 °C,

weight of solid α (composition Y_1)$\cdot Y_1 P$
$$= \text{weight of liquid (composition } X_1) \cdot PX_1$$

or $\qquad \dfrac{\text{weight of solid } \alpha \text{ (composition } Y_1)}{\text{weight of liquid (composition } X_1)} = \dfrac{PX_1}{Y_1 P}$

Similarly, when the temperature has fallen to 183 °C, we have α (now of composition B) and some remaining liquid (of composition E) in a ratio given by:

$$\frac{\text{weight of solid } \alpha \text{ (composition } B)}{\text{weight of liquid (composition } E)} = \frac{QE}{BQ}$$

At a temperature just below 183 °C, the remaining liquid solidifies as a eutectic, by depositing alternate layers of α (composition B) and β (composition F), the *overall* composition of this eutectic being given by E. Thus the structure, represented by a point just below Q, will consist of primary crystals of α of uniform composition B, surrounded by a eutectic mixture of α (composition B) and β (composition F).

9.52   In this diagram, we have two phase boundaries of a type not previously encountered in our studies; namely the *solvus* lines BC and FG. These boundary lines separate phase fields in which only solid phases exist, and therefore denote microstructural changes which occur *after an alloy is completely solid*. The slope of BC indicates that, as the temperature falls, the solubility of solid tin in solid lead will diminish from 19.5 per cent at 183 °C to about 2 per cent at 0 °C (point C); and similarly the slope of FG indicates that the solubility of solid lead in solid tin will fall from 2.6 per cent at 183 °C to less than 1 per cent at 0 °C (point G). Consequently, as our 70 per cent lead/30 per cent tin alloy cools slowly from 183 °C to room temperature, the composition of any α in the structure alters along BC, whilst the composition of any β will alter along FG.

In practice, such an alloy will never cool slowly enough for equilibrium to be reached at each stage of the process, and some coring will inevitably occur, particularly in the crystals of primary solid solution. Accordingly, the sketches representing microstructures in Fig. 9.7 assume that cooling has been fairly rapid, and that considerable coring has occurred as a result. Extremely slow cooling (as outlined above) or prolonged annealing eventually eliminates coring.

Precipitation from a solid solution

9.60   In the preceding section, the significance of the sloping solvus lines – BC and FG (Fig. 9.7) – was mentioned. It has been my experience that many students find it difficult to accept that changes in solid solubility can occur *after* an alloy has become completely solid. I would refer them to 8.35 for a suggestion of how this can take place. However, this change in solid solubility as temperature is raised or lowered is a very important feature involved in the heat-treatment of alloys so it will be discussed further here. Amongst other phenomena it explains largely how the mechanical properties of some aluminium alloys can be changed by 'precipitation hardening' (17.70). This was still very much a mystery (then known as 'age hardening') during my student days – and not only to us students!

9.61   Let us first consider a parallel case concerning the solubility of a salt in water. If some of the salt is put into water in a beaker, and stirred, much of the salt may dissolve; but some solid may remain at the bottom of the beaker. We thus have two phases in the beaker – a *saturated* solution, and some solid salt. If we now gently warm the solution, more and more salt will dissolve, until only solution remains. At a higher temperature still, the solution would dissolve more salt, if it were available in the beaker. The solution is therefore said to be *unsaturated*, and only a single phase remains in the beaker – the unsaturated solution.

9.62   It is quite easy to plot a curve showing the variation in the solubility of the salt with a rise in temperature. Such a curve is shown in Fig. 9.8. It indicates that, as the temperature increases, so does the solubility of salt in water.

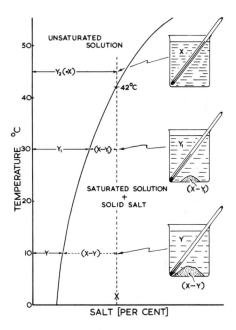

Fig. 9.8   *The solubility curve for salt in water. Solubility increases as the temperature increases.*

Suppose we add some salt to pure water, so that X denotes the total percentage of salt present. After mixing the two together at, say, 10 °C, we shall find that we still have a quantity of solid salt remaining. In fact, the solubility diagram tells us that Y per cent of salt has actually dissolved, giving a saturated solution at that temperature, and that $(X-Y)$ per cent salt remains at the bottom of the beaker. If we now warm the beaker slowly to 30 °C, we shall find that more salt dissolves, and the solubility diagram confirms that the amount in solution has increased to Y_1.

At, say, 40 °C, we would find that a very small quantity of solid salt remained, and at a slightly higher temperature (42 °C) this would just dissolve. If the temperature were raised to, say, 45 °C, the solution would then be unsaturated – that is, it would dissolve more salt at that temperature, if solid salt were added to the beaker.

9.63   This process of solution is reversible, and, if we allow the beaker to cool slowly, tiny crystals of salt will precipitate when the temperature has fallen a little below 42 °C. These crystals will increase in size as the temperature falls; and by the time 10 °C has been reached, we shall again have an amount Y left in solution, and $(X-Y)$ as solid crystals at the bottom of the beaker.

9.64   In the above case, we have been dealing with a *liquid* solution of salt in water, but exactly the same principles are involved if we consider instead a *solid* solution of, say, copper in aluminium. Naturally, in the case of a solid solution, the reversible process of solution and precipitation will take place much more slowly, since the individual atoms in a metallic structure are not able to move about as freely as the particles of salt and water in a beaker.

9.65   The aluminium-rich end of the aluminium-copper thermal equilibrium diagram is shown in Fig. 9.9. The sloping phase boundary AB shows that the solubility of *solid* copper in *solid* aluminium increases from 0.2 per cent at 0 °C (at A) to 5.7 per cent at 548 °C (at B). Any point to the left of AB will represent in composition and temperature an unsaturated solid solution (α) of copper in aluminium; whilst any point to the right of AB will represent in

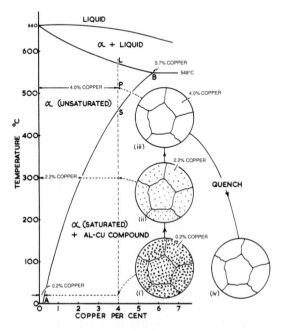

Fig. 9.9   *The solubility curve for copper in aluminium. This is the aluminium-rich end of the aluminium-copper thermal equilibrium diagram.*

composition and temperature a structure consisting of saturated solid solution α, along with some excess aluminium-copper compound.

9.66 We will consider an alloy containing 4 per cent copper, since this forms the basis of the well-known aluminium-copper alloy duralumin (17.70). If this has been allowed to cool very slowly to room temperature, its structure will have reached equilibrium, and is represented by diagram (i) (Fig. 9.9). This consists of solid solution α, which at room temperature contains only about 0.2 per cent of dissolved copper, the remainder of the 4 per cent copper being present as particles of the aluminium-copper intermetallic compound scattered throughout the structure.

Suppose this alloy is now heated slowly. As the temperature rises, the solid aluminium-copper compound gradually dissolves in the solid solution α, by means of a process of diffusion (8.35). At, say, 300 °C, the solid solution α will already have absorbed about 2.2 per cent copper, and for this reason there will be less of the intermetallic compound left in the structure (ii).

At about 460 °C (point S), the solution of the intermetallic compound will be complete (iii), the whole of the 4 per cent copper now being dissolved in the solid solution α. In practice, the alloy will be heated to about 500 °C (point P), in order to ensure that all of the intermetallic compound has been absorbed by the solid solution α. (Care must be taken not to heat the alloy above the point L, since at this point it would begin to melt).

The alloy is maintained at 500 °C for a short time, so that its solid solution structure can become uniform in composition. It is then quickly removed from the furnace, and immediately quenched in cold water. As a result of this treatment, *the rate of cooling will be so great that particles of the intermetallic compound will have no opportunity to be precipitated.* Therefore, at room temperature we shall have a uniform structure of α solid solution; though normally, and with slow cooling, an equilibrium structure consisting of almost pure aluminium (composition A) along with particles of the intermetallic compound would be formed (i). Quenching, however, has prevented equilibrium from being attained. Hence the quenched structure is *not* an equilibrium structure, and is in fact a *super-saturated* solid solution, since α contains much more dissolved copper than is normal at room temperature.

This treatment forms the basis of the first stage in the heat-treatment of duralumin-type alloys, and will be dealt with in detail in Chapter 17.

Ternary equilibrium diagrams

9.70 In this chapter, we have been dealing only with equilibrium diagrams which represent *binary* systems – that is, containing *two* metals. If an alloy contains *three* metals, this will introduce an additional variable quantity for our consideration, since the relative amounts of any two of the three metals can be altered independently. Temperature remains the other 'variable'. Thus we have a system with a total of three variables, and this can be represented graphically only by a three-dimensional system. Such a 'solid' diagram (Fig. 9.10) will consist of a base in the form of an equilateral triangle, each point of the triangle representing 100 per cent of one of the three metals (in this case cadmium, tin, and bismuth), whilst ordinates normal to this base represent temperature.

9.71 A solid diagram of this type is not of much practical use since phase boundaries 'inside' the model are not visible. In my student days research

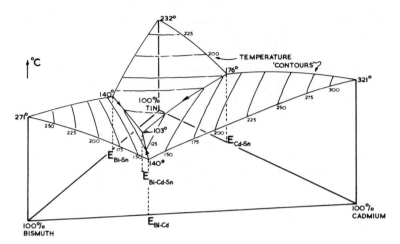

Fig. 9.10   *The bismuth-cadmium-tin thermal equilibrium diagram. This is of course a ternary diagram. The three 'valleys' drain down to the ternary (or 'triple') eutectic point at 103 °C. This alloy contains 53.9 per cent bismuth, 25.9 per cent tin, and 20.3 per cent cadmium. The 'temperature contours' are at 25 °C intervals. The ternary eutectic would melt at a temperature just above that of boiling water.*

workers at the University used to construct models representing such systems by employing bits of coloured plastics-coated electrical wire to indicate phase boundaries. With a complex system, the resulting model was quite fantastic, and resembled one of the more lurid examples of modern 'sculpture' – or possibly a parrot cage which had been designed by a committee.

However, for ternary (that is, three-metal) alloys we can still draw a useful *two-dimensional* constitutional diagram by fixing the quantities of *two* of the metals so that only the quantity of the third is variable. Of course we would then need a large number of separate sheets – each expressing two of the metals in different fixed amounts – to cover a ternary system comprehensively. It is often useful to do this as in the case of the diagram representing the structures of high-speed steel (Fig. 13.1). Here we have chosen a high-speed steel of standard composition, and have indicated the effects of variations in the carbon content and the temperature on the structure of this alloy. The amounts of the other alloy additions, viz. tungsten, chromium, and vanadium, are fixed at the values shown. The use of this diagram enables us to explain the basic principles of the heat-treatment of high-speed steel quite adequately. It is not a true equilibrium diagram, but is used like one, and is generally referred to either as a 'constitutional' or a 'pseudo-binary' diagram.

9.72   It is of course possible to make alloys containing yet more than three metals. Thus if we add lead to the bismuth-cadmium-tin alloy system described in the foregoing section a *quaternary* alloy series will result. However such a system could not be represented by an equilibrium diagram even of the 'parrot's cage' type simply because we have run out of dimensions – a point in space is represented by *three* coordinates in the Cartesian system.

In passing it is interesting to note that an alloy from the quaternary system mentioned here and containing 50Bi-24Pb-14Sn-12Cd melts at 71 °C. It was used in those frivolous unsophisticated days of my long-ago youth for the manufacture of tea spoons for practical jokers. Which is perhaps a note of light relief on which to end this serious and very important chapter.

CHAPTER 10

Practical microscopy

10.10 Examination of the microstructure of metals has been practised since it was developed by Professor Henry Sorby at Sheffield in the early 1860s and it is safe to say that of all the investigational tools available to him, the average materials scientist would least like to be without his microscope. With the aid of quite a modest instrument, a trained metallurgist can obtain an enormous amount of information from the microscopial examination of a metal or alloy. In addition to being able to find evidence of possible causes of failure of a material, he can often estimate its composition, as well as forecast what its mechanical properties are likely to be. Moreover, in the fields of pure metallurgical research, the microscope figures as the most frequently used piece of equipment. The primary object of this chapter is to help the engineering student to acquire some skill in the preparation and examination of a microsection, using a minimum of apparatus.

Selecting a specimen

10.20 Thought and care must be exercised in selecting a specimen from a mass of material, in order to ensure that the piece chosen is representative of the material as a whole. For example, free-cutting steels (7.55) contain a certain amount of slag (mainly manganese sulphide, MnS). This becomes elongated in the direction in which the material is rolled (Fig. 10.1). If only the cross-section (i) were examined the observer might be forgiven for assuming that the slag was present in free-cutting steel as more or less spherical globules instead of as elongated fibres. The latter fact could only be established if a longitudinal section (ii) were examined in addition to the cross-section (i).

10.21 In some materials both structure and composition may vary across the section. Thus, case-hardened steels (14.30) have a very different structure in the surface layers from that which is present in the core of the material. The same may be said of steels which have become decarburised at the surface due to faulty heat-treatment. Frequently it may be necessary to examine two or more specimens in order to obtain comprehensive information on the material.

10.22 A specimen approximately 20 mm in diameter or 20 mm square is convenient to handle. Smaller specimens are best mounted in one of the cold-setting plastics materials available for this purpose, since such a specimen may

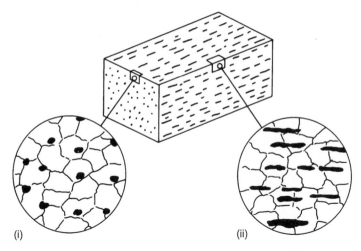

Fig. 10.1 *At least two specimens are necessary adequately to represent the microstructure of a free-cutting steel.*

rock during the grinding process, giving rise to a bevelled surface. Moreover, mounting in plastic affords a convenient way of protecting the edge of a specimen in cases where investigation of the edge is necessary, as, for example, in examining a section through a carburised surface.

The cold-setting plastics materials, available under a variety of trade names, consist of a white powder and a colourless liquid. When the powder is wetted with the liquid copolymerization (19.22) slowly takes place and a hard transparent solid is formed in about half an hour. A simple 'mould' (Fig. 10.2) is all that is required to mount specimens in such a material. The specimen is placed face-down on a sheet of glass which has first been lightly smeared with vaseline. The two L-shaped retaining pieces are then placed around the specimen, as shown, to give a mould of convenient size. The specimen is then covered with powder, which is in turn moistened with the liquid supplied. If a surplus of liquid is accidentally used, this can be absorbed by sprinkling a little more powder on the surface. In about thirty minutes the mass will have hardened, and the L-shaped members can be detached.

Specimens can also be mounted using thermoplastic or thermosetting materials in conjunction with a mould capable of being electrically heated and a suitable press. The advantage of using a thermosetting material such as bakelite is that it is less likely to be dissolved by organic degreasing agents such as acetone or warm alcohol.

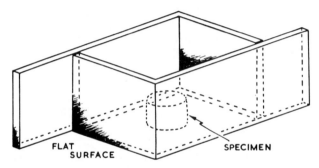

Fig. 10.2 *A method of mounting a specimen in a cold-setting plastics material. No pressure is required.*

Grinding and polishing the specimen

10.30 It is first necessary to obtain a flat surface on the specimen. This is best achieved by using a file held flat on the work-bench, and then rubbing the specimen on the file. It is much easier for an unskilled operator to produce a single flat surface on the specimen by using this technique, rather than by using a file in the orthodox manner. When the original hack-saw marks have been eliminated, the specimen should be rinsed in running water, to remove any coarse grit which may otherwise be carried over to the grinding papers.

10.31 Grinding is then carried out by using emery papers of successively finer grades. These papers must be of the best quality, particularly in respect of uniformity of particle size. For successful wet-grinding, at least four grades of paper are required ('220', '320', '400', and '600', from coarse to fine), and these must be of the type with a waterproof base. Special grinding tables can be purchased, in which the standard 300 mm × 50 mm strips of grinding papers can be clamped. The surface of the papers is flushed by a current of water, which serves not only as a lubricant in grinding, but also carries away coarse emery particles, which might otherwise scratch the surface of the specimen. If commercially produced grinding tables are not available – and certainly the prices of these simple pieces of apparatus seem to be unreasonably high – there is no reason why simple equipment should not be improvised, as indicated in Fig. 10.3. Here a sheet of 6 mm plate glass about 250 mm × 100 mm has a sheet of paper clamped to its surface by a pair of stout paper-clips. The paper should be folded round the edge of the glass plate, so that it will be held firmly. A suitable stream of water can be obtained by using a piece of rubber tubing attached to an ordinary tap, and the complete operation may be conducted in the laboratory sink. Alternatively, the apparatus can be contained in an old photographic developing dish, fitted with a suitable drain, in order to carry away the stream of water. The glass plate is tilted at one end, so that the water flows fairly rapidly over the grinding paper.

In busy laboratories rotating grinding tables are likely to be used. These are supplied with water which drips from a small reservoir above the rotating table.

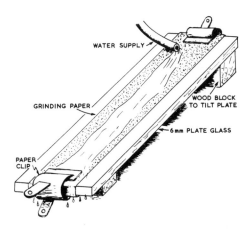

WATER SUPPLY

GRINDING PAPER

WOOD BLOCK TO TILT PLATE

6mm PLATE GLASS

PAPER CLIP

Fig. 10.3 *A simple grinding table adapted from odds and ends.*

10.32   The specimen is first ground on the '220' grade paper. Assuming that a stationary table is being used, this is achieved by rubbing it back and forth on the paper, in a direction which is roughly *at right angles* to the scratches left by the filing operation. In this way, it can easily be seen when the original deep scratches left by the file have been completely removed. If the specimen were ground so that the new scratches ran in the same direction as the old ones, it would be virtually impossible to see when the latter had been erased. With the primary grinding marks removed, the specimen is now washed free of '220' grit. Grinding is then continued on the '320' paper; again turning the specimen through 90°, and grinding until the previous scratch marks have been erased. This process is repeated with the '440' and '600' papers.

If circumstances demand that dry-grinding be used, complete cleanliness must be maintained at all stages, in order to avoid the carrying-over of coarse grit to the finer papers. After use, each paper should be shaken free of grit by smartly pulling it taut a number of times. Papers can be stored safely between the pages of a *glossy* magazine. Alternatively, a strip of each grade of paper can be permanently attached to its own polishing block. It is most important that the specimen be washed before passing from one grade of paper to the next, and particularly before transferring to the final polishing cloth.

Steels and the harder non-ferrous metals can be ground dry, provided that care is taken not to overheat them, since this may modify the microstructure. For the softer non-ferrous materials, such as aluminium alloys, and bearing metals, the paper should be moistened with a lubricant such as paraffin. A lighter pressure can then be used, and there will be much less risk of particles of grit becoming embedded in the soft metal surface. Modern wet-grinding processes are far more satisfactory for all materials, and have generally replaced dry-grinding methods.

10.33   Up to this stage, the process has been one of grinding, and each set of parallel 'furrows' has been replaced successively by a finer set. The final polishing operation is different in character, in that it removes the ridged surface layers by means of a burnishing process. Although the surface is made smooth by this operation, the structure still cannot be seen, because the nature of the polishing process is such that it leaves a 'flowed' or amorphous layer of metal on the surface of the specimen (Fig. 10.4). In

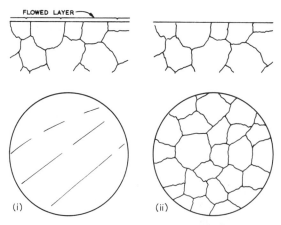

Fig. 10.4   *The 'flowed layer' on the surface of a polished microsection. (i) In the polished state the structure is hidden by the flowed layer – only a few polishing scratches are visible on the otherwise featureless surface. (ii) Etching removes the flowed layer, thus revealing the crystal structure beneath.*

order that the structure can be seen, this flowed layer must be dissolved – or 'etched' away – by a suitable chemical reagent.

10.34   Iron or steel specimens are polished by means of a rotating cloth pad impregnated with a suitable polishing medium. 'Selvyt' cloth is probably the best-known material with which to cover the polishing wheel, though special cloths are now available for this purpose. The cloth is thoroughly wetted with *distilled* water, and a small quantity of the polishing powder is gently rubbed in with *clean* finger-tips. Possibly the most popular polishing medium is alumina (aluminium oxide), generally sold under the name of 'Gamma Alumina'. During the polishing process, water should be permitted to spot on to the pad, which should be run at a low speed until the operator has acquired the necessary manipulative technique. Light pressure should be used, since too heavy a pressure on the specimen may result in a torn polishing cloth. Moreover, the specimen is more likely to be scratched by grit particles embedded deep in the cloth if heavy pressure is applied.

A disadvantage of most of the water-lubricated polishing powders is that they tend to dry on the pad, which generally becomes hard and gritty as a result. If the pad is to be used only intermittently, it might be worthwhile to use one of the proprietary diamond-dust polishing compounds. In these materials, graded diamond particles are carried in a 'cream' base which is soluble both in water and in the special polishing fluid, a few spots of which are applied to the pad in order to lubricate the work and lead to even spreading of the polishing compound. If properly treated, the pad remains in good condition until it wears out; so, although these diamond dust materials are more expensive than other polishing media, a saving may result in the long term, as polishing cloths will need to be changed less frequently.

10.35   Since non-ferrous metals and alloys are much softer than steels, they are best polished by hand, on a small piece of 'Selvyt' cloth wetted with 'Silvo'. During polishing, a circular sweep of the hand should be used, rather than the back-and-forth motion used in grinding.

When the surface appears free from scratches, it is cleaned thoroughly, dried, and then examined under the microscope, using a magnification between 50 and 100. If reasonably free from scratches, the specimen can at this stage be examined for inclusions, such as those of manganese sulphide (in steel), slag fibres (in wrought iron), or globules of lead (in free-cutting brasses). Such inclusions would be less obvious were the specimen etched *before* this primary examination.

10.36   The most important points to be observed during the grinding and polishing processes are the following:

- Absolute cleanliness is necessary at each stage.
- Use very light pressure during both grinding and polishing.
- Deep scratches are often produced during the final stage of grinding. Do *not* attempt to remove these by prolonged polishing, as such scratches tend to be obliterated by the flowed layer, only to reappear on etching. Moreover, prolonged polishing of non-ferrous metals tends to produce a rippled surface. If deep scratches are formed, wash the specimen, and return to the last-but-one paper, remembering to grind in a direction at 90° to the scratches.

- Care should be taken not to overheat the specimen during preliminary filing or grinding. Hardened steels could be tempered by such treatment, particularly if a linishing machine is used.

Etching the specimen

10.40   Etching is generally the stage in preparing a microsection that the beginner finds most difficult. Often the first attempt at etching results in a badly stained or discoloured surface, and this is invariably due to inadequate cleaning and degreasing of the specimen before attempting to etch it.

The specimen should first be washed free of any adhering polishing compound. This can be rubbed from the *sides* of the specimen using the fingers, but great care must be taken in dealing with the polished face. The latter can be cleaned and degreased successfully by *very gently* smearing the surface with a *clean* finger-tip dipped in grit-free soap solution, followed by thorough rinsing under the tap. Even now, traces of grease may still be present, as shown by the fact that a film of water will not flow evenly over the surface, but instead remains as isolated droplets. The last traces of grease are best removed by immersing the specimen for a minute or so in boiling[1] alcohol ('white industrial methylated spirit').

From this stage onwards, the specimen should not be touched by the fingers, but be handled with a pair of nickel tongs. It is lifted from the alcohol, and cooled under the tap before being etched. Some thermoplastic mounting materials are dissolved by hot alcohol; in such cases, swabbing with a piece of cotton wool saturated with dilute caustic-soda solution may degrease the surface effectively.

10.41   When the specimen is clean and free of grease, it is etched by plunging it into the etching solution, and agitating it vigorously for a few seconds. The specimen is then *very quickly* transferred to running water, in order to wash away the etchant as rapidly and as evenly as possible. It is then examined with the naked eye, to see to what extent etching has taken place. If successfully etched, the highly polished surface will now appear dull, and, in the case of cast metals, individual crystals may be seen. A bright surface at this stage will usually indicate that further etching is necessary. The time required for etching varies with different alloys and etchants, and may be limited to a few seconds for a specimen of carbon steel etched in 2 per cent 'nital', or extend to as long as 30 minutes for a stainless steel etched in a mixture of concentrated acids.

After being etched, the specimen is washed in running water, and then quickly immersed in boiling alcohol, where it should remain for a minute. On withdrawal from the alcohol, the specimen is shaken with a flick of the tongs, to remove surplus alcohol so that it will dry almost instantaneously. With specimens mounted in a plastic likely to be affected by boiling alcohol, it is better to spot a few drops of cold alcohol on the surface of the specimen. The surplus is then shaken off, and the specimen is held in a current of hot air from a domestic hair-drier. Unless the specimen is dried *evenly and quickly*, it will stain.

A summary of the more popular etching reagents which can be used for most metals and alloys is given in Table 10.1.

---

[1]   On no account should alcohol be heated over a naked flame, as the vapour is highly inflammable. An electrically heated water-bath should be used – an electric kettle with the lid removed is serviceable.

Table 10.1 *Etching reagents.*

Type of etchant	Composition	Characteristics and uses
2% 'nital' – for iron, steel, and bearing metals	2 cm^3 nitric acid, 98 cm^3 alcohol ('white industrial methylated spirit')	The best general etching reagent for irons and steels, both in the normalized and heat-treated conditions. For pure iron and wrought iron, the quantity of nitric acid may be raised to 5 cm^3. Also suitable for most cast irons and for some alloys, such as bearing metals.
Alkaline sodium picrate – for steels	2 g picric acid, 25 g sodium hydroxide, 100 cm^3 water	The sodium hydroxide is dissolved in the water, and the picric acid is then added. The whole is heated on a boiling water-bath for 30 min, and the clear liquid is poured off. The specimen is etched for 5–15 min in the boiling solution. It is useful for distinguishing between ferrite and cementite; the latter is stained black, but ferrite is not attacked.
Ammonia/hydrogen peroxide – for copper, brasses, and bronzes	50 cm^3 ammonium hydroxide (0.880), 50 cm^3 water. Before use, add 20–50 cm^3 hydrogen peroxide (3%)	The best general etchant for copper, brasses, and bronzes. Used for swabbing or immersion. Must be freshly made, as the hydrogen peroxide decomposes. (The 50% ammonium hydroxide solution can be stored, however.)
Acid ferric chloride – for copper alloys	10 g ferric chloride, 30 cm^3 hydrochloric acid 120 cm^3 water	Produces a very contrasty etch on brasses and bronzes. Use at full strength for nickel-rich copper alloys, but dilute one part with two parts of water for brasses and bronzes.
Dilute hydrofluoric acid – for aluminium and its alloys	0.5 cm^3 hydrofluoric acid, 99.5 cm^3 water	A good general etchant for aluminium and most of its alloys. The specimen is best swabbed with cotton-wool soaked in the etchant. N.B. *On no account should hydrofluoric acid or its fumes be allowed to come into contact with the eyes or skin. Care must be taken with all concentrated acids.*

Details of etching reagents for other specific purposes will be found in *Engineering Metallurgy, Part 1* (Section 10.24), by this author. No mention is made here of specialized methods such as electrolytic etching. These methods are generally used only for materials which are difficult to etch using conventional chemical reagents.

The metallurgical microscope

10.50   The reader may have used a microscope during his school days, but the chances are that this would be an instrument designed for biological work. Biological specimens can generally be prepared as thin, transparent slices, mounted between thin sheets of glass, so that illumination can be arranged simply by placing a source of light *behind* the specimen. Since metals are opaque substances, which must be illuminated by frontal lighting, the source of light must be *inside* the microscope tube itself. This is generally accomplished as shown in Fig. 10.5, by placing a small, thin plain-glass reflector, R, inside the tube. Since it is necessary for the returning light to pass through R, the latter must be of unsilvered glass. This means that much of the total light available is lost, both by transmission through the glass when it first strikes the plate, and by reflection when the returning ray from the specimen strikes the plate again. Nevertheless, a small 6-volt bulb is generally sufficient as a source of illumination. The width of the beam is controlled by the iris diaphragm, D. This should be closed until the width of the beam of light is just sufficient to cover the rear component of the objective lens, O. Excess light, reflected within the microscope tube, would cause light-scatter and consequently 'glare' in the field of view, leading to a loss of contrast and definition in the image formed.

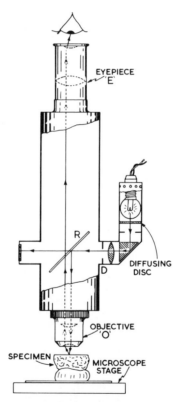

Fig. 10.5   *A metallurgical microscope. This illustrates a very basic 'student's model'. Most modern instruments are more streamlined and may be of inverted construction, i.e. the objective is at the upper end of the instrument so that the specimen is suspended above it. The basic method of illumination remains similar to that shown.*

10.51   The optical system of the microscope consists of two lenses: the objective, O, and the eyepiece, E. The former is the more important and expensive of the two lenses, since it must resolve fine detail of the object under examination. Good-quality objectives, like camera lenses, must be of compound construction. However, there is a limit to the degree of accuracy which is worthwhile in constructing an objective. At magnifications of ×1000 or so, one is dealing with dimensions comparable with the wavelength of light itself, and further improvements in the quality of the lens would produce no corresponding improvements in the 'sharpness' of the image. Thus, in purchasing a very expensive microscope, one may well be paying for extra refinements which may not be necessary. It is doubtful whether one will obtain higher optical quality than is offered in the 'standard' model of the same manufacturer's range of products. The same is not true of ordinary camera lenses, however; here, as a general rule, the more one is prepared to pay, the better the quality of the lens.

10.52 The magnification given by an objective depends upon its focal length – the shorter the focal length, the higher the magnification obtainable. As mentioned above, the resolving power of a lens is also important, and, within the limits stated, depends upon the quality of the lens. When working at high magnifications of ×1000 or more, it is generally useless to increase the size of the image either by extending the tube length or by using a higher-power eyepiece, as a point is reached where there is a falling-off in definition. A parallel example in photography is where a 12×8 inch enlargement taken from a 35 mm negative fails to show any more *detail* than it did in the original 6×4 inch print, but instead gives a rather blurred image.

10.53 The eyepiece is so called because it is the lens nearest the eye. Its purpose is to magnify the image formed by the objective. Eyepieces are made in a number of powers, generally ×6, ×8, ×10, and ×15.

10.54 The distance separating the objective and the eyepiece is known as the tube-length of the microscope, and is usually 200 mm for most instruments. The magnification of the system can then be calculated from:

$$\text{magnification} = \frac{\text{tube-length (mm)} \times \text{power of eyepiece}}{\text{focal length of objective (mm)}}$$

Thus, for a microscope having a tube-length of 200 mm, and using a 4 mm focal-length objective in conjunction with a × 8 eyepiece,

$$\text{magnification} = \frac{200 \times 8}{4} = 400$$

Using the microscope

10.60 The specimen must first be mounted so that its surface is normal to the axis of the instrument. This is best achieved by fixing the specimen to a microscope slide, by means of a pellet of 'plasticine', using a mounting-ring to ensure normality between the surface of the specimen and the axis of the microscope (Fig. 10.6). Obviously the mounting-ring must have perfectly parallel end-faces. If the specimen is in a plastics mount it can of course be placed direct on to the microscope stage provided that the end-faces of the mount are completely parallel.

10.61 The specimen is brought into focus by first using the coarse adjustment, and then the fine adjustment. Lenses are designed to work at an

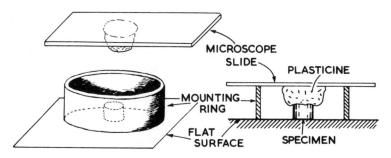

Fig. 10.6   *Mounting a specimen so that its surface is normal to the axis of the microscope.*

optimum tube-length (usually 200 mm), and give best results under these conditions. Hence the tube carrying the eyepiece should be drawn out to the appropriate mark (a scale is generally engraved on the side of the tube). Slight adjustments in tube-length should then be made to suit the individual eye. Finally, the iris, D, in the illumination system should be closed to the point where illumination begins to decrease. Glare due to internal reflection will then be at a minimum.

10.62   Invariably, the newcomer to the microscope selects the combination of objective and eyepiece which will give him the maximum magnification, but it is a mistake to assume that high magnifications in the region of ×1000 are necessarily the most useful. In fact, they may well give a misleading impression of the structure, by pin-pointing some very localized feature rather than giving a general picture of the microstructure of the material. Directional properties in wrought structures, or dendritic formations in cast structures, are best examined using low powers of between ×30 and ×100. Even at ×30, a single crystal of a cast structure may completely fill the field of view. The dendritic pattern, however, will be clearly apparent, whereas at ×500, only a small area between two dendrite arms would fill the field of view. In a similar way, a more representative impression of the lunar landscape may be obtained by using a pair of good binoculars than by using the very high-powered system of an astronomical telescope. Hence, as a matter of routine, a low-powered objective should *always* be used in the initial examination of a microstructure, before it is examined at a high magnification.

The care of the microscope

10.70   Care should be taken never to touch the surface of optical glass with the fingers, since even the most careful cleaning may damage the surface coating (most high-quality lenses are 'bloomed', that is, coated with magnesium fluoride to increase light transmission). In normal use, dust particles may settle on a lens, and these are best removed by sweeping gently with a high-quality camel-hair brush.

10.71   If a lens becomes finger-marked, this is best dealt with by wiping *gently* with a piece of soft, well-washed linen moistened with the solvent xylol. Note that the operative word is *wipe*, not *rub*. Excess xylol should be avoided, as it may penetrate into the mount of the lens, and soften the cement holding the glass components together so that they become detached. Such a mishap will involve an expensive re-assembly job!

High-power objectives of the oil-immersion type should be wiped free of cedar-wood oil before the latter has a chance to harden. If, due to careless neglect, the oil has hardened on a lens, then it will need to be removed with the minimum amount of xylol; but the use of the latter should be avoided whenever possible.

Soft, well-washed linen should always be used to clean lenses. It is superior to chamois leather, which is likely to absorb particles of grit, and to silk, which has a tendency to scratch the surface of soft optical glass.

The electron microscope

10.80   As the reader will have gathered, the greater proportion of routine microscopic examination of metals is carried out using magnifications in the region of ×100. It is often necessary, particularly in the field of research, to examine structures at much higher magnifications. Unfortunately, it is not practicable to use magnifications in excess of ×2000 with an ordinary optical microscope, since, as indicated earlier in this chapter (10.51), one is then dealing with an object size of roughly the same order as the wavelength of light itself. This leads to a loss of definition of the image at magnifications greater than ×2000.

For high-power microscopy – between ×2000 and about ×200 000 – an electron microscopic may be used. In this instrument, rays of light are replaced by a beam of electrons which have been made to travel at a suitable velocity. Since electrons are very tiny electrically charged particles (1.20) their path can be altered by an electromagnetic field. Consequently, the 'lenses' in an electron microscope consist of a system of coils which produce the necessary electromagnetic field to focus the electron beam. The other important difference between an optical and an electron microscope is that, whereas in the former light is *reflected* from the surface of the specimen, in the electron microscope the electron beam *passes through* the specimen, rather in the manner of light rays in a biological microscope. Hence, thin foil specimens must be used, or, alternatively, a very thin replica of the etched metallic surface may be produced in a suitable plastics material. This replica is then examined by the electron microscope.

10.81   The limitations of the optical microscope have already been mentioned, but it should be noted that the range of the ordinary electron microscope too is limited. Whilst it is true that magnification obtained with this instrument exceed a quarter of a million times, it should be appreciated that a magnification of something like twenty million times is required for us to be able to see an average-size atom. Magnifications of this order can now be obtained using a modern *field-ion* microscope, and the results so far obtained at least confirm the physicist's idea of the general form of atoms.

CHAPTER 11

Iron and steel

11.10 Since the onset of the Industrial Revolution, the material wealth and power of a nation has depended largely upon its ability to make steel. During the nineteenth century, Britain was prominent among steel-producing nations, and, towards the end of the Victorian era, was manufacturing a great proportion of the world's steel but exploitation of vast deposits of ore abroad changed the international situation so that by the middle of the twentieth century what we then referred to as the two 'Superpowers' – the USA and the USSR – owed their material power largely to the presence of high-grade ore within, or very near to, their own vast territories. Consequently they led the world in terms of the volume of steel produced annually. More recently rapid technological development in the Far East has meant that Japan is currently the world's premier steel producer followed closely by the People's Republic of China and the USA with Russia, Germany and the Korean Republic some way behind. The UK now occupies twelfth position in the international steel producers' 'league table'.

Low-grade ores mined in Britain now account for only a small part of the input to British blast-furnaces and most of the ore used must be imported (as high-grade concentrates) from Canada, Brazil, Australia, Norway, North Africa, Venezuela and Sweden.

11.11 Smelting of iron ore takes place in the blast-furnace (Fig. 11.1). A modern blast-furnace is something like 60 m high and 7.5 m in diameter at the base, and may produce from 2000 to 10 000 tonnes of iron per day. Since a refractory lining lasts for several *years*, it is only at the end of this period that the blast-furnace is shut down; otherwise it works a 365-day year. Processed ore, coke, and limestone are charged to the furnace through the double-bell gas-trap system, whilst a blast of heated air is blown in through the tuyères near the hearth of the furnace. At intervals of several hours, the furnace team opens both the slag hole and the tap hole, in order to run off first the slag, and then the molten iron. The holes are then plugged with clay.

11.12 The smelting operation involves two main reactions.

1. The chemical reduction of iron ore by carbon monoxide gas, CO, arising from the burning coke:

$$Fe_2O_3 + 3CO \longrightarrow 2Fe + 3CO_2$$
iron oxide   iron   carbon
(ore)   dioxide

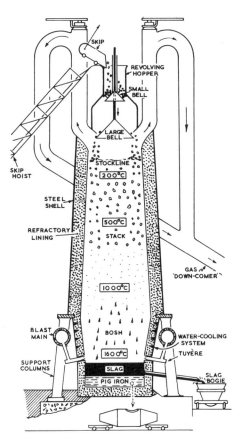

Fig. 11.1   *A modern blast-furnace.*

2. Lime (from limestone added with the furnace charge) combines with
    many of the impurities and also the otherwise infusible earthy waste
    (mainly silica, SiO_2) in the ore to form a fluid slag which will run from
    the furnace:

$$SiO_2 \quad + \quad 2CaO \longrightarrow 2CaO{\cdot}SiO_2$$

earthy waste   lime   slag

(mainly silica)   (calcium silicate)

The slag is broken up, and used for road-making, or as a concrete aggregate
(23.42). The molten iron is either cast into 'pigs', for subsequent use in an
iron foundry, or transferred, still molten, to the steel-making plant.

In the case of a large modern furnace, a daily output of 2000 tonnes of pig
iron would involve the following materials.

Charge	(tonnes)	Products	(tonnes)
ore (say 50% iron)	4 000	pig iron	2 000
limestone	800	slag	1 600
coke	1 800	dust	200
air	8 000	furnace gas	10 800
	14 600		14 600

11.13   One feature of the above table which may surprise the reader is the vast amount of furnace gas passing along the 'down-comer' each day. The gas contains a large amount of carbon monoxide, and therefore has a considerable calorific value. The secondary function of the blast-furnace is, in fact, to act as a large gas-producer. If the blast-furnace plant is part of an integrated steel works, then much of this vast quantity of gas will be used for raising electric power; but its major function is to be burned in the Cowper stoves, and so provide heat which in turn will heat the air blast to the furnace. Two such stoves are required for each blast-furnace. One is being heated by the burning gas whilst the other is heating the ingoing air.

In recent years much research has been conducted into steel production by the 'direct reduction' of iron ore but it seems that for some years yet the blast-furnace will survive as the major unit in iron production. The thermal efficiency of the blast-furnace is very high (7.61) and this is improved further in modern management by the injection of oil or pulverized low-cost coal at the tuyères in order to reduce the amount of expensive metallurgical coke consumed.

Be that as it may we must not lose sight of the fact that the blast-furnace we have been considering briefly here is responsible for releasing some 6600 *tonnes* of 'greenhouse gas' (carbon dioxide) into the atmosphere each day of the year.

Steel-making

11.20   Until Henry Bessemer introduced his process for the mass-production of steel, in 1856, all steel was made from wrought iron. Nowadays wrought iron is no longer produced, except perhaps in small quantities for decorative purposes (though much of the ornamental 'wrought iron' work is in fact 'mild steel'). The Bessemer process, too, is obsolete, and, as far as steel production in Great Britain is concerned, has been followed into obscurity by the open-hearth process, though the latter is still used in a few countries abroad. In Britain the bulk of steel is made either by one of the basic oxygen processes developed since 1952, or in the *electric-arc furnace*.

11.21   Basic oxygen steelmaking (BOS)   The process of steelmaking is mainly one involving oxidation of impurities present in the original charge, so that they form a slag which floats on the surface of the molten steel – or are lost as fume. In the Bessemer process impurities were removed from the charge of molten pig iron by blowing air through it. The impurities, mainly carbon, phosphorus, silicon, and manganese, acted as fuel; and so the range of compositions of pig iron was limited, because sufficient impurities were necessary in order that the charge did not 'blow cold' from lack of fuel. The oxidized impurities either volatilized or formed a slag on the surface of the charge.

Since the air blast contained only 20 per cent of oxygen by volume, much valuable heat was carried away from the charge by the 80 per cent nitrogen also present. Worse still, a small amount of this nitrogen dissolved in the charge, and, in the case of mild steel destined for deep-drawing operations, caused a deterioration in its mechanical properties. The new oxygen processes produce mild steel very low in nitrogen, so that its deep-drawing properties are superior to those of the old Bessemer steel. Improvements of this type are essential if mild steel is to survive the challenge of reinforced plastics such as ABS (19.31.9) in the field of automobile bodywork.

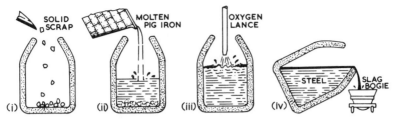

Fig. 11.2 *Stages in the manufacture of steel by the basic oxygen process. Steel scrap is added first (i), followed by molten pig iron (ii). At the end of the 'blow' (iii) the slag is run off first (iv), before 'teeming' the steel into a ladle.*

11.22 The earliest of these oxygen processes was the L-D process, so called because it originated in the Austrian industrial towns of Linz and Donawitz in 1952. It was made possible by the low-cost production of 'tonnage oxygen' and it is interesting to note that Bessemer had foreseen these possibilities almost a century earlier but of course did not have access to tonnage oxygen. Since 1952 a large number of variations of the original process have been developed but have become rationalized under the general heading of 'basic oxygen steelmaking' or 'BOS'. Generally the BOF (basic oxygen furnace) is a pear-shaped vessel of up to 400 tonnes capacity lined with basic refractories – magnesite bricks covered with a layer of dolomite. This lining must be basic to match the basic slag which is necessary for the removal of impurities from the charge. If the lining were chemically acid (silica bricks) it would be attacked by the basic slag and would quickly disintegrate.

In this process no heat is carried away by useless nitrogen (as was the case in the old Bessemer process) so a charge containing up to 40 per cent scrap can be used. This scrap is loaded to the converter first, followed by lime and molten pig iron. Oxygen is then blown at the surface of the molten charge from a water-cooled lance which is lowered through the mouth of the converter to within 0.5 m of the surface of the metal. Impurities in the charge are oxidized and form a slag on the surface. At the end of the blow this slag is run off first and any adjustments made to the carbon content of the charge which is then transferred to the ladle, preparatory to being cast as ingots, or, much more probably, fed to a continuous-casting unit (3.21) which, in the United Kingdom (and the European Union) now deals with some 90 per cent of the steel cast.

11.23 **Electric-arc steelmaking** is now the only alternative process in Britain to BOS, to which it is complementary rather than competitive. Originally electric-arc furnaces were used for the manufacture of high-grade tool and alloy steels but are now widely employed both in the treatment of 'hot metal' and of process scrap as well as scrap from other sources. The high cost of electricity is largely offset by the fact that cheap scrap can be processed economically to produce high-quality steel.

Since electricity is a perfectly 'clean' fuel no impurities are transmitted to the charge as was the case with producer gas used in the largely extinct open-hearth process. Moreover the chemical conditions within the electric-arc furnace can be varied at will to favour the successive removal of the various impurities present in the charge. Sulphur, which was virtually impossible to eliminate in either the Bessemer or open-hearth processes, can be effectively reduced to extremely low limits in the electric-arc process.

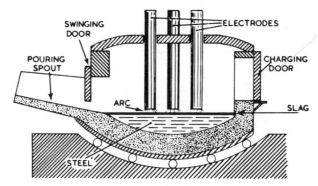

Fig. 11.3   *The principles of the electric-arc furnace for steelmaking. Modern furnaces have water-cooled panels built into the side walls to protect the refractory lining. Some furnaces lack the pouring spout and the charge is poured from a tap hole in the bottom of the furnace.*

The furnace (Fig. 11.3) employs carbon electrodes which strike an arc on to the charge. Lime and millscale are added in order to produce a slag which removes most of the carbon, silicon, manganese and phosphorus. This is run off and is often replaced by a slag containing lime and anthracite which effectively removes sulphur.

The composition of steel

11.30   Plain-carbon steels are those alloys of iron and carbon which contain up to 1.7 per cent carbon. In practice, most ordinary steels also contain up to 1.0 per cent manganese, which is left over from a deoxidation operation carried out at the end of the steelmaking process. This excess of manganese dissolves in the solid steel, slightly increasing its strength and hardness. It also helps to reduce the sulphur content of the steel. Both sulphur and phosphorus are extremely harmful impurities which give rise to brittleness in steels. Consequently most specifications allow no more than 0.05 per cent of either of these elements, whilst specifications for higher-quality steels limit the amount of each element to 0.04 per cent or less. In fact the quality, in respect of chemical composition, of mild steel is continually improving and it is common for specifications for steel used in gas and oil pipelines to demand sulphur contents as low as 0.002 per cent with phosphorus at 0.015 per cent maximum and carbon at 0.04 per cent.

Although the remainder of this chapter and most of the next are devoted to the heat-treatment of carbon steels used for constructional and tool purposes, we must not lose sight of the fact that by far the greatest quantity of steel produced is mild steel and low-carbon steel for structural work, none of which is heat-treated except for stress relief.

11.31   At ordinary temperatures, most of the carbon in a steel which has not been heat-treated is chemically combined with some of the iron, forming an extremely hard compound known by chemists as iron carbide, though metallurgists know it as 'cementite'. Since cementite is very hard, the hardness of ordinary carbon steel increases with the carbon content. Carbon steels can be classified into groups, as shown in Fig. 11.6 and in Table 12.2.

The structure of plain-carbon steels

11.40   In Chapter 2 (2.13) we saw that iron is what is called a polymorphic element; that is, an element which leads a sort of Jekyll-and-Hyde existence by appearing in more than one crystal form. Below 910 °C, pure iron has a body-centred cubic crystal structure; but on heating the metal to a temperature above 910 °C, its structure changes to one which is face-centred cubic. Now face-centred cubic iron will take quite a lot of carbon – up to 2.0 per cent in fact – into solid solution (8.34); whereas body-centred cubic iron will dissolve scarcely any – a maximum of only 0.02 per cent. Since the solid solubility of carbon in iron alters in this way, it follows that changes in the structure will also occur on heating or cooling through the polymorphic transformation temperature. Thus it is the polymorphic transformation, and the structural changes which accompany it, which cause the thermal equilibrium diagram (Fig. 11.4) to have a somewhat unusual shape as compared with those already dealt with in Chapter 9.

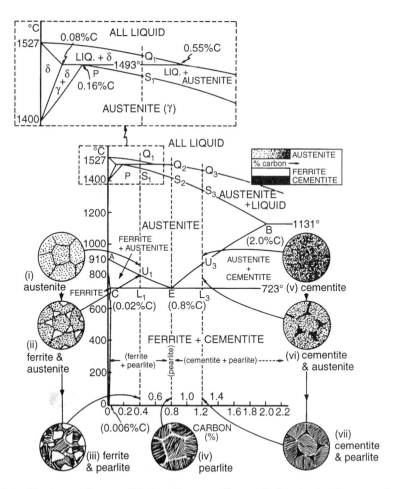

Fig. 11.4   *The iron-carbon equilibrium diagram. The small dots in the diagrams depicting structures containing austenite do not represent visible particles of cementite – they are meant to indicate the concentration of carbon atoms dissolved in the austenite, and in the real microstructures would of course be invisible.*
  The inset shows the 'peritectic part' of the diagram in greater detail.

11.41   Any solid solution of carbon up to a maximum of 2.0 per cent in face-centred cubic iron is called *austenite* (γ), whilst the very dilute solid solution formed when up to 0.02 per cent carbon dissolves in body-centred cubic iron is called *ferrite* (α). For all practical purposes, we can regard ferrite as being more or less pure iron, since less than 0.02 per cent carbon will have little effect on its properties. Thus, in carbon steel at, say, 1000 °C, all of the carbon present is dissolved in the solid austenite. When this steel cools, the austenite changes to ferrite, which will retain practically no carbon in solid solution. 'What happens to this carbon?', we may ask. The answer is that, assuming the cooling has taken place fairly slowly, the carbon will be precipitated as the hard compound *cementite*, referred to previously (11.31).

11.42   By referring to Fig. 11.4, let us consider what happens in the case of a steel containing 0.4 per cent carbon as it solidifies and cools to room temperature. It will begin to solidify at a temperature of about 1500 °C (Q_1) by forming dendrites of δ-iron (a body-centred cubic polymorph of iron) but as the temperature falls to 1493 °C the δ-dendrites react with the remaining liquid to form crystals of a new phase – γ-iron containing 0.16 per cent carbon (γ is the phase we call austenite). This process of change in structure takes place by what is termed a *peritectic reaction*. As the temperature continues to fall the remaining liquid solidifies as austenite, the composition of which changes along PS_1. The steel is completely solid at about 1450 °C (S_1). The structure at this stage will be uniform austenite – there will be no coring of the dendrites because the diffusion of the interstitially dissolved carbon atoms is very rapid (8.34) particularly at high temperatures in the region of 1400 °C. As this uniform austenite cools, nothing further will happen to its structure – except possibly grain growth – until it reaches the point U_1, which is known as the 'upper critical point' for this particular steel. Here austenite begins to change to ferrite, which will generally form as small new crystals at the grain boundaries of the austenite (Fig. 11.4 (ii)). Since ferrite contains very little carbon, it follows that at this stage the bulk of the carbon must remain behind in the shrinking crystals of austenite; and so the composition of the latter moves to the right. Thus, by the time the temperature has fallen to 723 °C, we shall have a mixture of ferrite and austenite crystals of compositions C and E respectively. The *overall* composition of the piece of steel is given by L_1, and so we can apply the lever rule:

weight of ferrite (composition C) $\cdot CL_1 =$
$$\text{weight of austenite (composition } E) \cdot L_1E$$

Since CL_1 and L_1E are of more or less equal length, it follows that the amounts of ferrite and austenite at this temperature (723 °C) are roughly equal for this particular composition of steel (0.4 per cent carbon).

The reader will recognize the point E as being similar to the eutectic points dealt with in Chapter 9 (Figs 9.6 and 9.7). In the present case, however, we are dealing with the transformation of a *solid* solution (austenite), instead of the solidification of a *liquid* solution. For this reason, we refer to E as a eutect*oid* point, instead of a eutec*tic* point.

11.43   As the temperature falls just below 723 °C (the 'lower critical temperature'), the austenite, now of composition E, transforms to a eutectoid (Fig. 11.4 (iii)) by forming alternate layers of ferrite (composition C) and the compound cementite (containing 6.69 per cent carbon). Clearly, since the austenite at this temperature was of composition E (0.8 per cent carbon), the *overall* composition of the eutectoid which forms from it will be of composition E (0.8 per cent carbon), even though the separate layers comprising it contain 0.02 per cent and 6.69 per cent carbon respectively. Since the specific gravities of ferrite and cementite are roughly the same this explains why the layers of ferrite are about seven times as thick as the layers of cementite (Fig. 11.7).

As the steel cools to room temperature no further important change will take place in the structure.

11.44   The austenite ⟶ pearlite transformation mentioned above begins at the grain boundaries of the austenite (Fig. 11.5). It is thought that carbon atoms congregate there in sufficient numbers to form cementite nuclei which grow inwards across the austenite grains. Since carbon atoms are removed from the austenite by this process the adjacent layer of austenite is left very low in carbon and so it transforms to produce a layer of ferrite which grows inwards following close behind the cementite. Beyond the new ferrite layer an increase in carbon atoms occurs so that further cementite nucleates and so on. In this way the structure builds up as alternate layers of cementite and ferrite.

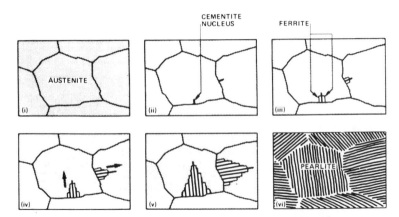

Fig. 11.5   *The transformation of austenite to pearlite (in this case in a 0.8 per cent carbon steel).*

11.45   In most cases, a eutectic or eutectoid in an alloy system is not given a separate name, since it is really a mixture of two phases. The iron-carbon system, however, is the most important of all alloy systems with which the metallurgist or engineer has to deal; so the eutectoid of ferrite and cementite referred to above is given the special name of *pearlite*. This name is derived from the fact that the etched surface of a high-carbon steel reflects an iridescent sheen like that from the surface of 'mother of pearl'. In both cases this is due to the diffraction of white light as it is 'unscrambled', by minute ridges (of cementite in the case of steel) protruding from the surface of the structure, into the colours of the spectrum.

These then are the main stages in the foregoing process of solidification and cooling of the 0.4 per cent carbon steel.

1. Solidification is complete at S_1, and the structure consists of uniform austenite.
2. This austenite begins to transform to ferrite at U_1, the upper critical temperature of this steel (about 825 °C).

3. At 723 °C (the lower critical temperature for all steels), formation of primary ferrite ceases, and, as the austenite is now saturated with carbon, the eutectoid pearlite is produced as alternate layers of ferrite and cementite.
4. Below 723 °C, there is no further significant change in the structure.

11.46   A steel which contains exactly 0.8 per cent carbon will begin to solidify as austenite at about 1490 °C (Q_2), and be completely solid at approximately 1410 °C (S_2). For a steel of this composition, the upper critical and lower critical temperatures coincide at E (723 °C), so that no change in the uniformly austenitic structure occurs until a temperature slightly below 723 °C is reached, when the austenite will transform to pearlite by precipitating alternate layers of ferrite and cementite. The final structure will be entirely pearlite (Fig. 11.4 (iv)).

11.47   Now let us consider the solidification and cooling of a steel containing, say, 1.2 per cent carbon. This alloy will begin to solidify at

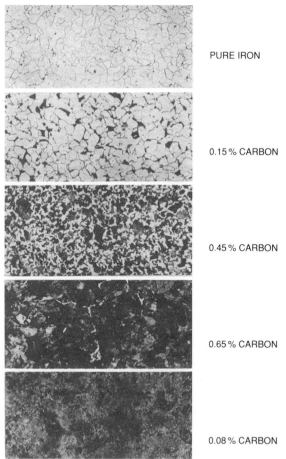

PURE IRON

0.15% CARBON

0.45% CARBON

0.65% CARBON

0.08% CARBON

Fig. 11.6   *This series of photomicrographs depicts steels of varying carbon content, in the normalized condition. As the carbon content increases, so does the relative proportion of pearlite (dark), until with 0.8 per cent carbon the structure is entirely pearlitic. The light areas consist of primary ferrite. The magnification (×80) is not high enough to reveal the laminated nature of the pearlite. In a similar way, craters on the surface of the moon are not visible unless the latter is viewed through a low-power telescope.*

approximately 1480 °C (Q_3), by depositing dendrites of austenite, and these will grow as the temperature falls, until at about 1350 °C (S_3) the structure will be uniform solid austenite. No further change in the structure occurs until the steel reaches its upper critical temperature, at about 880 °C (U_3). Then, needle-like crystals of cementite begin to form, mainly at the grain boundaries of the austenite (Fig. 11.4 (vi)). In this case, the remaining austenite becomes less rich in carbon, because the carbon-rich compound, cementite, has separated from it. This process continues, until at 723 °C the remaining austenite contains only 0.8 per cent carbon (E). This is, of course, the eutectoid composition; so, at a temperature just below 723 °C, the remaining austenite transforms to pearlite, as in the previous two cases.

Thus the structure of a carbon steel which has been allowed to cool fairly slowly from any temperature above its upper critical temperature will depend upon the carbon content as follows.

- Hypo-eutectoid steels; that is those containing *less than* 0.8 per cent carbon – primary ferrite and pearlite (Fig. 11.4 (iii)).
- Eutectoid steels, containing *exactly* 0.8 per cent carbon – completely pearlite (Fig. 11.4 (iv)).
- Hyper-eutectoid steels; that is those containing *more than* 0.8 per cent carbon – primary cementite and pearlite (Fig. 11.4 (vii)).

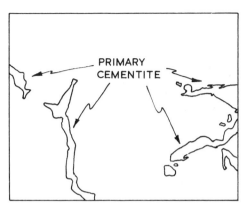

Fig. 11.7   *A high-carbon tool steel (1.2 per cent C) in the cast condition (×620). Since this steel contains more than 0.8 per cent carbon, its structure shows some primary cementite (indicated on the lower 'key' diagram). The remainder of the structure consists of typical lamellar pearlite, comprising layers of cementite sandwiched between layers of ferrite which have tended to join up and so form a continuous background or 'matrix'.*

11.48 Naturally, the proportion of primary ferrite to pearlite in a hypo-eutectoid steel, and also the proportion of primary cementite to pearlite in a hyper-eutectoid steel, will vary with the carbon content, as indicated in Fig. 11.8. This diagram summarizes the structures, mechanical properties, and uses of plain-carbon steels which have been allowed to cool slowly enough for equilibrium structures to be produced.

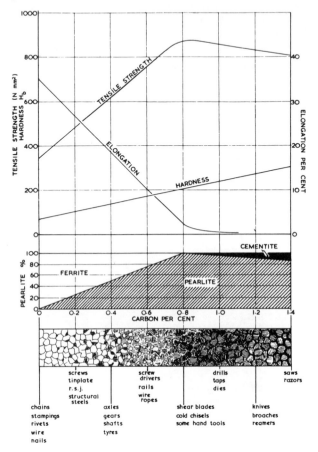

Fig. 11.8 *A diagram showing the relationship between carbon content, mechanical properties, microstructure, and uses of plain-carbon steels which have been slowly cooled from above their upper critical temperatures.*

The heat-treatment of steel

11.50 The exploitation of the properties of iron-carbon alloys is both a tribute to Man's ingenuity – or probably blind chance (12.10) – and the great diversity of the properties of the elements, in this case the polymorphism of iron. Depending upon its composition, steel has a wide range of properties such as are found in no other engineering alloy. Steel can be a soft, ductile material suitable for a wide range of forming processes, or it can be the hardest and strongest metallurgical material in use. This enormous range of properties is controlled by varying both the carbon content and the programme of heat-treatment. Structural effects of the type obtained by the heat-treatment of steel would not be possible were it not for the natural phenomenon of polymorphism (1.40) exhibited by the element iron. It is the transformation from a face-centred cubic structure to one which is basically body-centred cubic, occurring at 910 °C when iron cools, that makes it possible to heat-treat these iron-carbon alloys.

There is not just *one* heat-treatment process, but many which can be applied to steels. In the processes we shall deal with in this chapter, the object of the treatment is to obtain a pearlite type of structure; that is, one in which the steel has been allowed to reach structural equilibrium by employing a fairly slow rate of cooling following the heating process. In the next chapter, we shall deal with those processes where quenching is employed, to arrest the formation of pearlite, and, as a result, increase the hardness and strength of the steel. Within certain limits, *the properties of a steel are independent of the rate at which it has been heated, but are dependent on the rate at which it was cooled.*

11.51 Normalizing The main purpose in normalizing is to obtain a structure which is uniform throughout the work-piece, and which is free of any 'locked-up' stresses. For example, a forging may lack uniformity in structure, because its outer layers have received much more deformation than the core. Thicker sections, which have received little or no working, will be coarse-grained, whilst thin sections, which have undergone a large amount of working, will be fine-grained. Moreover, those thin sections may have cooled so rapidly that they were, in effect, cold-worked, and will consequently be suffering residual stress. If a forging were machined in this condition, its dimensions might well be unstable during subsequent heat-treatment.

Normalizing is a relatively simple heat-treatment process. It involves heating a piece of steel to just above its upper critical temperature, allowing it to remain at that temperature for only long enough for it to attain a uniform temperature throughout, then withdrawing it from the furnace, and allowing it to cool to room temperature in still air.

When, on heating, the work-piece reaches the lower critical temperature, L (Fig. 11.9), the pearlitic part of the structure changes to one of *fine-grained*

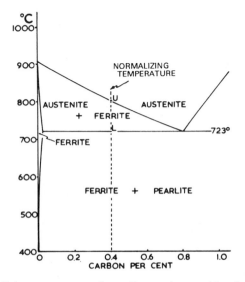

Fig. 11.9 *The normalizing temperature of a medium-carbon steel in relation to the equilibrium diagram.*

austenite; and as the temperature rises, the remaining primary ferrite will be absorbed by the new austenite crystals until, at the upper critical temperature, *U*, this process will be complete, and the whole structure will be of uniformly fine-grained austenite. In practice, a temperature about 30 °C above the upper critical temperature is used, to ensure that the whole structure has reached a temperature just above the upper critical. When the work-piece is withdrawn from the furnace, and allowed to cool, the uniformly fine-grained austenite changes back to a structure which is of uniformly fine-grained ferrite and pearlite. Naturally, the grain size in thin sections may be a little smaller than that in heavy sections, because of the relatively fast rate of cooling of thin sections.

11.52   Annealing A number of different heat-treatment processes are covered by the general description of annealing. These processes are applied to different steels of widely ranging carbon content. The three principal annealing processes will be discussed here.

11.52.1 *Annealing of castings* Sand-castings in steel commonly contain about 0.3 per cent carbon, so that a structure consisting of ferrite and pearlite is obtained. Such a casting, particularly if massive, will cool very slowly in the sand mould. Consequently, its grain size will be somewhat coarse, and it will suffer from brittleness because of the presence of what is known as a 'Widmanstätten' structure. This consists of a directional plate-like formation of primary ferrite grains along certain crystal planes in the original austenite (Fig. 11.10). Since fracture can easily pass along these ferrite plates, the whole structure is rendered brittle as a result.

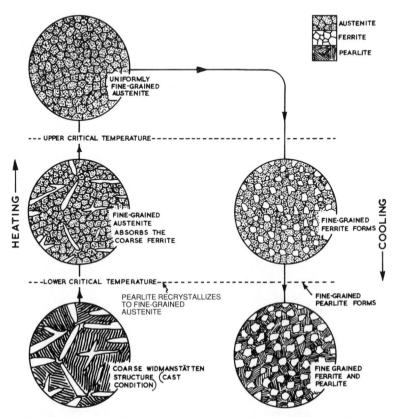

Fig. 11.10   *The refinement of grain in a steel casting during a suitable annealing process.*

The annealing process which is applied in order to refine such a structure is fundamentally similar to that described under 'normalizing' (11.51); that is, the casting is heated to just above its upper critical temperature, so that the coarse grain structure is replaced by one of fine-grained austenite. It is held at this temperature for a sufficient time for the temperature of the casting to become uniform throughout and for the recrystallization to fine-grained austenite to be complete. On cooling this gives rise to a structure of fine-grained ferrite and pearlite. It is in the cooling stage where the two processes differ. Whereas air-cooling is employed in normalizing, in this process the casting is allowed to cool with the furnace. This ensures complete removal of all casting stresses which might otherwise lead to distortion or cracking of the casting, without causing a substantial increase in grain size over that obtained by normalizing. Whilst the tensile strength is not greatly improved by this treatment, both toughness and ductility are considerably increased, so the casting becomes more resistant to mechanical shock.

An annealing temperature of 30–50 °C above the upper critical temperature for the casting is commonly used. If this temperature range is exceeded then the newly formed small crystals of austenite are likely to grow and result in a final structure almost as coarse-grained as the original cast structure. A prolonged holding time at the maximum temperature will also produce a coarse-grained structure. On the other hand annealing at too low a temperature (i.e. below the upper critical temperature) or for too short a period, may mean that the original coarse as-cast structure does not recrystallize completely so that some coarse-grained Widmanstätten plates of ferrite are retained from this original as-cast structure.

11.52.2 *Spheroidization annealing*   Although this may appear an onerous title, it is hoped that its meaning will become clear in the following paragraphs. Essentially it is an annealing process which is applied to high-carbon steels in order to improve their machinability, and, in some cases, to make them amenable to cold-drawing.

It is an annealing process which is carried out *below* the lower critical temperature of the steel; consequently, no phase change is involved, and we do not need to refer to the equilibrium diagram. The work-piece is held at a temperature between 650 and 700 °C for twenty-four hours or more. The pearlite, which of course is still present in the structure at this temperature, undergoes a physical change in pattern – due to a surface-tension[1] effect at the surface of the *cementite* layers within the pearlite. These layers break up into small plates, due to a tendency of the surface to shrink. This effect causes the plates to become gradually more spherical in form (Fig. 11.11) – that is, they spheroidize, or 'ball up'.

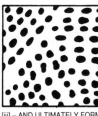
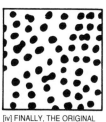

[i] THE NORMAL LAMELLAR FORM OF PEARLITE.   [ii] CEMENTITE LAYERS BEGIN TO BREAK UP –   [iii] – AND ULTIMATELY FORM GLOBULES.   [iv] FINALLY, THE ORIGINAL PATTERN OF THE PEARLITE IS LOST.

Fig. 11.11   *The spheroidization of pearlitic cementite during a subcritical annealing process.*

[1] In a similar manner, surface tension causes water to form rounded droplets when it is spilled on a very dusty floor.

When this condition has been reached, the charge is generally allowed to cool in the furnace. Steel is more easily machined in this state, since stresses set up by the pressure of the cutting edge cause minute chip cracks to form (7.51) in advance of the cutting edge. This is a standard method of improving machinability (7.53).

11.52.3   *Annealing of cold-worked steel*   Like the spheroidization treatment described above, this also is a subcritical annealing process. It is employed almost entirely for the softening of cold-worked mild steels, in order that they may receive further cold-work. Such cold-worked materials must be heated to a temperature above the minimum which will cause recrystallization to take place. Again, the equilibrium diagram is not involved, and the reader should *not* confuse this recrystallization temperature with the lower critical temperature. The latter is at 723 °C, whilst the recrystallization temperature varies according to the amount of previous cold-work the material has received, but is usually about 550 °C. Consequently, stress-relief annealing of mild steel usually involves heating the material at about 650 °C for one hour.

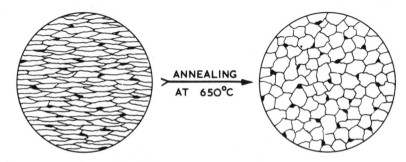

Fig. 11.12   *Annealing of cold-worked mild steel causes recrystallization of the distorted ferrite, so producing new ferrite crystals, which can again be cold-worked.*

This causes the distorted ferrite crystals to recrystallize (Fig. 11.12), so that the structure becomes soft again, and its capacity for cold-work is regained. Since the cold-working of mild steel is usually confined to the finishing stages of the product, any annealing is generally carried out in a controlled atmosphere, in order to avoid oxidation of the surface of the charge. The furnace used generally consists of an enclosed 'retort' through which an inert gas mixture passes whilst the charge is being heated. Such mixtures are based on 'burnt ammonia' or burnt town gas (6.71).

11.60   All of the foregoing heat-treatment processes produce a microstructure in the steel which is basically pearlitic. By this, metallurgists mean that the structure contains some pearlite (unless, of course, the steel is dead-mild). Thus, a hypo-eutectoid steel will contain ferrite and pearlite, a eutectoid steel all pearlite, and a hyper-eutectoid steel cementite and pearlite. Figure 11.13 summarizes the temperature ranges at which these treatments are carried out, and also indicates the carbon contents of steels most commonly involved in the respective processes.

11.70   Brittle fracture in steels   Fracture in metals generally takes place following a measurable amount of slip. Sometimes, however, fracture occurs with little or no such plastic extension. This is termed 'cleavage' or 'brittle' fracture. In plain-carbon steels the phenomenon is temperature dependent

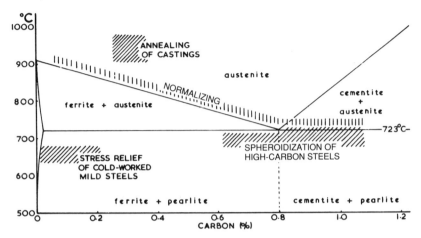

Fig. 11.13   *Temperature ranges of various annealing and normalizing treatments for carbon steels related to the carbon contents.*

and a low-carbon steel which at ambient temperature is tough and ductile will suddenly become extremely brittle at temperatures not far below 0 °C. This type of failure was experienced in the welded 'liberty ships' manufactured during World War II for carrying supplies from USA to Europe, particularly when these ships were used in the very cold North Atlantic.

11.71   Plastic flow depends upon the movement of dislocations during a finite time. As temperature decreases the movement of these dislocations becomes more sluggish so that when a force is increased very rapidly it is possible for stress to increase so quickly that it cannot be relieved by slip. A momentary increase in stress to a value above the yield stress will produce fracture. Such failure will be precipitated from faults like sharp corners or arc-weld spots. In the liberty ships a crack, once started, progressed right round the hull whereas in a normal riveted hull the crack would have been halted at an adjacent rivet hole.

11.72   Metals with a face-centred cubic (FCC) structure maintain ductility at low temperatures whilst some metals with structures other than FCC exhibit this type of brittleness. Body-centred cubic (BCC) ferrite is very susceptible to brittle fracture at low temperatures and the temperature at which brittleness suddenly increases is called the *transition temperature*. For applications involving near atmospheric temperatures the transition point can be depressed to a safe limit by increasing the manganese content to about 1.3 per cent. Where lower temperatures are involved it is better to use a low-nickel steel.

CHAPTER 12

The heat-treatment of plain-carbon steels

12.10   Some of the simpler heat-treatments applied to steel were described in the previous chapter. In the main, they were processes in which the structure either remained or became basically pearlitic as a result of the treatment. Here we shall deal with those processes which are perhaps better known because of their wider use, namely hardening and tempering.

Almost any schoolboy knows that a piece of carbon steel can be hardened by heating it to redness, and then plunging it into cold water. The Ancients knew this too, and no one can tell us who first hardened steel. Presumably such knowledge came by chance, as indeed did most knowledge in days before systematic research methods were instituted. One can well imagine that sooner or later some prehistoric metal-worker heating an iron tool among the glowing charcoal of his fire prior to forging it, and then, seeking to cool the tool quickly, would plunge it into water.

12.11   Although the fundamental *technology* of hardening steel has been well established for centuries, the *scientific principles* underlying the process were for long a subject of argument and conjecture. More than a century ago, Professor Henry Sorby of Sheffield began his examination of the microstructures of steel, but it was only during my undergraduate years that convincing explanations of the phenomenon of hardening were forthcoming.

Principles of hardening

12.20   If a piece of steel containing sufficient carbon is heated until its structure is austentic – that is, until its temperature is above the upper critical temperature – and is then quenched (cooled quickly), it becomes considerably harder than it would be were it cooled slowly.

Generally, when a metallic alloy is quenched, there is a tendency to suppress any change in structure which might otherwise take place if the alloy were cooled slowly. In other words, it is possible to 'trap' or 'freeze in' a metallic structure which existed at a higher temperature, and so preserve it for examination at room temperature. Metallurgists often use this technique when plotting their equilibrium diagrams, and it is also used industrially as, for example, in the solution treatment of some aluminium alloys (17.70).

12.21   Clearly, things do not happen in this way when we quench a steel. Austenite, which is the phase present in a steel above its upper critical temperature, is a soft, malleable material – which is why steel is generally shaped by hot-working processes. Yet when we quench austenite, instead of trapping the soft malleable structure, a very hard, brittle structure is produced, which is most unlike austenite. Under the microscope, this structure appears as a mass of uniform needle-shaped crystals, and is known as *martensite* (Fig. 12.1 (i)). Even at very high magnifications, no pearlite can be seen; so we must conclude that all of the cementite (which is one of the components of pearlite) is still dissolved in this martensitic structure. So far, this is what we would expect. However, investigations using X-ray methods tell us that, although the rapid cooling has prevented the formation of pearlite, *it has not arrested the polymorphic change from face-centred cubic to body-centred cubic.*

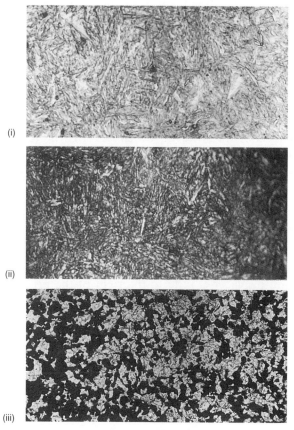

Fig. 12.1   *Representative structures of quenched and tempered specimens of a 0.5 per cent carbon steel: (i) water-quenched from 850 °C – martensite which appears as an irregular mass of needle-shaped crystals but what we see is a cross-section through roughly discus-shaped crystals (×700); (ii) water-quenched from 850 °C and tempered at 400°C – tempered martensite, the crystals of which have become darkened by precipitated particles of cementite (×700); (iii) oil-quenched from 850 °C – the slower cooling rate during quenching has allowed a mixture of bainite (dark) and martensite (light) to form (×100).*

Ferrite is fundamentally body-centred cubic iron which normally will dissolve no more than 0.006 per cent carbon at room temperature (see Fig. 11.4). Thus the structure of martensite is one which is essentially ferrite very much super-saturated with carbon (assuming that the steel we are dealing with contains about 0.5 per cent carbon). It is easy to imagine that this large amount of carbon remaining in *super-saturated* solid solution in the ferrite causes considerable distortion of the internal crystal structure of the latter. Such distortion will tend to prevent slip from taking place in the structure.

Consequently, large forces can be applied, and no slip will be produced. In other words, the steel is now hard and strong.

In order to obtain the hard martensitic structure in a steel, it must be cooled quickly enough; that is, at a speed which is at least as fast as *the critical cooling-rate*. If the steel is cooled at a rate slower than this, then the structure will be less hard, because some of the carbon has had the opportunity to precipitate as cementite. Under the microscope, some dark patches will be visible among the martensite needles. These are due to the precipitation of some tiny particles of cementite, and the structure so produced is called *bainite*, after Dr E. C. Bain – the American metallurgist who did much of the original research into the relationship between structure and rate of cooling of steels. Bainite is of course softer than martensite, but it is tougher and more ductile. Even slower rates of cooling will give structures of fine pearlite.

12.22   The ultimate structure obtained in a plain carbon steel is independent of the rate of heating, assuming that it is heated slowly enough to allow it to become completely austenitic before being quenched. It is the *rate* of subsequent cooling, however, which governs the resultant structure and hence the degree of hardness.

What are termed TTT (time-temperature-transformation) curves govern the relationship between the rate of cooling of a steel and its final microstructure and properties. Initially these curves were constructed, as a result of experimental techniques, for steels which had transformed *isothermally*, that is at a series of single fixed temperatures. Nevertheless in practice we are generally more interested in transformations which occur on falling temperature gradients such as prevail during the water quenching or oil quenching of steels. For this reason slightly modified TTT curves are generally used and these are displaced a little to the right of the original TTT curves. Modified TTT curves are used here (Fig. 12.2). It should be appreciated that a set of TTT curves relates to *one* particular steel of fixed composition.

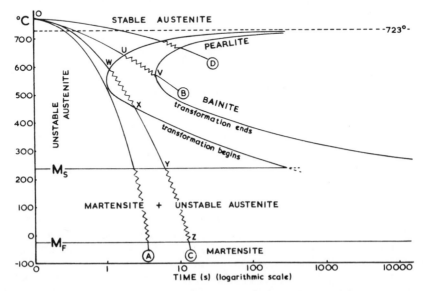

Fig. 12.2   *Modified TTT curves for a 0.8 per cent plain carbon steel. (Note that a logarithmic scale is used on the horizontal axis in order to compress the right-hand end of the diagram without at the same time cramping the important left-hand side.)*

12.23   A TTT diagram consists essentially of two C-shaped curves. The left-hand curve indicates the time interval which elapses at any particular temperature before a carbon steel (in this case one containing 0.8 per cent carbon) in its austenitic state begins to transform, whilst the right-hand curve shows the time which must elapse before this transformation is complete. The expected transformation product at that temperature is indicated on the diagram. The two parallel lines near the foot of the diagram are, strictly speaking, not part of the TTT curves but indicate the temperatures where austenite will start to transform to martensite (M_s) and where this transformation will finish (M_f).

It will be apparent that in order to obtain a completely martensitic structure the steel, previously heated to point O to render it completely austenitic, must be cooled at a rate at least as rapid as that indicated by curve Ⓐ (Fig. 12.2). This represents the *critical cooling rate* for the steel mentioned above. Thus curve Ⓐ just grazes the nose of the 'transformation begins' curve so that the austenitic structure is retained right down to about 180 °C (M_s) when this unstable austenite suddenly begins to change to martensite, this change being completed at about –40 °C (M_f). Since quenching media are at a temperature higher than –40 °C some 'retained austenite' may be present in the quenched component. This retained austenite usually transforms to martensite during subsequent low-temperature tempering processes.

12.24   It follows then that a 0.8 per cent carbon steel must be cooled very rapidly, i.e., from about 750 °C to 120 °C in little more than one second if it is to be completely martensitic (with possibly a little retained austenite as mentioned above). The situation is even more difficult for carbon steels with either more or less than 0.8 per cent carbon. For both hypo- and hyper-eutectoid steels the TTT curves are displaced further to the left making their critical cooling rates even faster. Fortunately the presence of alloying elements slows down transformation rates considerably so that the TTT curves are displaced to the right (Fig. 12.3 (ii)) giving much lower critical

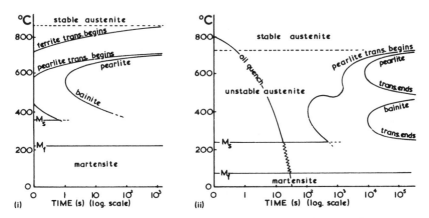

Fig. 12.3   *(i) The TTT curves for a 0.35 per cent carbon steel showing that it is virtually impossible to produce a completely martensitic structure by quenching, since however rapid the cooling rate, ferrite separation inevitably begins as the 'ferrite transformation begins' curve is cut. (ii) The TTT curves for an alloy steel containing 0.6 per cent C; 0.6 per cent Mn; 1.8 per cent Ni; 0.6 per cent Cr and 0.3 per cent Mo. This indicates the effects of alloying in slowing transformation rates so that the TTT curves are displaced far to the right. This steel can be oil quenched to give a martensitic structure.*

cooling rates so that oil- or even air-quenching is possible to give a completely martensitic structure. Even so-called 'plain-carbon steels' contain enough manganese, residual from de-oxidation processes (11.30), to give lower critical cooling rates than equivalent pure iron-carbon alloys.

In Fig. 12.2 curve Ⓑ illustrates the result of quenching a plain 0.8 per cent carbon steel in oil. Here transformation begins at U and is completed at V, the resultant structure being bainite. Curve Ⓒ indicates a rate of cooling intermediate between Ⓐ and Ⓑ. Here transformation to bainite begins at W but is interrupted at X and no further transformation takes place until the remaining austenite begins to change to martensite (at Y), this final transformation being complete at Z. Thus the resulting structure is a mixture of bainite and martensite.

The conditions prevailing during normalizing are indicated by curve Ⓓ. Here transformation to pearlite begins a few degrees below 723 °C (the lower-critical temperature) and is complete a few degrees lower still.

12.25 In practice, factors such as the composition, size, and shape of the component to be hardened govern the rate at which it shall be cooled. Generally, no attempt is made to harden plain-carbon steels, which contain less than 0.35 per cent carbon since the TTT curves for such a steel are displaced so far to the left (Fig. 12.3 (i)) that it is impossible to cool the steel rapidly enough to avoid the precipitation of large amounts of soft ferrite as the cooling curve inevitably cuts far into the nose of the 'transformation begins' curve. Large masses of steel of heavy section obviously cool more slowly than small components of thin section when quenched; so, whilst the outer skin may be martensitic, the inner core of a large component may contain bainite or even pearlite (Fig. 12.4). More important still, articles of heavy section will be more liable to suffer from quench-cracking. This is due to the fact that the outer skin changes to martensite a fraction of a second before layers just beneath the surface, which are still austenitic. Since sudden *expansion* takes place at the instant when face-centred cubic austenite changes to body-centred cubic martensite, considerable stress will be set up between the skin and the layers beneath it, and, as the skin is now hard and brittle due to martensite formation, cracks may develop in it.

Design also affects the susceptibility of a component to quench-cracking. Sharp variations in cross-section, and the presence of sharp angles, grooves, and notches are all likely to increase the possibility of quench-cracking, by causing uneven rates of cooling throughout the component.

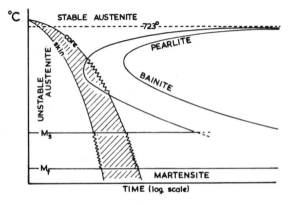

Fig. 12.4

12.26   The rate at which a quenched component cools is governed by the quenching medium and the amount of agitation it receives during quenching. The following media are commonly used, and are arranged in order of quenching speeds.

5% caustic soda solution
5–20% brine
cold water
warm water

mineral oil | synthetic polymer
animal oil | quenchants
vegetable oil

The very drastic quenching resulting from the use of caustic soda solution or brine is used only when extreme hardness is required in components of relatively simple shape. For more complex shapes, it would probably be better to use a low-alloy steel, which has a much lower critical cooling rate (12.24) and can therefore be hardened by quenching in oil. Mineral oils used for quenching are derived from petroleum, whilst vegetable oils include those from linseed and cottonseed. Animal oils are obtained from the blubber of seal and whale, though near extinction of the latter has led to the long overdue banning of whaling by most civilized countries. Synthetic polymer quenchants are now being developed to replace oils – which is good news for those gentle giants of the oceans and, incidentally, for the heat-treatment operatives since less fumes and offensive smells are generated by the 'synthetics'. These new quenchants consist of synthetic polyalkane glycols which can be mixed with water in varying proportions to give different quenching rates.

The hardening process

12.30   To harden a hypo-eutectoid steel component, it must be heated to a temperature of 30–50 °C above its upper critical temperature, and then quenched in some medium which will produce in it the required rate of cooling. The medium used will depend upon the composition of the steel, the size of the component, and the ultimate properties required in it. Symmetrically shaped components, such as axles, are best quenched 'end-on', and all components should be violently agitated in the medium during the quenching operation.

12.31   The procedure in hardening a hyper-eutectoid steel is slightly different. Here a quenching temperature about 30 °C above the *lower critical* temperature is generally used. In a hyper-eutectoid steel, primary cementite is present, and, on cooling from above the upper critical temperature, this primary cementite tends to precipitate as long, brittle needles along the grain boundaries of the austenite. This type of structure would be very unsatisfactory, so its formation is prevented by continuing to forge the steel whilst this primary cementite is being deposited – that is, between the upper and lower critical temperatures. In this way, the primary cementite is broken down into globules during the final stages of shaping the steel. During the subsequent heat-treatment, it must never be heated much more than 30 °C above the lower temperature, or there will be a tendency for primary cementite to be absorbed by the austenite, and then precipitated again as long brittle needles on cooling. When a hyper-eutectoid steel has been correctly

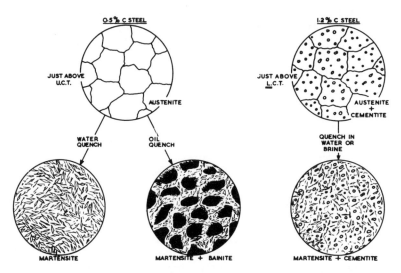

Fig. 12.5 *Typical microstructures produced when quenching both medium-carbon and tool steels in their appropriate media.*

hardened, its structure should consist of small near-spherical globules of very hard cementite (Fig. 12.5) in a matrix of hard, strong martensite.

Tempering

12.40   A fully hardened carbon steel is relatively brittle, and the presence of quenching stresses makes its use in this condition inadvisable unless extreme hardness is required. For these reasons, it is usual to reheat, or 'temper', the quenched component, so that stresses are relieved, and, at the same time, brittleness and extreme hardness are reduced.

As we have seen, the martensitic structure in hardened steel consists essentially of ferrite which is heavily super-saturated with carbon. By heating such a structure to a high enough temperature, we shall enable it to begin to return to equilibrium, by precipitating carbon in the form of tiny particles of cementite.

12.41   On heating the component up to 200 °C, no change in the microstructure occurs, though quenching stresses are relieved to some extent whilst hardness may even increase slightly as any retained austenite transforms to martensite. At about 230 °C, tiny particles of cementite are precipitated from the martensite, though these are so small that they are difficult to see with an ordinary microscope. Generally, the microstructure appears somewhat darker, but still retains the shape of the original martensite needles. This type of structure persists as the temperature is increased to about 400 °C (Fig. 12.1 (ii)) with more and more tiny cementite particles being precipitated, and the steel becoming progressively tougher – though softer than the original martensite. The structure so produced was commonly known as *troostite*.

12.42   Tempering at temperatures above 400 °C causes the cementite particles to coalesce (or fuse together) to such an extent that they can be seen clearly at magnifications of about ×500. At the same time, more cementite is precipitated. The structure, which is relatively granular in appearance, was known as *sorbite*. It must be emphasized that there is no fundamental difference between troosite and sorbite, since both are formed by precipitation of cementite from martensite; and there is no definite temperature where troosite formation ceases and formation of sorbite begins. Naturally, sorbite is softer and tougher than troosite, because still more carbon has been precipitated from the original martensite structure.

The names 'troosite' and 'sorbite' are long since obsolete and should not be used. The modern metallurgist describes these structures as 'tempered martensite' mentioning the temperature used during the tempering process.

Generally speaking, low temperatures (200–300 °C) are used for tempering various types of high-carbon steel tools where hardness is the prime consideration, higher temperatures (400–600 °C) being used for tempering stress-bearing medium-carbon constructional steels where strength, toughness, and general reliability are more important.

12.43   Furnaces used for tempering are usually of the batch type, in which the charge is carried in a wire basket through which hot air circulates. By this method, the necessarily accurate temperature can easily be maintained. The traditional method of treating tools is to 'temper by colour', and this still provides an accurate and reliable method of dealing with plain-carbon steels. After the tool has been quenched, its surface is first cleaned to expose bright metal. The tool is then slowly heated until the thin oxide skin which forms on the surface attains the correct colour (Table 12.1). It should be noted that this technique applies only to plain-carbon steels, since some of the alloy steels, particularly those containing chromium, do not oxidize readily. A summary of typical heat-treatment programmes, and uses of the complete range of plain-carbon steels, is given in Table 12.2.

Table 12.1   *Tempering colours for carbon steels.*

Temperature (°C)	Colour	Types of component
220	pale yellow	scrapers, hack-saws, light turning-tools
230	straw	hammer faces, screwing-dies for brass, planing- and slotting-tools, razor-blades
240	dark straw	shear blades, milling-cutters, drills, boring-cutters, reamers, rock-drills
250	light brown	penknife blades, taps, metal shears, punches, dies, woodworking tools for hardwood
260	purplish-brown	plane blades, stone-cutting tools, punches, reamers, twist-drills for wood
270	purple	axes, augers, gimlets, surgical tools, press-tools
280	deeper purple	cold-chisels (for steel and cast iron), chisels for wood, plane-cutters for softwood
290	bright blue	cold-chisels (for wrought iron), screwdrivers
300	darker blue	wood-saws, springs

Table 12.2 *Heat-treatments and typical uses of plain-carbon steels.*

Type of Steel	Carbon %	Heat-treatment			Typical uses
		Hardening temp. (°C)	Quenching medium	Tempering temp. (°C)	
dead-mild	Up to 0.15	These steels do not respond to heat-treatment, because of their low carbon-content			nails, chains, rivets, motor-car bodies
mild	0.15 to 0.25				structural steels (RSJ), screws, tinplate, drop-forgings, stampings, shafting, free-cutting steels
medium-carbon	0.25 to 0.35	880 to 850*	Oil or water, depending upon type of work		couplings, crankshafts, washers, steering arms, lugs, weldless steel tubes
	0.35 to 0.45	870 to 830*		Temper as required	crankshafts, rotor shafts, crank pins, axles, gears, forgings of many types
	0.45 to 0.60	850 to 800*	Oil, water, or brine, depending upon type of tool		hand-tools, pliers, screwdrivers, gears, die-blocks, rails, laminated springs, wire ropes
high-carbon tool	0.60 to 0.75	820 to 800*		275–300	hammers, dies, chisels, miners' tools, boilermakers' tools, set-screws
	0.75 to 0.90	800 to 820	Water or brine. Tools should not be allowed to cool below 100 °C before tempering	240–250	cold-chisels, blacksmiths' tools, cold-shear blades, heavy screwing dies, mining drills
	0.90 to 1.05	780 to 800		230–250	hot-shear blades, taps, reamers, threading and trimming dies, mill-picks
	1.05 to 1.20	760 to 780		230–250	taps, reamers, drills, punches, blanking-tools, large turning-tools
	1.20 to 1.35	760 to 780		240–250	lathe tools, small cold-chisels, cutters, drills, pincers, shear blades
	1.35 to 1.50	760 to 780		200–230	razors, wood-cutting tools, surgical instruments, drills, slotting-tools, small taps

*The *higher* temperature for the *lower* carbon-content.

Isothermal heat-treatments

12.50   A knowledge of isothermal transformations can be used in the practical heat-treatment of steels. The risk of cracking and/or distortion during the rather drastic water-quenching of carbon steels has already been mentioned (12.25) and such difficulties may be overcome in the case of suitably dimensioned work-pieces by both *martempering* and *austempering*.[1]

12.51   The principles of martempering are indicated in Fig. 12.6 (i). Here a carbon steel component has been quenched into a bath (either of hot oil or molten low-melting point alloy) held at a temperature just above M_s. The component is allowed to remain there for a time interval sufficient for the whole component to have *reached a uniform temperature throughout*. It is then removed from the bath and allowed to cool *very slowly* in warm air. (Note that in Fig. 12.6 (i) the cooling curve is foreshortened by the use of the logarithmic scale. Thus several minutes will elapse before the steel, at the quench-bath temperature, would begin to transform, giving ample time for uniformity of temperature to be attained in the work-piece). Since under these conditions both skin and core of the component pass through the M_s and M_f lines almost simultaneously there is little chance of those stresses being set up which may induce either distortion or cracking in the hard martensitic structure which results.

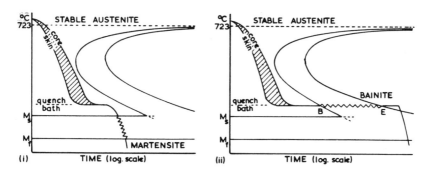

Fig. 12.6   *The isothermal treatments of (i) martempering and (ii) austempering.*

12.52   Austempering (Fig. 12.6 (ii)) is a means of obtaining a tempered type of structure without the necessity of a preliminary drastic water quench which is involved in traditional methods of heat-treatment. Again the work-piece is quenched from the austenitic state into a bath held at some suitable temperature above M_s but in this case it remains there long enough for transformation to occur to completion between B and E on the two TTT curves, yielding a structure of bainite which will be similar in properties to those structures of traditionally tempered martensite. The rate at which the work-piece is finally cooled is not important since transformation is already complete at E.

12.53   Although austempering and martempering would seem to provide enormous advantages in that risks of distortion and cracking of work-pieces

[1]   These two processes are now stuck with these rather ugly and clumsy titles, which unfortunately do not accurately describe the principles involved.

are largely removed there is one very obvious drawback to the wide application of such processes, namely that their use is limited generally to components of thin section since the *whole* of the cross-section of the work-piece must be capable of being cooled rapidly enough to miss the nose of the 'transformation-begins' curve of the TTT diagram appropriate to the composition of the steel being used. Thin-sectioned components which are austempered include steel toe-caps of industrial boots, whilst garden spades and forks were similarly heat-treated long before the fundamental theory of the process was investigated by Davenport and Bain in the nineteen-thirties. Similarly the *patenting* of high-tensile steel wire was achieved by winding it through an austenitizing furnace (at 970 °C) followed by a bath of molten lead (at 500 °C) where the structure transformed directly from austenite to bainite. In this condition the wire can be further hardened by cold-drawing.

Mass effect and hardenability

12.60   In order to harden a piece of steel completely, it must be cooled quickly from its austenitic state. The rate of cooling must be as great as, or greater than, the 'critical rate', otherwise the section will not be completely martensitic. Clearly, this will not be possible for a work-piece of very heavy section; for, whilst the outer skin will cool at a speed greater than the critical rate, the core will not. Consequently, whilst the outer shell may be of hard martensite, the core may be of bainite, or even fine pearlite.

12.61   This phenomenon is generally referred to as the 'mass effect' of heat-treatment, and plain-carbon steel is said to have a 'shallow depth of hardening', or, alternatively, 'a poor hardenability'. Whilst a rod of, say, 12 mm diameter in plain-carbon steel can be water-quenched and have a martensitic structure throughout, one of, say, 30 mm diameter will have a core consisting of bainite, and only the outer shell will be martensitic. The influence of sectional thickness and quenching medium used on the structure produced in a low-alloy steel is illustrated in Fig. 12.7.

12.62   For some applications, this variation in structure across the section will not matter, since the outer skin will be very hard, and the core reasonably tough; but in cases where a component is to be quenched and then tempered for use as a stress-bearing member, it is essential that the structure obtained by quenching is uniform throughout.

Ruling section

12.70   Fortunately, the addition of alloying elements to a steel reduces its critical rate so that such a steel can be oil-quenched in thin sections, or, alternatively, much heavier sections can be water-quenched. Thus an alloy steel generally has a much greater depth of hardening than has a plain-carbon steel of similar carbon content.

However, it would be a mistake to assume that *any* alloy steel in *any* thickness will harden right through when oil-quenched from above its upper critical temperature. The low-alloy steel used as an example in Fig. 12.7 illustrates this point. This steel has a critical cooling rate just a little lower than that of a plain-carbon steel; but, as the diagram shows, sections much over 14 mm in diameter will not harden completely unless quenched in brine, and the maximum diameter which can be hardened completely – even with this drastic treatment – is only about 30 mm.

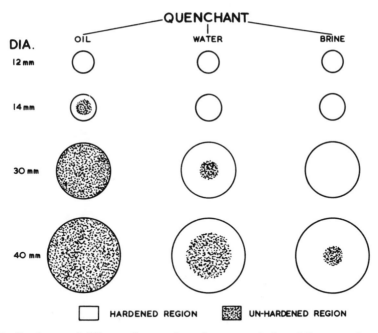

Fig. 12.7   *Specimens of different diameter have been quenched in different media, and the depth of hardening assessed in each case. A low-alloy steel containing 0.25 per cent C, 0.6 per cent Mn, 0.2 per cent Ni, and 0.2 per cent Mo was used.*

12.71   In order to prevent the misuse of steels by those who imagine that an alloy steel can be hardened to almost any depth, both the British Standards Institution and manufacturers now specify limiting ruling sections for each particular composition of steel. The limiting ruling section is quoted as the *maximum diameter* which can be heat-treated (under conditions of quenching and tempering suggested by the manufacturer) for the stated mechanical properties to be obtained. As an example, the following table shows a set of ruling sections for a low-alloy steel (BS 970/530M40), along with a manufacturer's suggested heat-treatments, and the corresponding BS specifications in respect of tensile strength.

Limiting ruling section (mm)	*Suggested heat-treatment*	*Tensile strength* $(N\,mm^{-2})$
100	oil harden	695
62.5	from 850 °C	770
29	and temper at 650 °C	850

In BS 970, formulae are given for deriving the equivalent diameters of rectangular and other bars, so that information as to the ruling section of the material can be correctly applied.

The Jominy end-quench test

12.80   This test is of considerable value in assessing the hardenability of a steel. A standard test-piece (Fig. 12.8 (ii)) is heated to above the upper

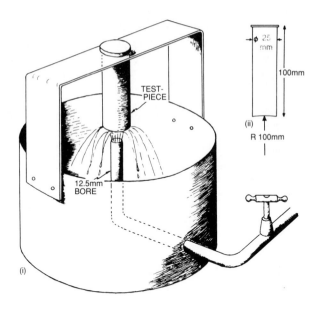

Fig. 12.8   *The Jominy end-quench test. (i) A simple type of apparatus in which to conduct the test. The end of the water pipe is 12 mm below the bottom of the test-piece, but the 'free height' of the water jet is 63 mm. (ii) A typical test-piece.*

critical temperature of the steel; that is, until it becomes completely austenitic. It is then very quickly transferred from the furnace, and immediately dropped into position in the frame of the apparatus shown in Fig. 12.8 (i)). Here it is quenched at one end only, by a standard jet of water at 25 °C; thus, different rates of cooling are obtained along the length of the test-piece. When the test-piece has cooled, a 'flat' approximately 0.4 mm deep is ground along the length of the bar, and hardness determinations are made every millimetre along the length, from the quenched end. The results are then plotted (Fig. 12.9).

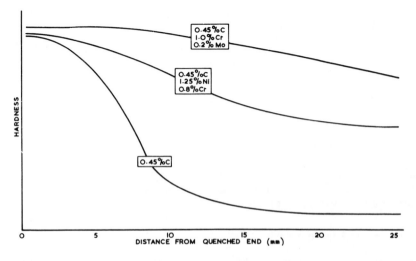

Fig. 12.9   *The depth of hardening of various steels of similar carbon content, as shown by the Jominy test.*

12.81   These curves show that a low nickel-chromium steel hardens to a greater depth than a plain-carbon steel of similar carbon content, whilst a chromium-molybdenum steel hardens to an even greater depth.

Whilst the Jominy test gives a good indication as to how deeply a steel will harden, there is no simple mathematical relationship between the results of this test and the ruling section of a steel. Using the results of the Jominy test as a basis, it is often more satisfactory to find the ruling section by trial and error. Hardenability curves of the type shown in Fig. 12.9 are included for a number of engineering steels in BS 970: Part 1.

Furnaces and furnace atmospheres used in heat-treatment of steels

12.90   It was mentioned earlier (12.43) that hardened steel is generally tempered in some form of furnace in which hot air is circulated over the charge. Since the temperatures involved are relatively low little or no change in composition occurs at the surface of the steel. Hardening temperatures however are much higher so that both decarburisation and oxidation of the surface can occur unless preventative measures are taken. Baths of electrically heated molten salt are sometimes used and these not only maintain an accurate quenching temperature due to the high heat capacity of the salt, but also provide protection of the surface from decarburisation and oxidation.

12.91   When atmospheres derived from hydrocarbons (6.72) are used both carbon dioxide and water vapour produced can act as decarburisers and oxidizers. Thus carbon (in the surface of the steel) will combine with carbon dioxide (in the furnace atmosphere) to form carbon monoxide:

$$C + CO_2 \rightleftharpoons 2CO$$

whilst iron itself reacts with carbon dioxide to form iron oxide (scale) and carbon monoxide:

$$Fe + CO_2 \rightleftharpoons FeO + CO$$

Water vapour can react with iron to form iron oxide and hydrogen:

$$Fe + H_2O \rightleftharpoons FeO + H_2$$

For these reasons the compositions of both exothermic and endothermic gases (6.73) must be controlled to reduce the quantities of carbon dioxide and water vapour they contain when used as furnace atmospheres in the high-temperature heat-treatment of steels.

CHAPTER 13

Alloy steels

13.10 The first deliberate attempt to develop an alloy steel was made by Sir Robert Hadfield during the 1880s but most of these materials are products of the twentieth century. Possibly the greatest advance in this field was made in 1900, when Taylor and White, two engineers with the Bethlehem Steel Company in the USA, introduced the first high-speed steel, and so helped to increase the momentum of the Industrial Revolution and, in particular, what has since been known as 'mass-production'.

So-called plain-carbon steels contain up to 1.0 per cent manganese, which is the residue of that added to deoxidize and desulphurize the steel just before casting. Consequently a steel used not to be classified as an 'alloy steel' unless it contained not less than 2 per cent manganese and/or other elements (nickel, chromium, molybdenum, vanadium, tungsten, etc.) in amounts of each between 0.1 per cent and 0.5 per cent minimum. Now that steels of very high purity are commonly produced commercially alloy additions of as little as 0.005 per cent can significantly influence mechanical properties and a range of 'micro-alloyed' or 'high-strength low-alloy' (HSLA) steel was introduced. This renders the above definition of an 'alloy steel' somewhat obsolete.

Although there are alloy steels with special properties, such as the stainless steels, and heat-resisting steels, the main purpose of alloying is to improve the existing properties of carbon steels, making them more adaptable and easier to heat-treat successfully. In fact, one of the most important and useful effects of alloying was mentioned in the previous chapter (12.70) – the improvement in 'hardenability'. Thus an alloy steel can be successfully hardened by quenching in oil, or even in an air blast, with less risk of distortion or cracking of the component than is associated with water-quenching. Moreover, suitable alloy steels containing as little as 0.2 per cent carbon can be hardened successfully because of the considerable slowing down of transformation rates imparted by the alloying elements (Fig. 12.3(ii)).

13.11 Alloying elements (or 'alloys' as they are often called) can be divided into two main groups.

1. Those which strengthen and toughen the steel by dissolving in the ferrite. These elements are used mainly in constructional steels, and include nickel, manganese, small amounts of chromium, and even smaller amounts of molybdenum.
2. Alloying elements which combine chemically with some of the carbon in

the steel, to form carbides which are much harder than iron carbide (cementite). These elements are used mainly in tool steels, die steels, and the like. They include chromium, tungsten, molybdenum, and vanadium.

13.12 Other alloying elements which are added in small amounts and for special purposes include titanium, niobium, aluminium, copper, boron and silicon. Even sulphur, normally regarded as the steel-maker's greatest enemy, is utilized in free-cutting and 'bright-drawn' steels (7.55 and 7.57).

Alloy steels may be classified into three main groups:

1. constructional steels, which are generally used for machine parts highly stressed in tension or compression;
2. tool steels, requiring great hardness and, in some cases, resistance to softening by heat;
3. special steels; for example, stainless steels and heat-resisting steels.

Constructional steels

13.20 Whilst the 'nickel-chrome' steels are the best known in this group, other alloy steels containing the elements nickel and chromium singly are also important.

13.21 **Nickel steels** Nickel increases the strength of a steel by dissolving in the ferrite. Its main effect, however, is to increase toughness by limiting grain-growth during heat-treatment processes. For this reason, up to 5.0 per cent nickel is present in some of the better quality steels used for case-hardening.

Unfortunately, nickel does not combine chemically with carbon, and, worse still, tends to make iron carbide (cementite) decompose and so release free graphite. Consequently, nickel steels are always low-carbon steels, or, alternatively, medium-carbon steels with very small amounts of nickel. However, because of their shortcomings in respect of carbide instability, they have been almost entirely replaced in recent years by other low-alloy steels.

13.22 **Chromium steels** When chromium is added to a steel, some of it dissolves in the ferrite (which is strengthened as a result), but the remainder forms chromium carbide. Since chromium carbide is harder than ordinary iron carbide (cementite), the hardness of the steel is increased. Because chromium forms stable carbides, these steels may contain 1.0 per cent or even more of carbon.

The main disadvantage of chromium as an alloying element is that, unlike nickel, it increases grain growth during heat-treatment. Thus, unless care is taken to limit both the temperature and the time of such treatment, brittleness may arise from the coarse grain produced.

As indicated by the uses mentioned in Table 13.2, these low-chromium steels are important because of their increased hardness and wear-resistance.

13.23 **Nickel-chromium steels** In the foregoing sections dealing with nickel and chromium, it was seen that in some respects the two metals have opposite effects on the properties of a steel. Thus, whilst nickel is a grain-refiner, chromium tends to cause grain-growth. On the other hand, whilst

Table 13.1   Nickel steels.

BS specification 970:	Composition (%)	Typical mechanical properties			Heat-treatment	Uses
		Yield point (N mm⁻²)	Tensile strength (N mm⁻²)	Izod (J)		
–	0.4 C 1.0 Mn 1.0 Ni	500	700	96	Oil-quench from 850°C, temper between 550°C and 650°C	Crankshafts, axles, other parts in the motor-car industry and in general engineering
–	0.12 C 0.45 Mn 3.0 Ni	510	775	86	After carburizing: refine grain by an oil-quench from 860°C, then harden by a water-quench from 770°C	A *cast-hardening steel*: crown-wheels, differential pinions, cam-shafts

Table 13.2   Chromium steels.

BS specification 970:	Composition (%)	Typical mechanical properties			Heat-treatment	Uses
		Yield point (N mm⁻²)	Tensile strength (N mm⁻²)	Hardness (Brinell)		
530M40	0.45 C 0.9 Mn 1.0 Cr	880	990	–	Oil-quench from 860°C, temper at 550–700°C	Agricultural machine parts, machine-tool components, parts for concrete and tar mixers, excavator teeth, automobile axles, connecting rods and steering arms, spanners
535A99	1.0 C 0.45 Mn 1.4 Cr	–	–	850	Oil-quench from 810°C, temper at 150°C	Ball- and roller-bearings, roller- and ball-races, cams, small rolls

chromium is a carbide-stabilizer, nickel tends to cause carbides to break down, releasing graphite. Fortunately, the beneficial effects of one metal are stronger than the adverse effects of the other, and so it is advantageous to add these metals together to a steel. Generally about two parts of nickel to one of chromium is found to be the best proportion.

In other respects, the two metals, as it were, work together, and so the hardenability is increased to the extent that, with 4.25 per cent nickel and 1.25 per cent chromium, an *air-hardening* steel is produced; that is, one which can be 'quenched' in an air blast, thus making cracking or distortion even less likely than if the steel were oil-quenched. However, for air-hardening, a ruling section of 62.5 mm diameter must be observed, and for greater diameters than this the steel must be oil-quenched if the stated properties are to be obtained.

Unfortunately, these straight nickel-chromium steels suffer from a defect known as 'temper brittleness' (described in the next section) and for this reason straight nickel-chromium steels have been almost entirely replaced by nickel-chromium-molybdenum steels.

13.24 Nickel-chromium-molybdenum steels As mentioned above, a severe drawback in the use of straight nickel-chromium steels is that they suffer from a defect known as 'temper brittleness'. This is shown by a serious decrease in toughness (as indicated by a low Izod or Charpy impact value) when a quenched steel is subsequently tempered in the range 250–580 °C. Further, if such a steel is tempered at 650 °C, it must be cooled quickly through the 'dangerous range' by quenching it in oil, following the tempering process. Although incorrect heat-treatment may lead to this disastrous reduction in impact toughness, the tensile strength and percentage elongation may not be seriously affected. Consequently, a tensile test alone would not reveal the shortcomings of such a steel, and the importance of impact testing in cases like this is obvious.

Fortunately, temper brittleness can be largely eliminated by adding about 0.3 per cent molybdenum to the steel, thus establishing the well-known range of 'nickel-chrome-moly' steels.

13.25 Manganese steels Most steels contain some manganese remaining from the deoxidation and desulphurization processes, but it is only when the manganese content exceeds 1.0 per cent that it is regarded as an alloying element. Manganese increases the strength and toughness of a steel, but less effectively than does nickel. Like all elements, it increases the depth of hardening. Consequently, low-manganese steels are used as substitutes for other, more expensive, low-alloy steels.

Manganese is a metal with a structure somewhat similar to that of austenite at ordinary temperatures; therefore, when added to a steel in sufficient quantities, it tends to stabilize the FCC (austenitic) structure of iron at lower temperatures than is normal for austenite. In fact, if 12.0 per cent manganese is added to a steel containing 1.0 per cent carbon, the structure remains austenitic even after the steel has been *slowly cooled* to room temperature. The curious – and useful – fact about this steel is that, if the surface suffers any sort of mechanical disturbance it immediately becomes extremely hard. Some suggest that this is due to spontaneous martensite formation but others think it is simply due to work-hardening. Whatever the reasons the result is a soft but tough austenitic core with a hard wear-resistant shell which is useful in conditions where both mechanical shock and severe abrasion prevail, as in dredging, earth-moving,

Table 13.3 *Nickel-chromium-molybdenum steels.*

BS specification 970:	Composition (%)	Typical mechanical properties			Heat-treatment	Uses
		Yield point (N mm^{-2})	Tensile strength (N mm^{-2})	Izod (J)		
817M40	0.4 C 0.55 Mn 1.5 Ni 1.1 Cr 0.3 Mo	990	1010	72	Oil-quench from 840 °C, temper at 600 °C	Differential shafts, crank-shafts and other high-stressed parts. (If tempered at 200 °C, it is suitable for machine-tool and automobile gears)
826M40	0.4 C 0.65 Mn 2.5 Ni 0.65 Cr 0.55 Mo	850	1000	45	Oil-quench from 830 °C, temper at 600 °C	Thin sections where maximum shock resistance and ductility are required, e.g. connecting rods, inlet-valves, cylinder-studs, valve-rockers
835M30	0.3 M 0.5 Mn 4.25 Ni 1.25 Cr 0.3 Mo	1450	1700	37	Air-harden from 830 °C, temper at 150–200 °C	An air-hardening steel for aero-engine connecting rods, valve-mechanisms, gears, differential shafts, etc. Suitable for other highly stressed parts

Table 13.4 *Manganese steels.*

BS specification 970:	Composition (%)	Typical mechanical properties				Heat-treatment	Uses
		Yield point (N mm⁻²)	Tensile strength (N mm⁻²)	Izod (J)	Brinell		
150M36	0.35 C 1.5 Mn	510	710	71	–	Oil-quench from 850 °C, temper at 600 °C	Automobile and general engineering, as a cheaper substitute for the more expensive nickel-chromium steels
605M36	0.36 C 1.6 Mn 0.3 Mo	1000	1130	72	–	Oil-quench from 850 °C, temper at 600 °C	
– (Hadfield steel)	1.2 C 12.5 Mn	–	–	–	Case-550 Core-200	Finish by quenching from 1050 °C to keep carbides in solution – quenching does *not* harden the steel, however	Rock-crushing equipment, buckets, heel-plates and bucket-lips for dredging equipment, earth-moving equipment, trackway crossings and points

and rock-crushing equipment. A further point of interest regarding this steel is that it was one of the very first alloy steels to be developed – by Sir Robert Hadfield in 1882 – though little use was made of it until the early days of the twentieth century.

13.26   Boron steels Boron is a non-metal generally familiar in the form of its compound 'borax'. The pure element is a hard grey solid with a melting point of 2300 °C. In recent years it has been developed as an alloying element in some steels, particularly in the USA.

Extremely small amounts – 0.0005 to 0.005 per cent – added to fully deoxidized steels are effective in reducing the austenite → ferrite + pearlite transformation rates in those steels containing between 0.2 and 0.5 per cent carbon, i.e. the TTT curves are displaced appreciably to the right so that these low-carbon steels can then be effectively hardened. Moreover in some low-alloy steels the amounts of other expensive elements like nickel, chromium and molybdenum can be reduced by as much as half if small amounts of boron are included. Low-carbon, manganese steels containing boron are used in Britain for high-tensile bolts and thread-rolled wood screws (see Table 13.5).

Table 13.5   *Some steels containing boron.*

Typical compositions (%)						Typical mechanical properties			Uses
						Yield point	Tensile strength	Elong.	
C	Mn	Ni	Cr	Mo	B	$(N\,mm^{-2})$	$(N\,mm^{-2})$	(%)	
0.2	1.5	–	–	–	0.0005	690	790	18	Some structural steels
0.17	0.6	–	–	0.6	0.003	690	860	18	High-tensile constructional steels
0.45	0.8	0.3	0.4	0.12	0.002	700	1000	15	Automobile engineering

13.27   Maraging steels are a group of very high-strength alloys which were used in aerospace projects such as the Lunar Rover Vehicle, but in general engineering have found a wide variety of uses such as the flexible drive shafts for helicopters, barrels for rapid-firing guns, die-casting dies and extrusion rams. On examining the composition of such alloys (Table 13.6) one notices that the amount of carbon present is very small. Indeed carbon plays no part in developing the high strength and is only residual from the manufacturing process, so that these alloys should be thought of as high-strength alloys rather than as steels, in the sense that steels normally depend for their properties on the presence of carbon.

Cobalt and nickel are essential constituents of maraging 'steels'. If such an alloy is solution treated (9.60) at 820 °C to absorb precipitated intermetallic compounds uniform austenite is formed. On cooling in air an iron-nickel variety of martensite is produced due to the retardation of transformation rates caused by the large amounts of alloying elements present. This form of 'martensite' however is softer and tougher than ordinary martensite based on the presence of carbon. If the alloy is now

Table 13.6 Typical maraging 'steels'.

Composition (%)						Heat-treatment	Typical mechanical properties			
Ni	Co	Mo	Ti	Al	C		0.2% proof stress (N mm−2)	Tensile strength (N mm−2)	Elongation (%)	Impact (Charpy) (J)
18	8.5	3	0.2	0.1	0.01	Solution treated at 820°C for one hour, air-cooled and age hardened at 480°C for three hours	1430	1565	9	52
18	9	5	0.5	0.1	0.01		1930	1965	7.5	21
17.5	12.5	3.75	1.6	0.15	0.01	Solution treated at 820°C for one hour, air-cooled and age hardened at 480°C for 12 hours	2390	2460	8	11

'age-hardened' (17.73) at 480 °C for three hours or more, coherent precipitates of intermetallic compounds ($TiNi_3$, $MoNi_3$ or $AlNi_3$) are formed making slip along crystal planes more difficult so that high tensile strengths up to $2400 \, N \, mm^{-2}$ result.

The main function of cobalt seems to be in providing more lattice positions where the coherent precipitates are able to form. These alloys combine considerable toughness with high strength when heat-treated and are far superior – but also very much more expensive – than conventional alloy constructional steels. Heat treatment however is relatively uncomplicated since there can be no decarburization and no water quench is required. They are also very suitable for surface hardening by nitriding (14.50).

13.28   High-strength low-alloy (HSLA) steels   Now that modern methods of steelmaking can produce dead-mild steel of *very high purity* it is possible to make 'micro' additions of some alloying elements which – in the absence of impurities that might otherwise 'interfere' – produce considerable improvements in mechanical properties. Quite small amounts of niobium, vanadium, titanium and aluminium are used in HSLA steels for the construction of buildings, bridges and pipelines. Increase in strength is due to the formation of tiny particles of titanium and niobium carbides and nitrides. These hard particles reduce grain growth and also oppose slip thus increasing strength. A typical off-shore pipeline alloy contains: C – 0.09; Al – 0.03; Mn – 1.35; Nb – 0.035; and V – 0.07 per cent.

In a different class, some micro-alloy steels are made from 'vacuum degassed' dead-mild steel to which small quantities of titanium or niobium are added. These elements combine with the very small amounts of carbon or nitrogen which may be present so that the latter are not available to form interstitial solid solutions in the iron. The absence of interstitially dissolved carbon or nitrogen results in a material with a very low yield stress but a very high ductility – ideal for cold-forming processes. It is used for auto-body pressings.

Tool steels and die steels

13.30   The primary requirement of a tool or die steel is that it shall have considerable hardness and wear-resistance, combined with reasonable mechanical strength and toughness. A plain high-carbon tool steel possesses these properties, but unfortunately its cutting edge softens easily on becoming over-heated during a high-speed cutting process. Similarly, dies which are to be used for hot-forging or extrusion operations cannot be made from plain-carbon steel, which, in the heat-treated state, begins to soften if heated to about 220 °C. Consequently, tool steels which work at high speeds, or die steels which work at high temperatures, are generally alloy steels containing one or more of those elements which form very hard carbides – chromium, tungsten, molybdenum, or vanadium. Of these elements, tungsten and molybdenum also cause the steel, once hardened, to develop a resistance to tempering influences, whether from contact with a hot work-piece, or from frictional heat. Thus, either tungsten or molybdenum is present in all high-speed steels, and in most high-temperature die steels.

13.31   Die steels As mentioned above, these materials will contain at least one of the four metals which form hard carbides; whilst hot-working dies

Table 13.7   *Tool steels and die steels (other than high-speed steels).*

BS specifi-cation 4659	Composition (%)	Hardness (VPN)	Heat-treatment	Uses
	0.6 C 0.65 Mn 0.65 Cr	700	Oil-quench from 830 °C. Temper: (i) for cold-working tools at 200–300 °C (ii) for hot-working tools at 400–600 °C	Blacksmiths' and boiler-makers' tools and chisels, swages, builders', masons' and miners' tools, chuck- and vice-jaws, hot-stamping and forging dies
BD 3	2.1 C 12.5 Cr	850	Heat slowly to 800 °C, then raise to 980 °C, and oil-quench. Temper between 150 °C and 400 °C for 30–60 minutes	Blanking punches, dies, shear blades for hard thin materials, dies for moulding ceramics and other abrasive powders, master-gauges, thread-rolling dies
BH 11	0.35 C 5.0 Cr 1.25 Mo 0.3 V 1.0 Si	Tempered at 550 °C: 600 Tempered at 650 °C: 375	Pre-heat to 850 °C, then heat to 1000 °C. Soak for 10–30 minutes, and air-harden. Temper at 550–650 °C for two hours	Hot-forging dies for steel and copper alloys where excessive temperatures are not encountered, extrusion dies for aluminium alloys, pressure and gravity dies for casting aluminium
BD 2A	1.6 C 13.0 Cr 0.8 Mo 0.5 V	Tempered at 200 °C: 800 Tempered at 400 °C: 700	Pre-heat to 850 °C, and then heat to 1000 °C. Soak for 15–45 minutes, and quench in oil or air. Temper at 200–400 °C for 30–60 minutes	Fine press-tools, deep-drawing and forming dies for sheet metal, wire-drawing dies, blanking dies, punches and shear blades for hard metals
BH 12	0.35 C 1.0 Si 5.0 Cr 1.5 Mo 0.45 V 1.35 W	–	Pre-heat to 800 °C. Soak, and then heat quickly to 1020 °C. Air-quench, and then temper for 90 minutes at 540–620 °C	Extrusion dies for aluminium and copper alloys, hot-forming, piercing-, and heading-tools, brass-forging dies
BH 21	0.3 C 2.85 Cr 0.35 V 10.0 W	–	Pre-heat to 850 °C, and then heat rapidly to 1200 °C. Oil-quench (or air-quench thin sections). Temper at 600–700 °C for 2–3 hours	Hot-forging dies and punches for making bolts, rivets, etc. where tools reach high temperatures, hot-forging dies, extrusion dies and die-casting dies for copper alloys, pressure die-casting dies for aluminium alloys

will in any case contain either tungsten or molybdenum, to provide resistance to tempering and, hence, the necessary strength and hardness at high temperatures.

The heat-treatment of these steels resembles that for high-speed steels, which will be described in the next section.

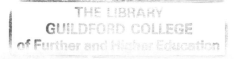

13.32   High-speed steel, as we know it, was first shown to an amazed public at the Paris Exposition of 1900. A tool was exhibited cutting at a speed of some 0.3 m s^{-1}, with its tip heated to redness. Soon after this, it was found that the maximum cutting efficiency was attained with a composition of 18 per cent tungsten, 4 per cent chromium, 1 per cent vanadium, and 0.75 per cent carbon, and this remains possibly the best-known general-purpose high-speed steel to this day.

13.33   Since high-speed steel is a complex alloy, containing at least five different elements, it cannot be represented by an ordinary equilibrium diagram (9.70); however, by grouping all the alloying elements together, under the title 'complex carbides', a simplified two-dimensional diagram (Fig. 13.1) can be used to explain the heat-treatment of this material.

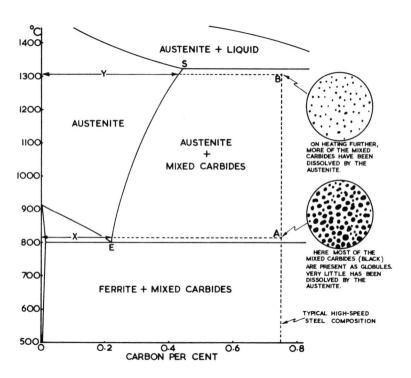

Fig. 13.1   *A modified equilibrium diagram for high-speed steel. If the 'typical high-speed steel composition' is heated to 'A', only an amount of mixed carbides equivalent to 'X' is dissolved by the austenite. Hence, on quenching, the steel would be soft, and would not resist tempering influences. The maximum amount of mixed carbides which can be safely dissolved (without beginning to melt the tool) is shown by 'Y', and this involves heating the tool to 1300 °C – just short of its melting-point.*

It will be seen that this diagram still resembles the ordinary iron-carbon diagram in general shape. The main difference is that the lower critical temperature has been raised (alloying elements usually raise or lower this temperature), and the eutectoid point E is now at only 0.25 per cent carbon (instead of 0.8 per cent). All alloying elements cause a shift of the eutectoid point to the left, for which reason alloy steels generally contain less carbon than the equivalent plain-carbon steels.

13.34 In the normalized condition, a typical high-speed steel contains massive globules of carbide in a matrix (or background) of ferrite. If this is now heated to just above the lower critical temperature (*A* in Fig. 13.1) the ferrite will change to austenite, and begin to dissolve the carbide globules. If the steel were quenched from this point, in the manner of a plain-carbon tool steel, the resultant structure would lack hardness, since only 0.25 per cent carbon, or thereabouts, would be dissolved in the martensite so produced. Moreover, it would not resist tempering influences, as little tungsten would be dissolved in the martensite and it is the presence of *dissolved* tungsten which provides resistance to tempering. It is therefore necessary to ensure that the maximum amount of tungsten carbide is dissolved in the austenite before the steel is quenched.

The slope of the boundary line *ES* shows that, as the temperature rises, the amount of carbides dissolved in the austenite increases to a maximum at *S*, where the steel begins to melt (approximately 1320 °C). Hence, to make sure that the maximum amount of carbide is dissolved before the steel is quenched, a high quenching-temperature in the region of 1300 °C is necessary. Since this is just short of the temperature at which melting begins, grain growth will proceed rather quickly. For this reason, a special heat-treatment furnace must be used. This consists of a lower chamber, usually heated by gas, and running at the quenching-temperature of 1300 °C, and

Table 13.8 *High-speed steels.*

BS specification 4659	Composition (%)	Heat-treatment		Hardness (VPN)	Uses
		Quench in oil or air from	Secondary-hardening treatment		
BT 21	0.65 C 4.0 Cr 14.0 W 0.5 V	1300 °C	Double temper at 565 °C for one hour. (The double-temper treatment gives extra hardness, as more austenite transforms to martensite.)	860	General shop practice – all-round work, moderate duties, also for blanking-tools and shear blades
BT 1	0.8 C 4.5 Cr 18.0 W 1.2 V	1310 °C		890	Lathe-, planer-, and shaping-tools, millers and gear-cutters, reamers, broaches, taps, dies, drills, hacksaws, roller-bearings for gas turbines
BT 6	0.8 C 4.25 Cr 20.0 W 1.5 V 0.5 Mo 12.0 Co	1320 °C		950	Lathe-, planer-, and shaper-tools, milling-cutters, drills for very hard materials. So-called 'super high-speed' steel, which has maximum hardness and toughness
BM 1	0.8 C 3.75 Cr 1.6 W 1.25 V 9.0 Mo	1230 °C		900	A general-purpose molybdenum-type high-speed steel for drills, taps, reamers, cutters. Susceptible to decarburization during heat-treatment, which therefore requires careful control

above this is a preheater chamber, maintained at about 850 °C by the exhaust gases which have already circulated around the high-temperature chamber. The tool is first preheated to 850 °C, and then transferred to the high-temperature compartment, where it will reach the quenching-temperature in a few minutes. In this way, the time of contact between tool and high-temperature conditions is reduced below that which would be necessary were the tool not preheated. At such high temperatures, decarburization of the tool surface would be serious, so a controlled non-oxidizing atmosphere is generally used in the furnace chamber.

13.35   As soon as it has reached the quenching-temperature, the tool is quenched in oil or in an air blast (depending upon its size and composition). The resultant structure contains some martensite, but also some soft austenite, because the high alloy content considerably reduces the rate of transformation. Hence the steel is heated to about 550 °C, to promote transformation of this austenite to martensite. This process is known as *secondary hardening*, and gives an increase in hardness from about 700 to over 800 VPN. 'Super high-speed' steels contain up to 12 per cent cobalt, and are harder than the ordinary tungsten types.

13.36   Since molybdenum is now cheaper than tungsten, many modern high-speed steels contain large amounts of molybdenum to replace much of the tungsten. These molybdenum-type steels are reputed to be more difficult to heat-treat successfully, and, whilst they are widely used in the USA, they are less popular in Britain.

Stainless steels

13.40   Although Michael Faraday had attempted to produce stainless steel as long ago as 1822, it was not until 1912 that Brearley discovered the rust-resisting properties of high-chromium steel.

Table 13.9   *Stainless steels.*

BS specification 970:	Composition (%)	Heat-treatment	Uses
403S17	0.04 C 0.45 Mn 14.0 Cr	Non-hardening, except by cold-work	'Stainless iron' – domestic articles such as forks and spoons. Can be pressed, drawn, and spun
420S45	0.3 C 0.5 Mn 13.0 Cr	Oil- or water-quench (or air-cool) from 960 °C. Temper (for cutting) 150–180 °C, temper (for springs) 400–450 °C	Specially for cutlery and sharp-edged tools, springs, circlips
302S25	0.1 C 0.8 Mn 8.5 Ni 18.0 Cr	Non-hardening except by cold-work. (Cool quickly from 1050 °C, to keep carbides dissolved)	Particularly suitable for domestic and decorative purposes
347S17	0.05 C 0.8 Mn 10.0 Ni 18.0 Cr 1.0 Nb		Weld-decay proofed by the presence of Nb. Used in welded plant where corrosive conditions are severe, e.g. nitric-acid plant

Chromium imparts the 'stainless' properties to these steels by coating the surface with a thin but extremely dense film of chromium oxide, which effectively protects the surface from further attack. Ordinary steel, on the other hand, becomes coated with a loose, porous layer of rust, through which the atmosphere can pass and cause further corrosion. For this reason, ordinary steel rusts quickly, the top flakes of rust being pushed off by new layers forming beneath.

13.41 Much corrosion in metals is of the 'electrolytic' type (25.40). Readers will be familiar with the working of a simple cell, in which a copper plate and a zinc plate are immersed in dilute sulphuric acid (called the 'electrolyte'). As soon as the plates are connected, a current flows, and the zinc plate dissolves ('corrodes') rapidly. In many alloys containing crystals of two different compositions, corrosion of one type of crystal will occur in this electrolytic manner when the surface of the alloy is coated with an electrolyte – which, incidentally, may be rain water. In stainless steels, however, the structure is a uniform solid solution. Since all of the crystals within a piece of the alloy are of the same composition, electrolytic action cannot take place. There are two main types of stainless steel:

1. The straight chromium alloys, which contain 13 per cent or more of chromium. These steels, provided they contain sufficient carbon, can be heat-treated to give a hard martensitic structure. Stainless cutlery steel is of this type. Some of these steel, however, contain little or no carbon, and are pressed and deep-drawn to produce such articles as domestic kitchen-sinks, refrigerator parts, beer-barrels, and tableware.
2. The '18/8' chromium/nickel steels, which are austenitic even after being cooled slowly to room temperature. This type of steel cannot be hardened (except of course by cold-work), and is used solely for constructional and ornamental work. Much of it is used in chemical plant, where acid-resisting properties are required, whilst the cheaper grades are widely used in tableware and kitchen equipment.

13.42 Although these austenitic stainless steels cannot be hardened by heat-treatment, they are usually 'finished' by quenching from 1050 °C. The purpose of this treatment is to prevent the precipitation of particles of chromium carbide, which would occur if the steel were allowed to cool slowly to room temperature.

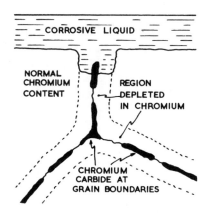

Fig. 13.2   *The effect of carbide precipitation on the resistance to corrosion.*

The precipitation of chromium carbide particles would draw out chromium from the surrounding structure, leaving it almost free of chromium (Fig. 13.2) so that rusting would occur in that region. Such corrosion would be due to a combination of electrolytic action and direct attack.

Because of the risk of precipitation of chromium carbide, these steels are unsuitable for welding, and suffer from a defect known as 'weld-decay'.

13.43   During welding, some regions of the metal near to the weld will be maintained between 650 and 800 °C long enough for chromium carbide to precipitate there (Fig. 13.4). Subsequently, corrosion will occur in this area near to the weld. The fault may largely be overcome by adding about 1 per cent of either titanium or niobium. These metals have a great affinity for carbon, which therefore combines with them in preference to chromium. Thus chromium is not drawn out of the structure, which, as a result, remains uniform.

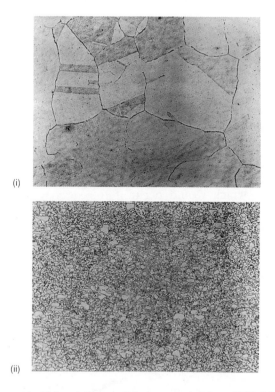

(i)

(ii)

Fig. 13.3   *(i) Austenitic 18.8 stainless steel, furnace-cooled from 1100 °C. Note the annealing 'twins' (left-hand side of picture) which are common in FCC structures which have been cold-worked and then annealed. The grain boundaries appear thick because some chromium carbide is precipitated there (see also Fig. 13.2) due to the slow cooling process. (ii) High-speed steel (18 per cent tungsten, 4 per cent chromium, 1 per cent vanadium) annealed at 900 °C. Rounded particles of tungsten carbide (light) in a ferrite-type matrix.*

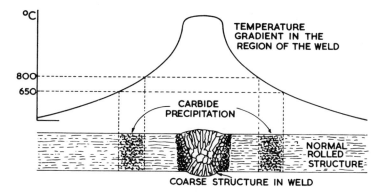

°C

800
650

TEMPERATURE GRADIENT IN THE REGION OF THE WELD

CARBIDE PRECIPITATION

NORMAL ROLLED STRUCTURE

COARSE STRUCTURE IN WELD

Fig. 13.4 *Microstructural changes during welding which lead to subsequent corrosion ('weld decay') in some stainless steels.*

Heat-resisting steels

13.50 The main requirement of a steel to be used at high temperatures are:

- it must resist oxidation and also attack by other gases in the working atmosphere,
- it must be strong enough at the working temperature.

Table 13.10   *Heat-resisting steels.*

Composition (%)	Heat-treatment	Maximum working temperature (°C)	Uses
0.4 C 0.2 Si 1.4 Mn 10.0 Cr 36.0 Ni	–	600	Steam-turbine blades and other fittings
0.1 C 0.7 Mn 12.0 Cr 2.5 Ni 1.8 Mo 0.35 V	Hardened from 1050 °C and tempered at 650 °C	600	Turbine blades and discs, bolts, some gas-turbine components
0.35 C 1.5 Si 21.0 Cr 7.0 Ni 4.0 W	–	950	Resists a high concentration of sulphurous gases
0.15 C 1.5 Si 25.0 Cr 19.0 Ni	–	1100	Heat-treatment pots and muffles, aircraft-engine manifolds, boiler and super-heater parts
0.35 C 0.6 Si 28.0 Cr	–	1150	Furnace parts, automatic stokers, retorts. Resistant to sulphurous gases

13.51 Resistance to oxidation is effected by adding chromium and sometimes small amounts of silicon. Both of these elements coat the surface with a tenacious layer of oxide, which protects the metal beneath from further attack. Nickel toughens the alloy by restricting grain-growth, but increased strength at high temperatures is achieved by adding small amounts of tungsten, titanium, or niobium. These form small particles of carbide, which raise the limiting creep stress (4.51) at the working temperature. Such steels are used for exhaust valves of internal-combustion engines, conveyor chains and other furnace parts, racks for enamelling stoves, annealing-boxes, rotors for steam and gas turbines, and retorts.

Magnet alloys

13.60 Magnetic fields are generated by the spin of electrons within the orbits of atoms. Most elements produce only extremely weak magnetic properties because the electrons are in pairs in their orbits and since they spin in opposite directions the magnetic fields so produced cancel each other. Only iron – along with nickel, cobalt and the rare earth metal, gadolinium – are strongly magnetic, or, *ferromagnetic*. In these metals unpaired electrons produce resultant magnetic fields. Even so the metal generally shows no resultant field in the unmagnetized condition because individual atoms are oriented at random and again magnetic fields cancel. The process of magnetization aligns atoms so that a strong resultant field is produced in one direction (Fig. 13.5).

Fig. 13.5 *The principles of ferromagnetism. (i) Random orientation of unit fields before magnetization. (ii) Alignment of unit fields to produce a resultant field following magnetization.*

13.61 Suppose a piece of magnetic material is placed in a solenoid through which an electric current is passing. This current will produce a magnetizing field (H) in the solenoid and a 'magnetic flux' (B) will be *induced* in the magnetic material. If the current is progressively increased H will increase and so will B until a point P is reached (Fig. 13.6(i)) where the magnet becomes 'saturated' with 'magnetic lines of force', i.e. any further increase in H will not produce any further increase in B (described loosely as the 'strength' of the magnet). If the current in the solenoid is now reduced H will of course decrease, reaching zero when the current has reached zero (point O). B on the other hand will decrease less rapidly and when H has reached zero B will still retain the value R. That is, a *magnetic flux density R* remains in the magnet. This value R is termed the *remanence* – or residual magnetism.

If the current in the solenoid is now increased in a reverse direction a magnetizing field ($-H$) is generated in a reverse direction and RUQ represents the change in the induced magnetism in the opposite direction in the magnet ($-B$). The value UO is the strength of the reverse magnetic field ($-H$) which is required to *demagnetize* the magnet completely. It is called the *coercive force* of the material and represents its *resistance to demagnetization* by opposing magnetic fields. If the magnetizing/

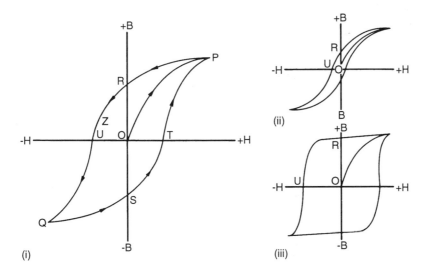

Fig. 13.6 *Magnetic hysteresis: (i) the derivation of a hysteresis 'loop'; (ii) typical hysteresis loop for a magnetically 'soft' alloy; (iii) typical hysteresis loop for a magnetically 'hard' (permanent magnet) alloy.*

demagnetizing cycle is completed by again reversing the field H from S then a *hysteresis loop* is formed. The area occupied by the loop is a measure of the magnetic efficiency of a material. An alloy suitable for use as a *permanent* magnet will require a high remanence (B_{rem}) and/or a high coercive force (H_c). The ultimate standard by which a permanent magnet material is judged is the maximum product of B and H obtained by a point Z on the hysteresis curve (Fig. 13.6(i)). This value BH_{max} – the product of B and H at Z – corresponds to the maximum energy the magnet can provide to an external circuit.

13.62 Magnetic materials used in engineering fall into two groups:

1. Magnetically *soft* materials which are those displaying a high *magnetic permeability*, that is, they provide a ready path for magnetic lines of force. At the same time they must have very *low* values of B_{rem} and H_c thus producing a very 'narrow' hysteresis loop (Fig. 13.6(ii)). These materials are used in transformer cores and dynamo pole-pieces where it is necessary for the induced magnetic field to fluctuate with the applied electromagnetic field. Pure soft iron and iron-silicon alloys (4 per cent Si) containing no carbon are generally used, but a 'metallic glass' (22.70) containing 80 Fe – 20 B is much 'softer' magnetically and gives up to 25 per cent savings on 'heat loss' in power transmission transformers.
2. Magnetically *hard* materials are those which possess very *high* values of B_{rem} and H_c and are therefore used for permanent magnets. The earliest of these included tungsten steels and cobalt steels (up to 35 Co), both of which are now obsolete. Alnico (16 Ni; 12 Co; 9 Al; 5 Cu; 1 Ti; bal – Fe) and Hycol 3 (34 Co; 15 Ni; 7 Al; 5 Ti; 4 Cu; bal – Fe) are representative of a range of cast and sintered alloys with very high remanence and coercive force, now used in the electrical, communications and general

engineering industries. A number of sintered ceramic materials are also important. These are based mainly on mixtures of barium oxide and iron oxide. Such a substance is Feroba, used extensively in loudspeakers, door catches and computer devices. In powder form it can be incorporated in a rubber matrix for use on display boards.

Recently magnetic alloys containing the 'rare earth' metals[1] samarium and neodymium have been developed for use in stepper motors for watches, camera apertures, micro-switches, servomotors, audiovisual and similar consumer products. Such alloys have very high values of B_{rem} and H_c.

Summary of the principal effects of the main alloying elements

13.70   The more important effects of the main alloying elements added to steels can be summarized as follows:

Element	Chemical symbol	Principal effects when added to steel (More important effects in italics)
Manganese	Mn	*Acts as a deoxidizer and a desulphurizer.* Stabilizes carbides.
Nickel	Ni	*Toughens steel by refining grain.* Strengthens ferrite. Causes cementite to decompose – hence used by itself only in low-carbon steel.
Chromium	Cr	Stabilizes carbides, and forms hard chromium carbide – hence *increases hardness of steel.* Promotes grain-growth, and so causes brittleness. *Increases resistance to corrosion.*
Molybdenum	Mo	*Reduces 'temper brittleness' in nickel-chromium steels.* Stabilizes carbides. Improves high-temperature strength.
Vanadium	V	Stabilizes carbides. *Raises softening temperature of hardened steels* (as in high-speed steels).
Tungsten	W	*Forms very hard stable carbides. Raises the softening temperature, and renders transformations very sluggish* (in high-speed steels). Reduces grain growth. Raises the limited creep stress at high temperatures.

13.71   Remember that *all* of these elements increase the depth of hardening of a steel, so an alloy steel has a bigger 'ruling section' than a plain-carbon steel. This is because alloying elements *slow down* the austenite $\longrightarrow$ martensite transformation rates, so it is possible to oil-harden or, in some cases, air-harden a suitable steel (Fig. 12.3(ii)).

---

[1]   A group of 14 closely related elements more correctly known as the 'lanthanides'.

CHAPTER 14

The surface-hardening of steels

14.10 Many metal components require a combination of mechanical properties which at first sight seems impossible to attain. Thus, bearing metals (18.60) must be both hard and, at the same time, ductile, whilst many steel components, like cams and gears, need to be strong and shock-resistant, yet also hard and wear-resistant. In ordinary carbon steels, these two different sets of properties are found only in materials of different carbon content. Thus, a steel with about 0.1 per cent carbon will be tough, whilst one with 0.9 per cent carbon will be very hard when suitably heat-treated.

The problem can be overcome in two different ways:

1. by employing a tough, low-carbon steel, and altering the composition of its surface, either by case-hardening or by nitriding;
2. by using a steel of uniform composition throughout, but containing at least 0.4 per cent carbon, and heat-treating the surface differently from the core, as in flame- and induction-hardening.

In the first case, it is the hardening material which is localized, whilst in the second case it is the heat-treatment which is localized.

Case-hardening

14.20 This process makes use of the fact that carbon will dissolve in appreciable amounts in *solid* iron, provided that the latter is in the face-centred

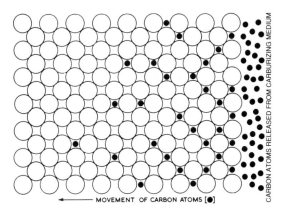

Fig. 14.1 *An impression of the penetration by carbon atoms into the lattice structure of FCC iron (austenite).*

cubic crystal form. This is due to the fact that carbon dissolves interstitially in iron (8.34) – the carbon atoms are small enough to infiltrate between the larger iron atoms; so solid iron can absorb carbon in much the same way that water is soaked up by a sponge. Since only face-centred cubic iron will dissolve carbon in this way, it follows that steel must be carburized at a temperature *above* the upper critical temperature. As it is generally low-carbon steel which is carburized, this involves using a temperature in the region of 900–950 °C. Thus carburizing consists of surrounding mild-steel components with some carbon-rich material, and heating them above their upper critical temperature for long enough to produce a carbon-rich surface layer of sufficient depth.

Solid, liquid, and gaseous carburizing materials are used, and the quantity of output required largely governs the method employed.

14.21   Carburizing in solid media   So-called 'pack-carburizing' is probably the process with which the reader is most likely to be familiar. Components to be treated are packed into steel boxes, along with the carburizing material, so that a space of roughly 50 mm exists between them. Lids are then fixed on the boxes, which are then slowly heated to the carburizing temperature (900–950 °C). They are maintained at this temperature for up to six hours[1] according to the depth of case required (Fig. 14.2). Carburizing mixtures vary in composition, but consist essentially of some carbon-rich material, such as charcoal or charred leather, along with an energizer which may account for about 40 per cent of the total. This energizer is generally a mixture of sodium carbonate ('soda ash') and barium carbonate. Its function is to accelerate the solution of carbon by taking part in a chemical reaction which causes single carbon atoms to be released at the surface of the steel.

If it is necessary to prevent any parts of the surface of the component from becoming carburized, this can be achieved by electroplating these areas with copper, to a thickness of 0.07 to 0.10 mm, since carbon does not dissolve in solid copper. In small-scale treatment, the same objective can be achieved by coating the necessary areas of the components with a paste of fireclay

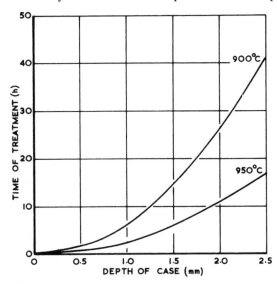

Fig. 14.2   *The relationship between time of treatment, temperature, and depth of case in a carburizing process using solid media (0.15 per cent plain-carbon steel).*

[1]   Much longer periods are sometimes necessary when deep cases are to be produced.

and ignited asbestos mixed with water. This is allowed to dry on the surface, before the components are loaded into the carburizing box.

When carburizing is complete, the charge is either quenched or allowed to cool slowly in the box, depending on the subsequent heat-treatment it will receive.

14.22   Carburizing in liquid media   Liquid-carburizing – or cyanide-hardening, as it is usually called – is carried out in baths of molten salt which contain 20 to 50 per cent sodium cyanide, together with as much as 40 per cent sodium carbonate, and varying quantities of sodium or barium chloride. The cyanide-rich mixture is heated in iron pots to a temperature of 870 to 950 °C, and the work, which is carried in wire baskets, is immersed for periods of about five minutes upwards, according to the depth of case required. The process is particularly suitable for producing shallow cases of 0.1 to 0.25 mm.

Carburizing takes place due to the decomposition of sodium cyanide at the surface of the steel. Atoms of both carbon and nitrogen are released, so cyanide-hardening is due to the absorption of nitrogen, as well as of carbon.

The main advantages of cyanide hardening are:

- the temperature of a liquid salt bath is uniform throughout, and can be controlled accurately by pyrometers;
- the basket of work can be quenched direct from the bath;
- the surface of the work remains clean.

Many readers will be aware of the fact that all cyanides are extremely poisonous chemicals. However, since *sodium cyanide is one of the most deadly poisonous materials* in common use industrially, it might be well to stress the following points, which should be observed by the reader should he find himself involved in the use of cyanides.

1. Every pot should be fitted with an efficient fume-extraction system.
2. The consumption of food by operators whilst working in a shop containing cyanide should be *absolutely forbidden.*
3. Cyanide-rich salts should never be allowed to come into contact with an open wound.
4. Advice should be sought before disposing of any waste hardening salts. They should *never* be tipped into canals or rivers.

14.23   Carburizing by gaseous media   Gas-carburizing is carried out in both continuous and batch-type furnaces. Whichever is used, the components are heated at about 900 °C for three hours or more in an atmosphere containing gases which will deposit carbon atoms at the surface of the components. The gases generally used are the hydrocarbons methane ('natural gas'–1.23) and propane (a by-product of petroleum production). These should be of high purity otherwise oily soot may be deposited on the work-pieces. The hydrocarbon is usually mixed with a 'carrier' gas (generally a mixture of nitrogen, hydrogen and carbon monoxide) which allows better gas circulation and hence greater uniformity of treatment.

Gas-carburizing is now used for most large-scale treatment, particularly for the mass production of thin cases. Its main advantages as compared with other methods of carburizing are:

- the surface of the work is clean after treatment;
- the necessary plant is more compact for a given output;
- the carbon content of the surface layers can be more accurately controlled by this method.

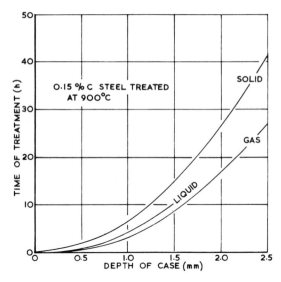

Fig. 14.3 *The relationship between time of treatment and depth of case produced when carburizing in solid, liquid, and gaseous media (0.15 per cent plain-carbon steel).*

Heat-treatment after carburizing

14.30   If the carburizing process has been successful, the core will still have a low carbon content (0.1 to 0.3 per cent carbon), whilst the case should have a maximum carbon-content of 0.8 per cent carbon (the eutectoid composition). Unfortunately, prolonged heating in the austenitic range will have caused the formation of coarse grain, and further heat-treatment is desirable if optimum properties are to be obtained.

The most common method of producing a fine-grained structure in steel is by normalizing it. This involves heating the steel to just above its upper critical temperature, followed by cooling it in air (11.51). The need for such treatment poses a problem here, since core and case are of widely different carbon contents, and therefore have different upper critical temperatures. Thus, if the best mechanical properties are to be obtained in both core and case, a double heat-treatment is necessary.

14.31   Refining the core   The component is first heat-treated to refine the grain of the core, and so toughen it. This is done by heating the component to a temperature just above the upper critical temperature for the core (point *A* in Fig. 14.4), so that the coarse ferrite/pearlite will be replaced by fine-grained austenite. The component is then generally water-quenched, so that a mixture of fine-grained ferrite and a little martensite is produced. The temperature of this treatment is high above the upper critical temperature for the case (723 °C), so at this stage the case will be of coarse-grained martensite (because the steel was quenched). Further heat-treatment is therefore necessary to refine the grain of the case.

14.32   Refining the case   The component is now heated to 760 °C (point *B* in Fig. 14.4), so that the structure of the case changes to fine-grained austenite. Quenching then gives a hard case of fine-grained martensite.

At the same time, any brittle martensite present in the core as a result of the *first* quenching process will be tempered to some extent by the second

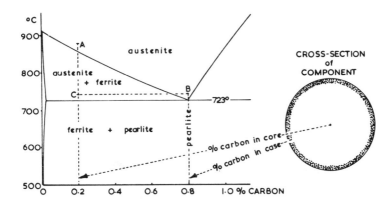

Fig. 14.4 *Heat-treatment of a carburized component in relation to the equilibrium diagram.*

heating-operation (point *C* in Fig. 14.4). Finally, the component is tempered at 200 °C, to relieve any quenching-stresses present in the case.

The above heat-treatment processes can be regarded to some extent as the counsel of perfection, and the needs of economy often demand that such treatments may be replaced by a single operation. Often the work may be 'pot-quenched'; that is, quenched direct from the carburizing process, followed by a low-temperature tempering process to relieve any quenching stresses.

Alternatively, the work may be cooled slowly from the carburizing temperature, to give maximum ductility to the core. It is then reheated to 760 °C, and water-quenched. This treatment leaves the core quite soft, but hardens the case, which will be fine-grained, due to the low quenching temperature.

Case-hardening steels

14.40 Plain-carbon and low-alloy steels are used for case-hardening, but, in either type, the carbon content should not be more than 0.2 per cent if a really tough core is to be obtained. Manganese may be present in amounts

Table 14.1 *Case-hardening steels.*

Composition (%)					*Characteristics and uses*
C	Mn	Ni	Cr	Mo	
0.15	0.7	–	–	–	Machine parts requiring a hard surface and a tough core, e.g. gears, shafts, cams.
0.15	1.3	–	–	–	A carbon-manganese steel giving high surface hardness where severe shock is unlikely.
0.13	0.5	3.25	0.85	–	High surface hardness combined with core toughness – high-duty gears, worm gears, crown wheels, clutch gears.
0.17	0.5	1.75	–	0.25	High hardness and severe shock resistance – automobile parts (steering worms, overhead valve mechanisms).
0.15	0.4	4.0	1.2	0.2	Best combination of surface hardness, core strength and shock resistance – crown wheels, bevel pins, intricate sections which need to be *air-hardened* (13.23).

up to 1.4 per cent, since it stabilizes cementite, and increases the depth of hardening. Unfortunately, it is also liable to increase the tendency of a steel to crack during quenching.

Alloy steels used for case-hardening contain up to 4.0 per cent nickel, since this increases the strength of the core, and retards grain-growth during the carburizing process. This often means that the core-refining heat-treatment can be omitted. Chromium is sometimes added to increase hardness and wear-resistance of the case, but it must be present only in small quantities, as it tends to promote grain-growth (13.22).

Nitriding

14.50 Nitriding and case-hardening have one factor in common – both processes involve heating the steel for a considerable time in the hardening-medium, but, whilst in case-hardening the medium contains carbon, in nitriding it contains gaseous nitrogen. Special steels – 'Nitralloy' steels – are necessary for the nitriding process, since hardening depends upon the formation of very hard compounds of nitrogen and such metals as aluminium, chromium and vanadium present in the steel.

Ordinary plain-carbon steels cannot be nitrided, since any compounds of iron and nitrogen which form will diffuse into the core, so that the increase in hardness of the surface is lost. The hard compounds formed by aluminium, chromium and vanadium, however, remain near to the surface, providing an extremely hard skin.

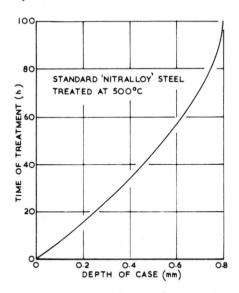

Fig. 14.5 *The relationship between time of treatment and depth of case produced in the nitriding process.*

14.51 Nitriding is carried out at the relatively low temperature of 500 °C. Consequently, it is made the *final* operation in the manufacture of the component, all machining and core heat-treatments having been carried out previously. The work is maintained at 500 °C for between 40 and 100 hours, according to the depth of case required, though treatment for 90 hours is general. The treatment takes place in a gas-tight chamber through which ammonia gas is allowed to circulate. Some of the ammonia decomposes, re-leasing single atoms of nitrogen, which are at once absorbed by the surface of the steel:

$$NH_3 = 3H + N(atom)$$

Ordinary 'atmospheric' nitrogen is not suitable since it exists in the form of molecules (N_2) which would not be absorbed by the steel.

14.52 Nitralloy steels containing aluminium are hardest, since aluminium forms very hard compounds with nitrogen. Unfortunately, aluminium tends to affect the core-strength adversely, and is replaced by chromium, vanadium, and molybdenum in those Nitralloy steels in which high strength and toughness of the core are important. Compositions and uses of some nitriding steels are given in Table 14.2.

Table 14.2   *Nitriding steels.*

Composition (%)					Typical mechanical properties		Characteristics and uses
C	Cr	Mo	V	Al	Tensile strength (N mm⁻²)	VPN	
0.5	1.5	0.2	–	1.1	1200	1075	Where maximum surface hardness, coupled with high core-strength is essential
0.2	1.5	0.2	–	1.1	600	1075	For maximum surface hardness, combined with ease of machining before hardening
0.4	3.0	1.0	0.2	–	1400	875	Ball-races, etc., where high core-strength is necessary
0.3	3.0	0.4	–	–	1000	875	Aero crankshafts, air-screw shafts, aero cylinders, crank-pins, and journals

14.53 Prior to being nitrided, the work-pieces are heat-treated, to produce the required properties in the core. Since greater scope is possible in this heat-treatment than is feasible in that associated with case-hardening, Nitralloy steels often have higher carbon contents, allowing high core-strengths to be developed. The normal sequence of operations will be:

1. oil-quenching from 850–900 °C, followed by tempering at between 600 and 700 °C;
2. rough machining, followed by a stabilizing anneal at 550 °C for five hours, to remove internal stresses;
3. finish-machining, followed by nitriding.

Any areas of the surface which are required soft are protected by coating with solder or pure tin, by nickel-plating, or by painting with a mixture of whiting and sodium silicate.

14.54   Advantages of nitriding over case-hardening are as follows.

- Since no quenching is required *after* nitriding, cracking or distortion is unlikely, and components can be machine-finished before treatment.
- An extremely high surface hardness of up to 1150 VPN is attainable with the aluminium-type Nitralloy steels.
- Resistance to corrosion is good, if the nitrided surface is left unpolished.
- Hardness is retained up to 500 °C, whereas a case-hardened component begins to soften at about 200 °C.
- The process is clean, and simple to operate.
- It is cheap if large numbers of components are to be treated.

14.55   Disadvantages of nitriding as compared with case-hardening are as follows.

- The initial outlay for nitriding plant is higher than that associated with solid- or liquid-medium carburizing; so nitriding is only economical when large numbers of components are to be treated.
- If a nitrided component is accidentally overheated, the loss of surface hardness is permanent, unless the component can be nitrided again. A case-hardened component would need only to be heat-treated, assuming that it had not been so grossly overheated as to decarburise it.

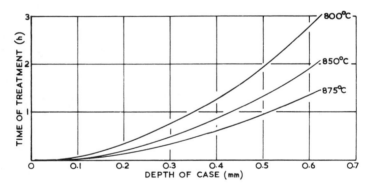

Fig. 14.6   *The relationship between time of treatment, temperature, and depth of case produced by the carbonitriding process.*

14.56   Carbonitriding is a surface-hardening process which makes use of a mixture of hydrocarbons and ammonia. It is therefore a gas treatment, and is sometimes known as 'dry-cyaniding' – a reference to the fact that a mixed carbide-nitride case is produced, as in ordinary liquid-bath cyanide processes (14.22).

14.57   Furnaces used for carbonitriding are generally of the continuous type, as the work is nearly always directly quenched in oil from the carbonitriding atmosphere. If 'stopping off' is necessary for any areas required soft, then good-quality copper-plating is recommended.

   Carbonitriding is an ideal process for hardening small components where great resistance to wear is necessary.

Ionitriding

14.60   More correctly termed *ion-nitriding*, this process is also known as *plasma nitriding* and *ion implantation*. The work load is made the cathode in a sealed chamber containing nitrogen under near-vacuum conditions. Under a potential difference approaching 1000 volts (d.c.) the low-pressure nitrogen ionizes. That is, the nitrogen atoms lose outer-shell electrons (1.20) and so become *positively* charged ions so that they are attracted and so, accelerated towards, the *negatively* charged cathode, i.e. the work load. They strike this at a very high velocity and so penetrate the surface, the kinetic energy lost on impact being converted into heat so that the surface temperature of the work load is raised to the nitriding temperature (400–600 °C). The work load is surrounded by a glow of ionized nitrogen and treatment time is between 10 minutes and 30 hours depending upon the type of steel and the depth of case required. Maximum hardness is achieved with Nitralloy-type steels (Table 14.2) – the higher the alloy content, the thinner and harder the case.

   The process is being used to nitride components weighing several tonnes down to the tiny balls of ball-point pens. Automobile parts, hot- and cold-working dies and tools are now ion-nitrided.

Flame-hardening

14.70   In this process, the work-piece is of uniform composition throughout, and it is the type of structure which varies across the section, because the surface layers receive extra heat-treatment as compared with the core material.

14.71   The surface is heated to a temperature above its upper critical temperature, by means of a travelling oxyacetylene torch (Fig. 14.7), and is immediately quenched by a jet of water issuing from a supply built into the torch-assembly. Symmetrical components, such as gears and spindles, are conveniently treated by this process, since they can be spun between centres, the whole circumference being treated simultaneously. Only steels with a sufficiently high carbon content – at least 0.4 per cent – can be hardened effectively in this way. Alloy steels containing up to 4.0 per cent nickel and 1.0 per cent chromium respond well to such treatment. Before

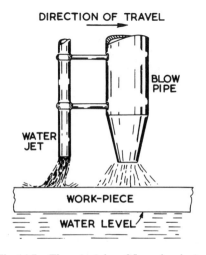

Fig. 14.7   *The principles of flame-hardening.*

being hardened, the components are generally normalized, so that the final structure consists of a martensitic case some 4 mm deep, and a tough ferrite-pearlite core. Core and case are usually separated by a layer of bainite, which helps to prevent the hard case from cracking away from the core material. Should a final tempering process be necessary, this can also be carried out by flame-heating, though furnace treatment is also possible, since such low-temperature treatment will have no effect on the core, particularly if it has been normalized.

Induction-hardening

14.80   This process is similar in principle to flame-hardening, except that the component is usually held stationary whilst the whole circumference is heated simultaneously by means of an induction-coil. This coil carries a high-frequency current, which produces eddy currents in the surface of the component, thus raising its temperature. The depth to which heating occurs varies inversely as the square root of the frequency, so that the higher the frequency used, the shallower the depth of heating. Typical frequencies used are:

3000 Hz for depths of 3 to 6 mm,
9600 Hz for depths of 2 to 3 mm.

As soon as the surface of the component has reached the necessary quenching temperature, the current is switched off, and the surface is simultaneously quenched by pressure jets of water, which pass through holes in the induction-block (Fig. 14.8).

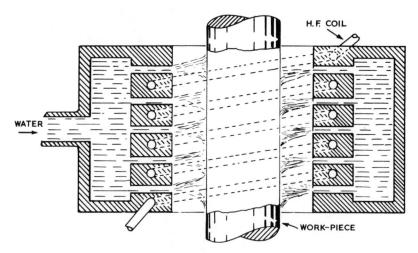

Fig. 14.8   *The principles of induction-hardening.*

This process lends itself to mechanization, so that selected regions of a symmetrical component can be hardened, whilst others are left soft. As in flame-hardening, the induction process makes use of the existing carbon content – which consequently must be at least 0.4 per cent – whilst in case-hardening, nitriding, and carbonitriding, an alteration in the composition of the surface layers takes place.

Summary of surface-hardening processes

14.90   The characteristics and uses of the processes dealt with in this chapter are summarized in Table 14.3.

Table 14.3   *Summary of surface-hardening processes.*

Process	Type of work	Characteristics
Case-hardening (solid and gas)	Gears, king-pins, ball- and roller-bearings, rocker-arms, gauges	A wide variety of low-carbon and low-alloy steels can be treated. Local soft surfaces are easily retained. Gas carburizing is a rapid process.
Case-hardening (liquid cyanide)	Used mainly for light cases	The case tends to be of poorer quality, but thin cases can be produced quickly.
Nitriding	Crankshafts, cam-shafts, gears requiring high core-strength	A very high surface hardness, combined with a high core-strength when required. Surface will withstand tempering influences up to 500 °C. Less suitable than other methods if surface has to withstand very high pressure, e.g. gear-teeth.
Carbonitriding	Particularly useful for treating small components	Safe, clean, and easy to operate, applicable to mass-production methods.
Ionitriding	Crankshafts; many other components in various industries.	A high degree of control and uniformity is possible.
Flame- and induction-hardening	Tappets, cam-shafts, gears where high core-strength is required	Particularly useful where high core-strength is necessary, since a high-carbon steel can be used and heat-treated accordingly. Rapid output possible, but equipment often needs to be designed for a particular job; hence suitable mainly for long runs.

CHAPTER 15

Cast iron

15.10 The Victorian era may well be remembered by the cast-iron monstrosities which it produced. Street lamps, domestic fireplaces, and railings were typical cast-iron products of that period. Most of these relics are gone – the railings fell victim of the need for steel during the Second World War – but many an industrial town still boasts an ornamental drinking-fountain in its local park, or a cast-iron clock presiding over the public-conveniences (of similar period) in the town square, whilst it seems that the once despised blackleaded fireplaces of my childhood are now regarded as being 'eminently collectable'.

During the nineteenth century, much cast iron was also used for engineering purposes. Today, the whole production of cast iron is directed towards these purposes, and, as in other fields of metallurgical technology, considerable progress has been made during the twentieth century. Special high-duty and alloy compositions have made cast iron an extremely important engineering material, which is suitable for the manufacture of crankshafts, connecting rods, and axles – components which were formerly made from forged steel.

15.11 Ordinary cast iron is similar in composition to the crude pig iron produced by the blast-furnace. The pig iron is generally melted in a cupola, any necessary adjustments in composition being made during the melting process. At present, the high cost of metallurgical coke, coupled with the desire to produce high-grade material, has led the foundryman to look for other methods of melting cast iron; consequently, line-frequency induction furnaces are being used on an increasing scale.

15.12 The following features make cast iron an important material.

1. It is a cheap metallurgical substance, since it is produced by simple adjustments to the compositions of ordinary pig irons.
2. Mechanical rigidity and strength under compression are good.
3. It machines with ease when a suitable composition is selected.
4. Good fluidity in the molten state leads to the production of good casting-impressions.
5. High-duty cast irons can be produced by further treatment of irons of suitable composition; e.g. spheroidal-graphite irons are strong, whilst malleable irons are tough.

The composition of cast iron

15.20   Ordinary cast irons contain the following elements:

carbon	3.0–4.0%	sulphur	up to 0.1%
silicon	1.0–3.0%	phosphorus	up to 1.0%
manganese	0.5–1.0%		

15.21   **Carbon** may be present in the structure either as flakes of graphite or as a network of hard, brittle iron carbide (or cementite). Naturally, if a cast iron contains much of this brittle cementite, its mechanical properties will be poor, and for most engineering purposes it is desirable for the carbon to be present as small flakes of graphite. Cementite is a silvery-white compound, and, if an iron containing much cementite is broken, the fractured surface will be silvery white, because the piece of iron breaks along the brittle cementite networks. Such an iron is termed a *white iron*. Conversely, if an iron contains much graphite, its fractured surface will be grey, due to the presence of graphite flakes in the structure and this iron would be described as a *grey iron*. Under the microscope these graphite flakes appear as two-dimensional strands but what we see is a cross-section through a micro-constituent shaped something like a breakfast cereal 'cornflake' (Fig. 15.1).

Fig. 15.1   *A three-dimensional impression of a cluster – or 'cell' – of graphite flakes in a grey cast iron. A micro-section of a high-silicon cast iron would show them as 'star-fish' shapes (Fig. 15.2 (iii)).*

15.22   **Silicon** to some extent governs the form in which carbon is present in cast iron. It causes the cementite to be unstable, so that it decomposes, thus releasing free graphite. Therefore a high-silicon iron tends to be a grey iron, whilst a low-silicon iron tends to be a white iron (Fig. 15.2).

15.23   **Sulphur** has the opposite effect on the structure; that is, it tends to stabilize cementite, and so helps to produce a white iron. However, sulphur causes excessive brittleness in cast iron (as it does in steel), and it is therefore always kept to the minimum amount which is economically possible. No self-respecting foundryman would think of altering the structure of an iron by adding sulphur to a cupola charge, any more than he would think of diluting his whisky with dirty washing-up water; instead, he obtains the desired microstructure by adjusting the silicon content of the iron.

During the melting of cast iron in a cupola, some silicon is inevitably burned away, whilst some sulphur will be absorbed from the coke. Both

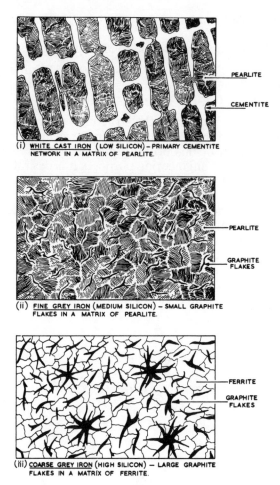

(i) <u>WHITE CAST IRON</u> (LOW SILICON) – PRIMARY CEMENTITE NETWORK IN A MATRIX OF PEARLITE.

(ii) <u>FINE GREY IRON</u> (MEDIUM SILICON) – SMALL GRAPHITE FLAKES IN A MATRIX OF PEARLITE.

(iii) <u>COARSE GREY IRON</u> (HIGH SILICON) – LARGE GRAPHITE FLAKES IN A MATRIX OF FERRITE.

Fig. 15.2 *The effects of silicon content on the structure of cast iron. The higher the silicon content, the more unstable the cementite becomes, until even the pearlitic cementite decomposes (iii). Magnifications – approx. ×100. (This illustration shows that a careful sketch of a microstructure can often reveal more than would a photomicrograph. Patience, assisted by a very sharp pencil, are necessary* and *some microstructures are more conducive to being sketched than others.)*

factors tend to make the iron 'whiter', so the foundryman begins with a charge richer in silicon than that with which he expects to finish.

15.24   Manganese toughens and strengthens an iron, partly because it neutralizes much of the unwelcome sulphur by forming a slag with it, and partly because some of the manganese dissolves in the ferrite.

15.25   Phosphorus forms a very brittle compound with some of the iron; it is therefore kept to a minimum amount in most engineering cast irons. However, like silicon, it increases fluidity, and considerably improves the casting qualities of irons which are to be cast in thin sections, assuming that components are involved in which mechanical properties are unimportant. Thus cast-iron water pipes contain up to 0.8 per cent phosphorus, whilst many of the old ornamental castings contained up to 1.0 per cent of the element.

The influence of cooling rate on the properties of a cast iron

15.30   When the presence of silicon in an iron tends to make cementite unstable, the latter does not break up or decompose instantaneously: this process of decomposition requires time. Consequently, if such an iron is cooled so that it solidifies rapidly, the carbon may well be 'trapped' in the form of hard cementite, and so give rise to a white iron. On the other hand, if this iron is allowed to cool and solidify slowly, the cementite has more opportunity to decompose, forming graphite, and so produce a grey iron.

15.31   This effect can be shown by casting a 'wedge-bar' in an iron of suitable composition (Fig. 15.3). If this bar is fractured, and hardness determinations are made at intervals along the centre line of the section, it will be found that the thin end of the wedge has cooled so quickly that decomposition of the cementite has not been possible. This is indicated by the white fracture and the high hardness in that region. The thick end of the wedge, however, has cooled slowly, and is graphitic, because cementite has had more opportunity to break up. Here the structure is softer.

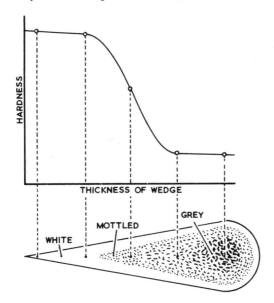

Fig. 15.3   *The effect of sectional thickness on the depth of chilling of a grey iron.*

15.32   If the reader has had experience in machining cast iron, he will know that such a casting has a hard surface skin, but that, once this is removed, the material beneath is easy to machine. This hard skin consists largely of cementite which has been prevented from decomposing by the chilling action of the mould. Iron beneath the surface has cooled more slowly, so that cementite has decomposed, releasing graphite.

15.33   To summarize: the engineer requires a cast iron in which carbon is present in the form of small flakes of graphite. The form in which the carbon is present depends upon:

- the silicon content of the iron, and
- the rate at which the iron solidifies and cools, which in turn, depends upon the cross-sectional thickness of the casting.

Thus the foundryman must strike a balance between the silicon content of the iron and the rate at which it cools.

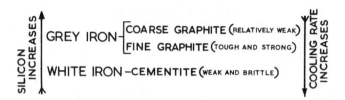

When casting *thin* sections, it will be necessary for the foundryman to choose an iron which has a coarser grey fracture in the pig form than that which is required in the finished casting. Thus the iron must have a higher silicon content than that used for the manufacture of castings of heavier section, which will consequently cool more slowly.

15.34   Sometimes it is necessary to have a hard-wearing surface of white iron at some point in a casting which otherwise requires a tough grey-iron structure. This can be achieved by incorporating 'chills' at appropriate points in the sand mould. The 'chill' usually consists of a metal block, which will cause the molten iron in that region to cool so quickly that a layer of hard cementite is retained adjacent to the chill.

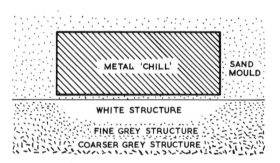

Fig. 15.4   *The use of 'chills' in iron-founding.*

'Growth' in cast irons

15.40   An engineering cast iron contains come cementite as a constituent of the pearlitic areas of the structure. If such an iron is heated for a prolonged period in the region of 700 °C or above, this cementite decomposes to form graphite and iron. Since the graphite and iron so formed occupy more space than did the original cementite, the volume of the heated region increases, and this expansion leads to warping of the casting, and the formation of cracks at the surface. Hot gases penetrate these cracks, gradually oxidizing both the graphite and the iron, so that the surface ultimately disintegrates. Fire bars in an ordinary domestic grate often break up in this way.

15.41   The best way to prevent 'growth' in cast irons which are to be used at high temperatures is to ensure that they contain *no* cementite in the first place. This can be achieved by using a high-silicon-content iron, in which decomposition of all the cementite will take place during the actual solidification of the casting. Thus, 'Silal' contains 5.0 per cent silicon, with relatively low carbon, so that the latter is present in the finished casting entirely as graphite. Unfortunately, Silal is rather brittle, so, when the cost is justified, the alloy cast iron 'Nicrosilal' (Table 15.3) may be used.

Ordinary cast irons

15.50   Ordinary cast irons fall into two main groups:

15.51   Engineering irons which must possess reasonable strength and toughness, generally coupled with good machinability (7.54). The silicon content of such an iron will be chosen in accordance with the cross-sectional thickness of the casting to be produced. It may be as much as 2.5 per cent for castings of thin section, but as low as 1.2 per cent for bulky castings of heavy section. This relationship between silicon content and sectional thickness of a casting is illustrated by the three irons specified for light, medium, and heavy machine castings given in Table 15.1. Amounts of sulphur and phosphorus generally are kept low, since both elements cause brittleness, though some castings contain as much as 1.0 per cent phosphorus, to give fluidity to the molten iron.

Table 15.1   *Compositions and uses of some ordinary cast irons.*

Composition %					Uses
C	Si	Mn	S	P	
3.30	1.90	0.65	0.08	0.15	Motor brake-drums
3.25	2.25	0.65	0.10	0.15	Motor cylinders and pistons
3.25	2.25	0.50	0.10	0.35	Light machine-castings
3.25	1.75	0.50	0.10	0.35	Medium machine-castings
3.25	1.25	0.50	0.10	0.35	Heavy machine-castings
3.60	1.75	0.50	0.10	0.80	Water pipes
3.50	2.75	0.50	0.10	0.90	Low-strength ornamental castings of yesteryear

The principal engineering grey irons are covered by BS 1452: 1990 which deals with eight different grades each with its minimum acceptable tensile strength. Grades 100 and 150 are fluid irons possibly containing phosphorus and which will only be used where low strength is acceptable, e.g. domestic pipes and gutters. Grade 180 is a high-silicon iron in which most of the carbon is present as coarse graphite flakes in a matrix mainly of ferrite. At the other extreme grade 350 is a low-silicon iron containing fine graphite flakes in a pearlite matrix. The properties of flake graphite (grey) iron depend on the amount and form of graphite and on the matrix structure. This structure may be controlled by a variety of factors such as production conditions, chemical composition, rate of solidification and rate of cooling *after* solidification as well as complexity of design. BS 1452 therefore specifies only tensile strength.

Table 15.2 *Typical properties of engineering grey irons in accordance with BS 1452: 1990. It will be noted that the Grade Number indicates the minimum acceptable tensile strength.*

Grade	Tensile strength (N mm-2) (minimum)	Brinell hardness at thickness of section (mm)				
		2.5	5	10	20	80
100	100	210	–	–	–	150
150	150	260	–	–	–	–
180	180	–	–	–	–	–
200	200	280	–	–	–	120
220	220	–	–	–	–	–
250	250	–	280	–	–	145
300	300	–	–	280	–	165
350	350	–	–	–	280	185

15.52　Fluid irons, in which mechanical strength is of secondary importance, were at one time widely used in the manufacture of railings, lamp-posts, and fireplaces. High fluidity was necessary in order that the iron should fill intricate mould impressions, and this was achieved by using a high silicon content of 2.5 to 3.5 per cent, as well as a high phosphorus content of up to 1.5 per cent. Cast iron has been replaced by other materials for the purposes mentioned above. As a small boy I was walking along the local High Street when a heavy lorry careered across the road crashing into the ornamental cast iron 'skirting' of a lampstandard on the pavement in front of me. The skirting – presumably of brittle high-phosphorus iron – fragmented on impact sending a hail of 'shrapnel' across the pavement. Shop windows were broken and some pedestrians received nasty wounds from the razor-sharp shards. Much scared I scuttled home unscathed! The local authority soon began to remove the cast iron skirtings which were ultimately replaced by concrete supports. The menace of HGVs had begun even that early in the twentieth century.

High-duty cast irons

15.60　As described earlier (15.21) what we see under the microscope as a thread-like inclusion with sharp-pointed ends is actually a cross-section through a graphite flake in shape very like a 'cornflake' with a very sharp-edged rim. Graphite has no appreciable strength and so these flakes have the same effect on mechanical properties as would have cavities in the structure. The sharp edges act as stress-raisers within the structure; that is, they give rise to an increase in local stress, in the same way that a sharp-cornered key-way tends to weaken a shaft. Hence, both the mechanical strength and the toughness of a cast iron can be improved by some treatment which disperses the graphite flakes, or, better still, which alters their shape to that of spherical globules.

15.61　Spheroidal-graphite (SG) cast iron, also known as 'nodular iron', or (in the USA) as 'ductile iron', contains its graphite in the form of rounded globules (Fig. 15.5). The sharp-edged, stress-raising flakes are thus eliminated, and the structure is made more continuous. Graphite is made to deposit in globular form by adding small amounts of either of the metals cerium or magnesium to the molten iron, just before casting. Magnesium is the more widely used, and is generally added as a magnesium-nickel alloy, in amounts to give a residual magnesium-content of 0.1 per cent in the iron. Such an iron may have a tensile strength of as much as $775\,\text{N}\,\text{mm}^{-2}$.

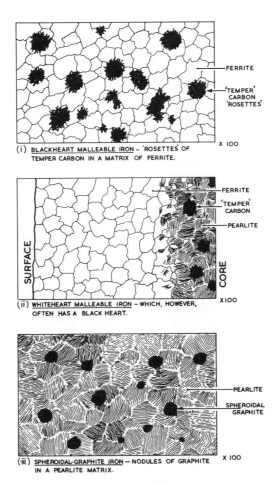

Fig. 15.5 *Structures of malleable and spheroidal-graphite cast irons. The structure of a whiteheart malleable iron (ii) is rarely uniform throughout, and often contains some carbon in the core. This has had insufficient time to diffuse outwards to the skin, and so be lost.*

15.62 Compacted-graphite (CG) cast irons are characterized by graphite structures – and consequently mechanical properties – intermediate between those of ordinary grey irons and those of SG irons. Before being cast, the molten iron is first desulphurized and then treated with an alloy containing magnesium, cerium and titanium, so that traces of these elements remain in the casting. Titanium prevents the graphite from being completely spherical as in SG iron. Instead the flakes are short and stubby and have *rounded* edges. These irons are used in vehicle brake parts, gear pumps, eccentric gears and fluid and air cylinders.

Malleable cast irons

15.70 These are irons of such a composition as will give, in the ordinary cast form, a white (cementitic) structure. However, they subsequently receive heat-treatment, the object of which is either to convert the cementite into small spherical particles of carbon (the 'black-heart' process), or, alternatively, to remove the carbon completely from the structure (the 'white-heart' process). In either process, the silicon content of the iron is usually less than 1.0 per cent, in order that the iron shall be 'white' in the cast condition. When the cementite has been either replaced by carbon or removed completely, a product which is both malleable and ductile is the result.

15.71   Blackheart malleable iron In this process, the low-silicon white-iron castings are heated at about 900 °C in a continuous-type furnace through which an oxygen-free atmosphere circulates. A moving hearth carries the castings slowly through the heating-zone, so that the total heating time is about 48 hours.

This prolonged annealing causes the cementite to break down; but, instead of coarse graphite flakes, the carbon deposits as small 'rosettes' of 'temper carbon'. A fractured surface appears dark, because of the presence of this carbon; hence the term 'blackheart'. Since the structure now consists entirely of temper carbon and ferrite, it is soft and ductile. Blackheart malleable castings are widely used in the automobile industries, because of their combination of castability, shock-resistance, and good machinability. Typical components include brake-shoes, pedals, levers, wheel-hubs, axle-housings, and door-hinges.

15.72   Whiteheart malleable iron Castings for this are also produced from a low-silicon white iron; but, in this process, the castings are heated at about 1000 °C for up to 100 hours, whilst in contact with some oxidizing material such as haematite ore.

During heating, carbon at the surface of the castings is oxidized by contact with the haematite ore, and is lost as carbon dioxide gas. Carbon then diffuses outwards from the core – rather like a carburizing process in reverse – and is in turn lost by oxidation. After this treatment is complete, *thin* sections will consist entirely of ferrite, and, on fracture, will give a steely-white appearance; hence the name 'whiteheart'. Thick sections may contain some particles of temper carbon at the core, because heating has not been sufficiently prolonged to allow all of the carbon to diffuse outwards from these thick sections (Fig. 15.5(ii)).

The whiteheart process is particularly suitable for the manufacture of thin sections which require high ductility. Pipe-fittings, parts for agricultural machinery, switchgear equipment, and fittings for bicycle and motorcycle frames are typical of whiteheart malleable castings.

15.73   Pearlitic malleable iron This is similar in its initial composition to the blackheart material. The castings are malleabilized, either fully or partially at about 950 °C, to cause the breakdown of the bulk of the cementite, and are then reheated to about 900 °C. This reheating process causes some of the carbon to dissolve in austenite which forms above the lower critical temperature; and, on cooling, a background of pearlite is produced, the mechanism of the process being something like that of normalizing in steels.

The choice between ordinary grey irons, spheroidal-graphite iron, and malleable iron is, as with most materials problems, governed by both economic and technical considerations. Grey iron is, of course, cheapest, and also the easiest in which to produce sound castings. For 'high-duty' purposes, the choice between SG iron and malleable iron may be less easy, and a number of factors may have to be considered in arriving at a decision, but in recent years malleable irons have lost ground to other forms of high-duty iron.

Table 15.3 *Properties and uses of some alloy cast irons.*

Name or type of iron	Composition %				Typical mechanical properties		Uses
	C	Si	Mn	Others	Tensile strength (N mm^{-2})	Brinell	
Chromidium	3.2	2.1	0.8	Cr – 0.3	270	230	Cylinder-blocks, brake-drums, clutch-casings, etc.
Ni-tensyl	2.8	1.4	–	Ni – 1.5	350	230	An 'inoculated' cast iron
Wear-resistant iron	3.6	2.8	0.6	V – 0.2	–	–	Piston-rings for aero, automobile and diesel engines. Possesses wear-resistance and long life
Ni-Cr-Mo iron	3.1	2.1	0.8	Ni – 0.5 Cr – 0.9 Mo – 0.9	360	300	Hard, strong, and tough – used for automobile crankshafts
Heat-resistant iron	3.4	2.0	0.6	Ni – 0.35 Cr – 0.65 Cu – 1.25	270	220	Good resistance to wear and to heat cracks – used for brake-drums and clutch-plates
Ni-hard	3.3	1.1	0.5	Ni – 4.5 Cr – 1.5	–	600	A 'martensitic' iron, due to the presence of nickel and chromium – used to resist severe abrasion – chute-plates in coke plant
Ni-resist	2.9	2.1	1.0	Ni – 15.0 Cr – 2.0 Cu – 6.0	210	130	Plant handling salt water – an austenitic iron
Nicrosilal	2.0	5.0	1.0	Ni – 18.0 Cr – 5.0	255	330	Also an austenitic corrosion- and heat-resistant iron
Silal	2.5	5.0	–	–	165	–	'Growth'-resistant at high temperatures

Alloy cast irons

15.80   Generally speaking, the effects which alloying elements have on the properties of cast iron are similar to the effects which the same elements have on steel. Alloying elements can therefore be used to improve the mechanical properties of an iron, by refining the grain size, stabilizing hard carbides, and, in some cases, producing cast irons with a martensitic or austenitic structure.

15.81   **Nickel**, like silicon, has a graphitizing effect on cementite, and so tends to produce a grey iron. At the same time, nickel has a grain-refining effect, which helps to prevent the formation of coarse grain in those heavy sections which cool slowly. It also toughens thin sections, which might otherwise be liable to crack.

15.82   **Chromium** is a carbide stabilizer, and forms chromium carbide, which is harder than ordinary cementite. It is therefore used in wear-resistant irons. Since chromium forms very stable carbides, irons which contain chromium are less susceptible to 'growth'.

15.83   **Molybdenum** increases the hardness of thick sections, and also improves toughness.

15.84   **Vanadium** increases both strength and hardness; but, more important still, promotes heat-resistance in cast irons, by stabilizing carbides so that they do not decompose on heating.

15.85   **Copper** dissolves in iron in only very small amounts, and has little effect on mechanical properties. It is added mainly to improve resistance to rusting.

Some typical alloy cast irons are shown in Table 15.3.

CHAPTER 16

Copper and its alloys

16.10 In our early history lessons we were told that the Stone Age was followed some 3500 years ago by the Bronze Age but it seems very likely that metallic copper was used in Egypt some 1500 years before bronze. Since copper is more easily corroded than is bronze only limited evidence of this early use of copper remains. Nevertheless we can regard copper as the most ancient metal of any engineering significance. Such brass – a copper-zinc alloy – as was used by the Ancients was made by smelting together ores of copper and zinc. The extraction of metallic zinc was not possible at that time. Later, the Romans made their brass by smelting copper along with the zinc ore calamine (zinc carbonate). In ancient times the World's output of copper – mainly for bronze manufacture – was ultimately outstripped by that of iron and, during the twentieth century, also by aluminium.

16.11 In former days, the bulk of the world's requirement of copper was smelted in Swansea, from ore mined in Cornwall, Wales, or Spain. Later, deposits of ore were discovered in the Americas and Australia, and shipped to Swansea to be smelted, but it was subsequently realized that it would be far more economical to smelt the ore at the mine. Thus, Britain ceased to be the centre of the copper industry. Today, the USA and the Russian Federation, along with Chile, Canada, Zambia, Zaire and Peru, are the leading producers of copper.

The extraction of copper

16.20 Copper is extracted almost entirely from ores based on copper pyrites (a mineral in which copper is chemically combined with iron and sulphur). The metallurgy of the process is rather complex, but is essentially as follows.

1. The ore is 'concentrated'; that is, it is treated by 'wet' processes to remove as much as possible of the earthy waste, or 'gangue'.
2. The concentrate is then heated in a current of air, to burn away much of the sulphur. At the same time, other impurities, such as iron and silicon, oxidize to form a slag which floats on top of the purified molten copper sulphide (called 'matte').
3. The molten matte is separated from the slag, and treated in a Pierce–Smith converter, the operation of which resembles to some extent that of the furnace used in steel-making by the 'oxygen process' (11.22).

Some of the copper sulphide is oxidized, and the copper oxide thus formed reacts chemically with the remainder of the sulphide, producing crude copper.

16.21   The crude copper is then refined by either

- remelting it in a furnace, so that impurities are oxidized, and are lost as a slag; or
- electrolysis, in which an ingot of crude copper is used as the anode, whilst a thin sheet of pure copper serves as the cathode. During electrolysis, the anode gradually dissolves, and high-purity copper is deposited on the cathode. 'Cathode copper' so formed is 99.97 per cent pure.

Properties of copper

16.30   The most important physical property of copper is its *very high electrical conductivity*. In this respect it is second only to silver; though, if we take the electrical conductivity of silver as being 100 units, then that of pure copper reaches 97 followed by gold (71) and aluminium (58). Consequently the greater part of the world's production of metallic copper is used in the electrical industries.

Much of the copper used for electrical purposes is of very high purity, as the presence of impurities reduces the electrical conductivity, often very seriously. Thus, the introduction of only 0.04 per cent phosphorus will reduce the electrical conductivity by almost 25 per cent. Other elements have less effect; for example, 1.0 per cent cadmium, added to copper used for telephone-wires, in order to strengthen them, has little effect on the conductivity.

16.31   The thermal conductivity and corrosion-resistance of copper are also high, making it a useful material for the manufacture of radiators, boilers, and other heating equipment. Since copper is also very malleable and ductile, it can be rolled, drawn, deep-drawn, and forged with ease.

In recent years, the cost of copper production has risen steeply; so for many purposes – electrical and otherwise – it has been replaced by aluminium, even though the electrical and thermal conductivities of the latter (17.32) are inferior to those of copper.

Commercial grades of copper

16.40   Copper is refined either electrolytically or by furnace-treatment, as outlined above, and both varieties are available commercially.

16.41   **Oxygen-free high-conductivity (OFHC) copper** is derived from the electrolytically refined variety. The cathodes are melted, cast, rolled, and then drawn to wire or strip for electrical purposes. This grade is usually 99.97 per cent pure, and is of the highest electrical conductivity.

16.42   **'Tough-pitch' copper** is a fire-refined variety which contains small amounts of copper oxide as the main impurity. Since this oxide is present in the microstructure as tiny globules which have little effect on the properties, it is suitable for purposes where maximum electrical conductivity is not required. It is unsuitable where gas welding processes are involved because reactions between the oxide globules and hydrogen in the welding gas cause extreme brittleness:

$$Cu_2O + H_2 = 2Cu + H_2O \text{ (steam)}$$

Steam is insoluble in solid copper and forms fissures at the crystal boundaries.

16.43 Deoxidized copper is made from the tough-pitch grade by treating it with a small amount of phosphorus just before casting, in order to remove the oxide globules. Whilst phosphorus-deoxidized copper may be valuable for processes where welding is involved, it is definitely not suitable for electrical purposes, because of the big reduction in electrical conductivity introduced by the presence of dissolved phosphorus.

The tensile strength of hard-rolled copper reaches about 375 N mm^{-2}; so, for most engineering purposes where greater strength is required, copper must be suitably alloyed.

16.44 Alloys of copper Alloys of copper are less widely used than they were. The growing price of copper relative to its increasing scarcity, coupled with the fact that the quality of cheaper alternatives has improved in recent years, has led to the replacement of copper alloys for many purposes. Moreover, improved shaping techniques have allowed less ductile materials to be employed; thus deep-drawing quality mild steel is now often used where ductile brass was once considered to be essential. Nevertheless, the following copper alloys are still of considerable importance.

The brasses

16.50 These are copper-base alloys containing up to 45 per cent zinc and, sometimes, small amounts of other metals, the chief of which are tin, lead, aluminium, manganese, and iron. The equilibrium diagram (Fig. 16.1) shows that plain copper-zinc alloys with up to approximately 37 per cent zinc have a structure consisting of a single phase – that labelled α. Phases occupying such a position on an equilibrium diagram are solid solutions (9.50). Solid solutions are invariably tough and ductile and this particular one is no exception, being the basis of one of the most malleable and

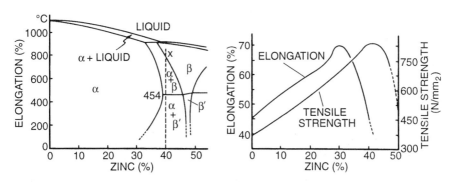

Fig. 16.1 *The section of the copper-zinc equilibrium diagram which covers brasses of engineering importance.*

ductile metallurgical materials in common use. Brasses containing between 10 and 35 per cent zinc are widely used for deep-drawing and general presswork – the maximum ductility being attained in the case of 70–30 brass, commonly known as 'cartridge metal', since it is used in the deep-drawing of cartridge- and shell-cases of all calibres. Many of these however are now produced in modern low-nitrogen, deep-drawing quality mild steel (11.21).

16.51 Brasses with more than 37 per cent zinc contain the phase β'. This is a hard, somewhat brittle substance; so a 60–40 brass lacks ductility. If such a brass is heated to 454 °C (Fig. 16.1), the phase β' changes to β, which is soft and malleable. As the temperature is increased further, the α phase present dissolves in β, until at X the structure is entirely malleable β. Therefore, 60–40 type brasses are best hot-worked at about 700 °C. This treatment also breaks up the coarse cast structure, and replaces it with a fine granular structure.

Thus, brasses with less than 37 per cent zinc are usually cold-working alloys, whilst those with more than 37 per cent zinc are hot-working alloys. A copper-zinc alloy containing more than 50 per cent zinc would be useless for engineering purposes, since it would contain the very brittle phase γ.

Up to 1 per cent *tin* is sometimes added to brasses, to improve their resistance to corrosion, particularly under marine conditions. *Lead* is insoluble in both molten and solid brass, and exists in the structure as tiny globules. About 2 per cent lead will improve the machinability (7.53) of brass. Small amounts of *arsenic* are said to improve corrosion-resistance and inhibit dezincification.

16.52 Manganese, iron, and aluminium all increase the tensile strength of a brass, and are therefore used in high-tensile brasses. These alloys are sometimes known, rather misleadingly, as 'manganese bronzes'. Additions of up to 2 per cent aluminium are also used to improve the corrosion-resistance of some brasses.

The more important brasses are shown in Table 16.1.

16.53 'Shape memory' alloys 'Shape memory' is a phenomenon associated with a limited number of alloys. The important characteristic of these alloys is the ability to exist in two distinct shapes or crystal structures, one above and one below a critical transformation temperature. As the temperature falls below this critical temperature a martensitic type of structure forms. As the temperature is raised again the martensite reverts to the original structure. This *reversible change* in structure is linked to a change in dimensions and the alloy thus exhibits a 'memory' of the high and low temperature shapes.

The best-known commercial alloys are the SME brasses[1] containing 55–80 per cent copper, 2–8 per cent aluminium and the balance zinc. By choosing a suitable composition, transition temperatures between −70 °C and +130 °C can be achieved. The force associated with the change in shape can be used to operate temperature-sensitive devices, the snap on/off positions often being obtained by using a compensating bias spring. Such devices are used in automatic greenhouse ventilators, thermostatic radiator valves, de-icing switches, electric kettle switches and valves in solar heating systems.

---

[1] 'Shape memory effect' brasses.

Table 16.1 *Brasses.*

BS 2870 designation	Composition %			Typical mechanical properties			Uses
	Cu	Zn	*Other elements*	*Condition*	*Tensile strength* (N mm^{-2})	*Elongation* (%)	
CZ 101	90	10	–	Soft Hard	280 510	55 4	*Gilding metal* – used for imitation jewellery, because of its gold-like colour, good ductility, and its ability to be brazed and enamelled.
CZ 106	70	30	–	Soft Hard	320 700	70 5	*Cartridge brass* – deep-drawing brass of maximum ductility. Used particularly for the manufacture of cartridge- and shell-cases.
CZ 107	65	35	–	Soft Hard	320 700	65 4	*Standard brass* – a good general-purpose cold-working alloy when the high ductility of 70–30 brass is not required. Widely used for press-work and limited deep-drawing.
CZ 108	63	37	–	Soft Hard	340 725	55 4	*Common brass* – a general-purpose alloy, suitable for limited cold-working operations.
CZ 123	60	40	–	Hot-rolled	370	40	*Yellow or Muntz metal* – hot-rolled plate. Can be cold-worked only to a limited extent. Also extruded as rods and tubes.
CZ 120 2Pb	59	39	Lead – 2	Extruded rod	450	30	*Free-cutting brass* – very suitable for high-speed machining, but can be deformed only slightly by cold-work.
CZ 112	62	37	Tin – 1	Extruded	420	35	*'Naval brass'* – structural uses, also forgings. Tin raises corrosion-resistance, especially in sea water.
–	58	Rem.	Manganese – 1.5 Aluminium – 1.5 Iron – 1 Lead – 1 Tin – 0.6	Extruded	500	15	*High-tensile brass* ('manganese bronze') – pump-rods, stampings, and pressings. Also as marine castings such as propellors, water-turbine runners, rudders, etc.
–	61.5	Rem.	Lead – 2.5 Arsenic – 0.1	Hot-stamping	380	25	*Dezincification-resistant brass* – water fittings where water supply dezincifies plain α/β' brasses. After stamping the brass is usually annealed at 525°C and water quenched to improve the dezincification resistance.

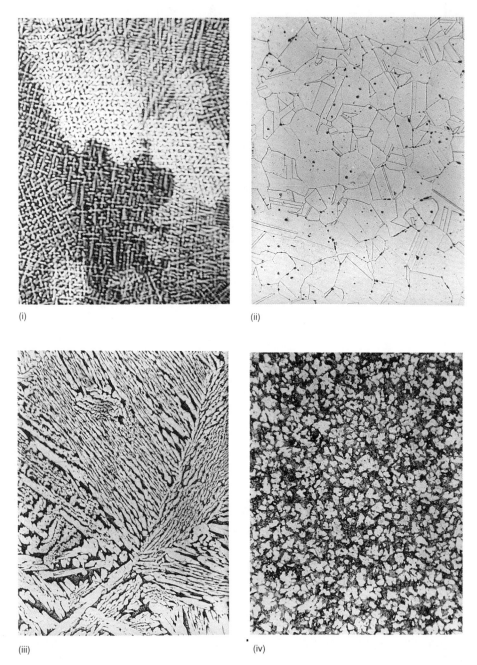

(i)

(ii)

(iii)

(iv)

Fig. 16.2   *Microstructures of some brasses. (i) 70–30 brass, as cast. Cored crystals of α solid solution (see also Fig. 8.3). (ii) 70–30 brass, cold-worked and then annealed at 600 °C. Coring of the original cast structure has been removed by this treatment and recrystallization has produced small crystals (twinned) of the solid solution α. (iii) 60–40 brass, as cast. This shows a typical Widmanstätten structure of α (light) and β (dark). (iv) 60–40 brass in hot worked (extruded) condition. Hot-working has broken up the coarse Widmanstätten structure and, on cooling, small α crystals (light) have precipitated from the β phase (dark).*

Tin bronzes

16.60 Tin bronze was almost certainly the first metallurgical *alloy* to be used by Man and it is a sobering thought that for roughly half of the 4000 years of history during which he has been using metallurgical alloys bronze was his sole material. Though tin bronzes have relatively limited uses these days – due in part to the high prices of both copper and tin – they still find application for special purposes. Tin bronzes contain up to 18 per cent tin, sometimes with smaller amounts of phosphorus, zinc or lead.

16.61 The complete copper-tin equilibrium diagram is a rather complex one and its interpretation is rendered more difficult by the fact that tin bronzes need to be cooled extremely slowly indeed – *far more slowly than is ever likely to prevail in a normal industrial casting process* – if microstructural equilibrium is to be attained. The reader may be forgiven for assuming, from a simple interpretation of the diagram, that a cast 10 per cent tin bronze which begins to solidify at X (about 1000 °C) will be completely solid at X_1 (about 850 °C) as a polycrystalline mass of α solid solution, which then cools without further change until at Z particles of the intermetallic compound ϵ begin to form within the α crystals. Sadly the transformations which take place are not nearly so simple.

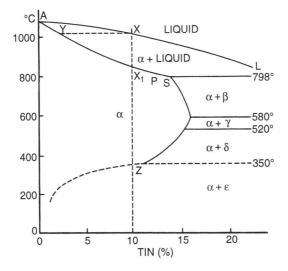

Fig. 16.3 *Part of the copper-tin equilibrium diagram. The phase δ is an intermetallic compound $Cu_{31}Sn_8$ whilst ϵ is the intermetallic compound Cu_3Sn but this latter will never be present in a bronze casting which solidifies and cools under normal industrial conditions.*

First, the wide range between the liquidus line AL and the solidus AS is an indication that extensive coring takes place in tin bronzes as they solidify. Thus the molten bronze containing 10 per cent tin begins to solidify at X by forming dendrites of composition Y (about 2 per cent tin). Since these dendrites contain only about 2 per cent tin this means that the remaining liquid has been left much richer in tin, i.e. its composition moves to the right along AL. Cooling is too rapid for the microstructure to attain equilibrium and solidification is not complete until the temperature has fallen to 798 °C when the remaining liquid will probably contain about 20 per cent tin. This will then solidify as a mixture of α and β as indicated by the diagram.

As the now solid alloy cools to 586 °C the $\alpha + \beta$ mixture transforms to $\alpha + \gamma$ which at 520 °C transforms further to $\alpha + \delta$. As the temperature falls lower transformations become increasingly sluggish and so the change to a structure of $\alpha + \epsilon$ at 350 °C does not occur at the speed at which the temperature is falling. So under industrial conditions of cooling we are left with a final microstructure of heavily cored α solid solution with a network and particles of the brittle intermetallic compound δ ($Cu_{31}Sn_8$) at the α dendrite boundaries.

Sorry if all of this sounds a bit of a fudge! It isn't really – it merely illustrates that in interpreting these equilibrium diagrams a number of complicating influences often have to be considered.

The networks of the δ phase at the crystal boundaries of the solid solution α make the 10 per cent tin bronze rather brittle under shock. Thus, commercial tin bronzes can be divided into two groups.

1. Wrought tin bronzes containing up to approximately 7 per cent tin. These alloys are generally supplied as rolled sheet and strip, or as drawn rod, wire, and turbine-blading.
2. Cast tin bronzes containing 10 to 18 per cent tin, used mainly for high-duty bearings.

During World War Two the company where I was employed as chief metallurgist accepted an order for slabs of hard-rolled 12 per cent tin bronze. Such material had never been produced commercially at that time but since – as it transpired later – the material was required for secret development of the jet engine it was presumably hoped that I could 'wave my magic wand'. After some weeks of laboratory trials I discovered that, as the copper-tin equilibrium diagram – though then rather incomplete – predicted, annealing the cast ingots of 12 per cent tin bronze at 740 °C for 12 hours was effective in absorbing all traces of the brittle δ phase and also eliminating almost completely all traces of coring. At 740 °C of course the structure of these $\alpha + \delta$ areas will have changed to $\alpha + \beta$ and since 740 °C is only just below the melting range, the movement of copper and tin atoms within the lattice structure will have accelerated – just as salt dissolves more quickly in hot water than in cold. Therefore uniform α (as indicated by point P in Fig. 16.3) was produced and equilibrium was finally reached.

My efforts bore fruit only temporarily. Not only was 12 per cent tin bronze expensive due to the high costs of tin and copper but production costs were high because of the long annealing process in a protective atmosphere and the fact that the alloy work-hardened very quickly during rolling and so required frequent inter-stage annealings. Very soon a cheaper material was sought. Often a R-and-D project serves only to prove that a suggested programme is not feasible!

16.62 Phosphor bronze Most of the tin bronzes mentioned above contain up to 0.05 per cent phosphorus, left over from the deoxidation process which is carried out before casting. Sometimes, however, phosphorus is added – in amounts up to 1.0 per cent – as a deliberate alloying element, and only then should the material be termed 'phosphor bronze'. The effect of phosphorus is to increase the tensile strength and corrosion-resistance, whilst, in the case of cast bearing alloys, reducing the coefficient of friction.

16.63 Gunmetal This contains 10 per cent tin and 2 per cent zinc, the latter acting as a deoxidizer, and also improving fluidity during casting. Since

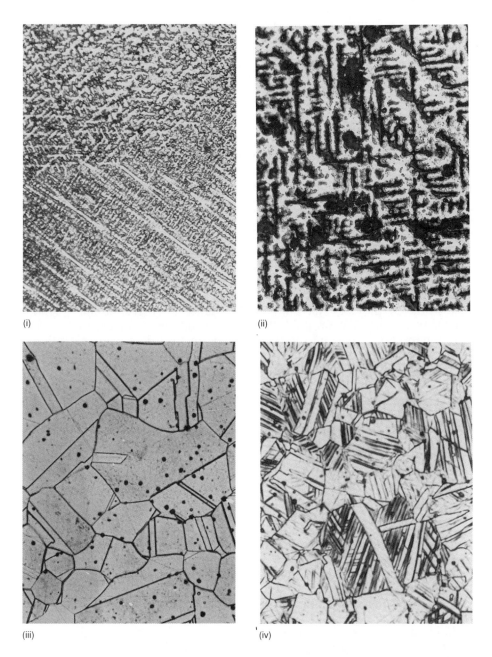

(i)

(ii)

(iii)

(iv)

Fig. 16.4   *Microstructure of some tin bronzes. (i) 95–5 tin bronze, as cast. Cored dendrites of α
solid solution. The black areas are voids, i.e. intercrystalline cavities probably caused by
shrinkage. (ii) 90–10 tin bronze, as cast. Cored α solid solution (dark) surrounded by δ which
has coagulated so that the original α + δ eutectoid structure has been lost. The large black areas
are again shrinkage cavities. (iii) 95–5 tin bronze, cold worked and annealed. 'Twinned' crystals
of α solid solution. Coring has been dispersed by the mechanical work and recrystallization
during annealing. The structure is soft. The dark spots are probably voids resulting from the
original shrinkage cavities. (iv) 95–5 tin bronze, cold-worked, annealed and cold-worked again.
Now the α solid-solution crystals show extensive strain bands (indicating the positions of the slip
planes in the crystals – locked-up strain energy there causes the region to etch more quickly).
The structure is much harder because of the cold-work it has received.*

zinc is considerably cheaper than tin, the total cost of the alloy is reduced. Gunmetal is no longer used for naval armaments, but it is used as a bearing alloy, and also where a strong, corrosion-resistant casting is required.

16.64 Coinage bronze   This is a wrought alloy containing 3 per cent tin and, in the interests of economy, 1.5 per cent zinc. In Britain even more of the tin has been replaced by zinc (see Table 16.2).

16.65 Leaded bronzes   Up to 2.0 per cent lead is sometimes added to bronzes, in order to improve machinability. Some special bearing bronzes contain up to 24 per cent lead, and will carry greater loads than will 'white metal' bearings. Since the thermal conductivity of these bronzes is also high, they can work at higher speeds, as heat is dissipated more quickly.

Aluminium bronzes

16.70   Like brasses, the aluminium bronzes can be divided into two groups: the cold-working alloys, and the hot-working alloys. Those alloys containing approximately 5 per cent aluminium are ductile and malleable, since they are completely solid solution in structure. They have a good capacity for cold-work. As they also have a good resistance to corrosion, and a colour similar to that of 22 carat gold, they were widely used for cheap jewellery and imitation wedding rings in those less permissive days when such things were deemed necessary; but for decorative purposes they have been replaced by coloured anodized aluminium and other cheaper materials.

The hot-working alloys contain in the region of 10 per cent aluminium, and, if allowed to cool slowly, the structure is brittle, due to the precipitation of a hard compound within it. When this structure is heated to approximately 800 °C, it changes to one which is completely solid solution, and hence malleable; so alloys of this composition can be hot-worked successfully. Similar alloys are also used for casting to shape by both sand- and die-casting methods. To prevent precipitation of the brittle compound mentioned above, castings are usually ejected from the mould as quickly as possible, so that they cool rapidly.

An interesting feature of the 10 per cent alloy is that it can be heat-treated in a manner similar to that for steel. A hard martensitic type of structure is produced on quenching from 900 °C, and its properties can be modified by tempering. Despite these apparently attractive possibilities, heat-treatment of aluminium bronze is not widely employed, and such of these alloys as are used find application mainly because of their good corrosion-resistance, retention of strength at high temperatures, and good wearing properties.

Aluminium bronze is a difficult alloy to cast successfully, because, at its casting temperature (above 1000 °C), aluminium oxidizes readily. This leads to aluminium oxide dross becoming entrapped in the casting, unless special casting techniques are employed, and an increase in the cost of the process inevitably results. Compositions and uses of some aluminium bronzes are given in Table 16.3.

Copper–nickel alloys

16.80   The metals copper and nickel 'mix' in all proportions in the solid state; that is, a copper-nickel alloy of any composition consists of only one

Table 16.2 Tin bronzes and phosphor bronzes.

BS specification number	Composition %			Condition	Typical mechanical properties		Uses
	Cu	Sn	Other elements		Tensile strength ($N\,mm^{-2}$)	Elongation (%)	
–	95.5	3	Zn – 1.5	Soft Hard	320 725	65 5	Coinage bronze – British 'copper' coinage now contains rather less tin (0.5%) and more zinc (2.5%).
2870/PB 101	96	3.75	P – 0.1	Soft Hard	340 740	65 15	Low-tin bronze – springs and instrument parts. Good elastic properties and corrosion-resistance.
2870/PB 102	94	5.5	P – 0.2	Soft Hard	350 700	65 15	Drawn phosphor bronze – generally used in the work-hardened condition; steam-turbine blading. Other components subjected to friction or corrosive conditions.
1400/PB1 – B	89	10	P – 0.5	Sand cast	280	15	Cast phosphor bronze – supplied as cast sticks for turning small bearings, etc.
–	81	18	P – 0.5	Sand cast	170	2	High-tin bronze – bearings subjected to heavy loads – bridge and turntable bearings.
1400/G1 – C	88	10	Zn – 2 Ni – 2 (max.)	Sand cast	290	16	Admiralty gunmetal – pumps, valves, and miscellaneous castings (mainly for marine work, because of its high corrosion resistance); also for statuary, because of good casting properties.
1400/LG2 – A	85	5	Zn – 5 Pb – 5 Ni – 2 (max.)	Sand cast	220	13	Leaded gunmetal (or 'red brass') – a substitute for Admiralty gunmetal; also where pressure tightness is required.
1400/LB5 – B	75	5	Pb – 20 Ni – 2 (max.)	Sand cast	160	6	Leaded bronze – a bearing alloy; can be bonded to steel shells for added strength.

Table 16.3 *Aluminium bronzes.*

BS specification number	Composition %			Condition	Typical mechanical properties		Uses
	Cu	Al	Other elements		Tensile strength (N mm⁻²)	Elongation (%)	
–	Remainder	7.5	Fe, Mn, and Ni up to 2.5 total	Hot-worked	430	45	Chemical engineering, particularly at fairly high temperatures.
2872/CA 104	80	10	Fe – 5 Ni – 5	Forged	725	20	Forged propellor-shafts, spindles, etc. for marine work. Can be heat-treated by quenching and tempering.
1.400/AB1 – B	Remainder	9.5	Fe – 2.5 Ni and Mn up to 1.0 each (optional)	Cast	520	30	The most widely used aluminium bronze for both die- and sand-casting. Used in chemical plant and marine conditions – pump-casings, valve-parts, gears, propellors, etc.

Table 16.4 *Cupro-nickels and nickel-silvers.*

BS specification number	Composition %			Condition	Typical mechanical properties		Uses
	Cu	Ni	Other elements		Tensile strength $(N\,mm^{-2})$	Elongation (%)	
2870/CN 104	80	20	Mn – 0.25	Soft Hard	340 540	45 5	Used for bullet-envelopes, because of its high ductility and corrosion-resistance.
2870/CN 105	75	25	Mn – 0.25	Soft Hard	350 600	45 5	Mainly for coinage – the current British 'silver' coinage.
3072/3076/NA13	29	68	Fe – 1.25 Mn – 1.25	Soft Hard	560 720	45 20	*Monel metal* – good mechanical properties and excellent corrosion-resistance. Chemical engineering plant, etc.
3072/3076/NA18	29	66	Al – 2.75 Fe – 1.0 Mn – 0.4 Ti – 0.6	Soft Hard Heat-treated	680 760 1060	40 25 22	*'K' Monel* – a heat-treatable alloy. Used for motor-boat propellor-shafts.
2870/NS 106	60	18	Zn – Bal. Mn – 0.4	–	–	–	*Nickel silver* – spoons, forks, etc.
2870/NS 111	60	10	Zn – Bal. Pb – 1.5 Mn – 0.25	–	–	–	*Leaded nickel silver* – Yale-type keys, etc.

phase – a uniform solid solution. For this reason, *all* copper-nickel alloys are relatively ductile and malleable, since there can never be any brittle phase present in the structure. In the cast state, a copper-nickel alloy may be cored (9.36), but this coring can never lead to the precipitation of a brittle phase. In other words, the metallurgy of these alloys is very simple – and not particularly interesting as a result.

Cupro-nickels may be either hot-worked or cold-worked, and are shaped by rolling, forging, pressing, drawing, and spinning. Their corrosion-resistance is high, and only the high cost of both metals limits the wider use of these alloys.

16.81   Nickel silvers contain from 10 to 30 per cent nickel, and 55 to 65 per cent copper, the balance being zinc. Like the cupro-nickels, they are uniform solid solutions. Consequently, they are ductile, like the high-copper brasses, but have a 'near white' colour, making them very suitable for the manufacture of forks, spoons, and other tableware, though in recent years these alloys have been generally replaced by the less attractive stainless steels. When used for such purposes these nickel silvers are usually silver-plated – the stamp 'EPNS' means 'electroplated nickel silver'. The machinability of these – as of all other copper alloys – can be improved by the addition of 2 per cent lead. Such alloys are easy to engrave, and are also useful for the manufacture of Yale-type keys, where the presence of lead makes it much easier to cut the blank to shape.

Other copper base alloys

16.90   A number of copper-base alloys contain small quantities of alloying elements which are generally added to increase the tensile strength.

16.91   Copper-beryllium (or beryllium bronze) contains approximately 1.75 per cent beryllium and 0.2 per cent cobalt. It is a heat-treatable alloy which can be precipitation-hardened in a manner similar to that of some of the

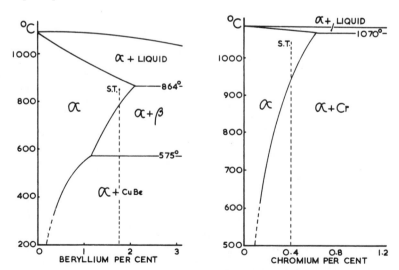

Fig. 16.5   *Copper-base alloys which can be precipitation hardened. In either case, the alloy is solution treated by heating it to the point ST, at which the structure becomes uniformly α. It is then quenched, to retain this structure, and then precipitation-hardened.*

aluminium alloys. A glance at the equilibrium diagrm (Fig. 16.5) will show why this is so. At room temperature, the slowly cooled structure will consist of the solid solution α (in this case, almost pure copper), along with particles of the compound CuBe. If this is heated – that is, solution-treated – at about 800 °C, the structure becomes completely α, as the CuBe is slowly dissolved by the solid solution. The alloy is then quenched, and in this condition is ductile, so that it can be cold-worked. If the alloy is now precipitation-hardened at 275 °C for an hour, a tensile strength of up to 1400 N mm^{-2} is obtained. The addition of 0.2 per cent cobalt restricts grain growth during the heat-treatment at 800 °C.

Since beryllium bronze is very hard in the cold-worked/heat-treated state, it is useful for the manufacture of non-sparking tools – chisels and hack-saw blades – for use in gas-works, 'dangerous' mines, explosives factories, or where inflammable vapours are encountered, as in paint and varnish works. Unfortunately, beryllium is a very scarce and expensive metal, and this limits its use in engineering materials.

16.92   Copper-chromium contains 0.5 per cent chromium, and is also a heat-treatable alloy, as is indicated by the equilibrium diagram (Fig. 16.5). It is used in some electrical industries and in spot-welding electrodes since it combines a high conductivity of about 80 per cent with a reasonable strength of 550 N mm^{-2}.

16.93   Copper-cadmium contains about 1 per cent cadmium, which raises the tensile strength of the hard-drawn alloy to 700 N mm^{-2}. Since the fall in electrical conductivity due to the cadmium is very small, this material is useful for telephone-wires, and other overhead lines.

16.94   Copper-tellurium contains 0.5 per cent tellurium, which, being insoluble in solid copper, exists as small globules in the structure. Since tellurium is insoluble, it has little effect on electrical conductivity, but gives a big improvement in machinability (resembling lead in this respect).

16.95   Arsenical copper contains 0.4 per cent arsenic, which increases from 200 to 550 °C the temperature at which cold-worked copper begins to soften when it is heated. This type of copper was widely used in steam-locomotive fireboxes and boiler-tubes, and still finds use in high-temperature steam plant. It is useless for electrical purposes, however, because of the great reduction in electrical conductivity caused by the presence of arsenic in solid solution.

CHAPTER 17

Aluminium and its alloys

17.10 Minerals containing aluminium are difficult to decompose because the metal is strongly electropositive, having a strong affinity for all of the non-metals. For this reason samples of the metal were not produced until in 1825 the Danish scientist H. C. Oersted used metallic potassium to chemically reduce aluminium from one of its compounds. Consequently, in those days aluminium cost about £250 per kg to produce, and was far more expensive than gold. It is reported that the more illustrious foreign visitors to the court of Napoleon III were privileged to use forks and spoons made from aluminium, whilst the French nobility had to be content with tableware in gold and silver. One still meets such cases of 'one-upmanship' even in the metallurgical world – quite recently I was told of a presentation beer tankard produced in the metal zirconium.

The extraction of aluminium

17.20 The modern electrolytic process for extracting aluminium was introduced simultaneously and independently in 1886 by Hall (in the USA) and Héroult (in France). Nevertheless, the metal remained little more than an expensive curiosity until the beginning of the twentieth century. Since then, the demands by both air- and land-transport vehicles for a light, strong material have led to the development of aluminium technology and an increase in the production of the metal, until now it is second only to iron in terms of annual world production.

17.21 The only important ore of aluminium is bauxite, which contains aluminium oxide (Al_2O_3). Unfortunately, this cannot be reduced to the metal by heating it with coke (as in the case of iron ore), because aluminium atoms are, so to speak, too firmly combined with oxygen atoms to be detached by carbon. For this reason, an expensive electrolytic process must be used to decompose the bauxite and release aluminium. Since each kg of aluminium requires about 91 megajoules of electrical energy, smelting plant must be located near to sources of cheap hydro-electric power (HEP), often at great distances from the ore supply, and from the subsequent markets. Consequently, most aluminium is produced in the USA, and in Canada, West Germany and Norway. HEP in the Western Highlands enables some aluminium to be smelted in Scotland, but the bulk of aluminium used in the UK is imported.

Crude pig iron can be purified (turned into steel) by blowing oxygen over it, to burn out the impurities (11.22), but this would not be possible in the case of aluminium, since the metal would burn away first, and leave us with the impurities. Instead, the crude bauxite ore is first purified by means of a chemical process, and the pure aluminium oxide is then decomposed by electrolysis. Since aluminium oxide has a very high melting-point, it is mixed with another aluminium mineral, cryolite, to form an electrolyte which will melt at a lower temperature.

17.22 The furnace used consists of a 'tank' some 2.5 m long to contain the molten electrolyte. It is lined with carbon and this constitutes the cathode of the electrolytic cell. Carbon rods dipping into the electrolyte form the anode. On dissolving in the molten electrolyte aluminium oxide, Al_2O_3, dissociates into ions:

$$Al_2O_3 \longrightarrow 2Al^{+++} + 3O^{--} \longrightarrow 2Al + 3O$$

The Al^{+++} ions are attracted to the cathode where they receive electrons to become atoms. This aluminium collects at the base of the 'tank' and is tapped off at intervals. The O^{--} ions are attracted to the anode where they discharge electrons to become atoms which immediately combine with the carbon electrodes. This causes the latter to burn away so that they need frequent replacement.

Properties of aluminium

17.30 Although aluminium has a high affinity for oxygen, and might therefore be expected to oxidize – or corrode – very easily, in practice it has an *excellent resistance to corrosion*. This is due largely to the thin but very dense film of oxide which forms on the surface of the metal and, since it is very tenacious, effectively protects it from further atmospheric attack. The reader will be familiar with the comparatively dull appearance of the surface of polished aluminium. This is due to the oxide film which immediately forms. The protective oxide skin can be artificially thickened by a process known as 'anodizing' in which the article is made the anode in an electrolyte of dilute sulphuric acid.[1] Oxygen atoms released at the anode combine with aluminium atoms at the surface to thicken the oxide film already present. Since aluminium oxide is extremely hard, anodizing makes the surface more wear-resistant. The anodized film is sufficiently porous to allow it to be dyed with either organic or inorganic dyes.

17.31 The fact that it has a *high thermal conductivity* and good corrosion-resistance, and that any corrosion products which are formed are non-poisonous, makes aluminium very suitable for the manufacture of domestic cooking-utensils such as kettles, saucepans, and frying-pans. In the form of disposable collapsible tubes it was used to contain a wide range of foodstuffs and toilet preparations ranging from caviare to toothpaste but for many such purposes it has been replaced by plastics materials. The high malleability of aluminium makes it possible to produce very thin foil which is excellent for the packaging of food and sweets. It is also widely used as 'cooking foil'.

---

[1] The function of the acid is merely to produce ionization of the water so that it is conductive.

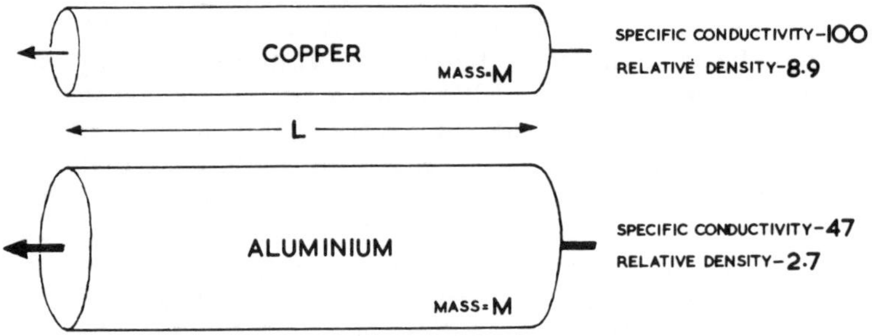

Fig. 17.1 *Conductors of copper and aluminium of equal length,* L, *and equal mass,* M. *Although copper has the better specific conductivity, the aluminium conductor passes a greater current under similar conditions, because of its greater cross-section.*

17.32 Aluminium has a *very good electrical conductivity*, which, though only about half that of copper, when considered weight for weight, can make aluminium a better proposition in some circumstances. Thus, if conductors of copper and aluminium of equal length and equal weight are taken, that of aluminium is a better overall conductor, because it has a greater cross-sectional area than that of the copper one. So, whilst aluminium may be unsuitable for use as windings of small components in the electronics industry because of the greater *volume* of wire needed to produce a unit of similar conductive capacity to that of copper, aluminium conductors are used in the grid system, generally strengthened by a steel core.

17.33 For use as a constructional material, pure aluminium lacks strength. In the 'soft' condition, its tensile strength is only 90 N mm^{-2}, whilst even in the work-hardened state it is no more than 135 N mm^{-2}. Hence, for most engineering purposes, aluminium is alloyed, in order to give a higher strength/weight ratio. Some of the high-strength alloys have a tensile strength in excess of 600 N mm^{-2} when suitably heat-treated.

Alloys of aluminium

17.40 Many aluminium alloys are used in the wrought form; that is, they are rolled to sheet, strip, or plate; drawn to wire; or extruded as rods or tubes. Other alloys are cast to shape by either a sand-casting or a die-casting process. In either case, some of the alloys may receive subsequent heat-treatment, in order further to improve their mechanical properties by inducing the phenomenon originally known as 'age-hardening', but now more properly termed 'precipitation-hardening'.

17.41 Thus the engineering alloys can be conveniently classified into four groups:

alloys which *are not* heat-treated   (1) wrought alloys   (2) cast alloys
alloys which *are* heat-treated   (3) wrought alloys   (4) cast alloys

In pre-war days, a bewildering assortment of aluminium alloys confronted the engineer. Worse still, they were covered by a rather untidy system of specification numbers, and each manufacturer used his own particular brand-name to describe an alloy. Since the Second World War, however, the number of useful alloys has been somewhat streamlined, and a systematic method of specification designation has been established internationally. In Britain this is operated by the British Standards Institution. In this system each alloy is identified by a number; whilst suffix letters are used to show what treatment (if any) it has received.

17.42 Wrought alloys are covered by BS 1470 (sheet and strip); 1471 (drawn tube); 1472 (forgings); 1473 (rivet, bolt and screw stock); 1474 (bars, extruded tube and sections) and 1475 (wire). Each alloy is identified by a four-digit number which itself carries some basic information about the alloy. Thus the first of the four digits indicates the *major* alloying element(s):

Commercial aluminium (more than 99.0 per cent Al):　1– – –

Aluminium alloy group by *major* alloying elements:

Copper	2– – –
Manganese	3– – –
Silicon	4– – –
Magnesium	5– – –
Magnesium and silicon	6– – –
Zinc	7– – –
Other elements.	8– – –
Unused series	9– – –

The second digit in the alloy designation denotes the number of modifications which have been made to the alloy specification. (If the second digit is zero it indicates the original alloy – integers 1 to 9 indicate alloy modifications.) The last two digits in the alloy designation have no significance other than to identify different alloys in the group. National variations of alloys registered by another country are identified by a serial letter immediately *after* the numerical designation, i.e. A, etc.

The condition in which the alloy is supplied is indicated by suffix letters:

O – anneald to its lowest strength
M – material 'as manufactured', e.g. rolled, drawn, extruded, etc.
D – material solution-treated and then drawn
H1, H2, H3, H4, Strain hardened – material subjected to cold work after
H5, H6, H7, H8. annealing or to a combination of cold work and partial annealing. Designations are in order of ascending tensile strength.
TB – material solution-treated and aged naturally
TD – material solution-treated, cold worked and aged naturally
TE – material cooled from an elevated-temperature shaping process and precipitation treated
TF – material solution treated and precipitation treated
TH – material solution treated, cold-worked and then precipitation treated.

Example: BS 1471/6082-TF represents the original specification of an aluminium alloy containing magnesium and silicon, in the tube form which has been solution treated and precipitation hardened after manufacture.

17.43 Cast alloys are covered by BS 1490, and specification numbers give information as to:

(a) Form of material:

Ingots – the alloy has *no* suffix letter
Castings – the suffix letter denotes the condition of the material.

(b) Condition:

    M – 'as cast' with no further treatment
    TS – stress-relieved only
    TE – precipitation treated
    TB – solution treated
    TB7 – solution treated and stabilized
    TF – solution treated and precipitation treated
    TF7 – full treatment, plus stabilization.

All of the alloy numbers representing cast alloys are prefixed by the letters LM. Thus LM 10 denotes alloy no. 10 in the ingot form, whilst LM 10-TF represents a casting of the same alloy in the fully heat-treated condition.

Wrought alloys which are not heat-treated

17.50   These are all materials in which the alloying elements form a solid solution in the aluminium, and this is a factor which contributes to their good corrosion resistance, high ductility and of course increased strength, the latter arising from the resistance to slip provided by the solute atoms present. Since these alloys are not heat-treatable the necessary strength and rigidity can be obtained only by controlling the amount of cold-work in the final shaping process. They are available in various stages of hardness between 'soft' (0) and 'full hard' (H).

17.51   The commercial grades of aluminium (1200) are sufficiently strong and rigid for some purposes, and the addition of up to 1.5 per cent manganese (3103) will produce a slightly stronger alloy. The aluminium-magnesium alloys have a very good resistance to corrosion, and this corrosion-resistance increases with the magnesium content, making them particularly suitable for use in marine conditions (Table 17.1).

Cast alloys which are not heat-treated

17.60   This group consists of alloys which are widely used for both sand-casting and die-casting. Rigidity, good corrosion-resistance, and fluidity during casting are their most useful properties.

17.61   The most important alloys in this group are those containing between 10 and 13 per cent silicon. Alloys of this composition are approximately eutectic in structure (Fig. 17.2). This makes them particularly useful for die-casting, since their freezing-range will be short, so that the casting will solidify quickly in the mould, making rapid ejection and hence a short production cycle possible. In the ordinary cast condition (Fig. 17.2) the eutectic structure is rather coarse, a factor which causes the alloy to be rather weak and brittle. However, the structure – and hence the mechanical properties – can be improved by a process known as 'modification'. This involves adding a small amount of metallic sodium (only 0.01 per cent by weight of the total charge) to the molten alloy, just before casting. The resultant casting has a very fine grain, and the mechanical properties are improved as a result (Table 17.2). The eutectic point is also displaced due to modification; so a 13 per cent silicon alloy, which would contain a little

Table 17.1 *Wrought aluminium alloys – not heat-treated.*

BS specification	Composition % Mn	Si	Mg	Others	Condition	Typical mechanical properties Tensile strength (N mm^{-2})	Elongation (%)	Use
1470: 1471: } 1200 1474:	99% aluminium 0.1 Maximum values	–	–	Fe + Si –1.0	O H8	90 155	35 5	Panelling and moulding, hollow-ware, electrical conductors, equipment for chemical, food- and brewing-plant, packaging
1470: 1475: } 3103	1.2	–	–	–	O H8	105 200	34 4	Metal boxes, bottle-caps, food-containers, cooking utensils, panelling of land-transport vehicles
1470: 1471: 1472: } 5251 1474: 1475:	–	–	2.25	–	O H28	185 255	20 2	Marine superstructures, panelling exposed to marine atmospheres, chemical plant, panelling for road and rail vehicles, fencing-wire
1470: 1471: 1472: 1473: } 5154A 1474: 1475:	–	–	3.5	–	O H24	220 275	18 6	Ship-building, deep-pressing for car bodies
1473: 1475: } 5083	–	–	5.0	–	O H24	275 350	16 6	Ship-building, and applications requiring high strength and corrosion-resistance, rivets

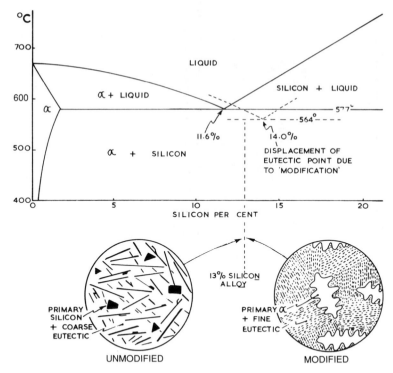

Fig. 17.2   *The aluminium-silicon thermal equilibrium diagram. The effects of 'modification' on both the position of the eutectic point and the structure are also shown. (See also Fig. 17.3.)*

brittle primary silicon in the unmodified condition now contains tough primary α solid solution instead, since the composition of the alloy is now to the *left* of the *new* eutectic point.

17.62   High casting-fluidity and low shrinkage make the aluminium-silicon alloys very suitable for presence die-casting. (Alloys of high shrinkage cannot be used, because they would crack on contracting in a rigid metal mould.) Since they are also very corrosion-resistant, the aluminium-silicon alloys are useful for marine work.

17.63   The most important property of the aluminium-magnesium-manganese alloys is their good corrosion-resistance, which enables them to receive a high polish. These alloys are noted for high corrosion-resistance, rigidity, and toughness, making them very suitable for use as moderately stressed parts working under marine conditions. Other casting alloys containing up to 10 per cent of either copper or zinc are now little used, because of their poor resistance to corrosion.

Wrought alloys which are heat-treated

17.70   In 1906, a German research metallurgist, Dr Alfred Wilm, was investigating the effects of quenching on the mechanical properties of some aluminium alloys containing small amounts of magnesium, silicon, and copper. To his surprise, he found that, if quenched test-pieces were allowed to remain at room temperature for a few days, without further heat-treatment, a

Table 17.2 *Cast aluminium alloys – not heat-treated.*

BS 1490 designation	Composition %					Condition	Typical mechanical properties		Uses
	Si	Cu	Mg	Mn	Ni		Tensile strength (N mm^{-2})	Elongation (%)	
LM 2M	10	1.6	–	–	–	Chill cast	250	3	General purposes, particularly pressure die-castings. Moderate strength alloy.
LM 4M	5	3	–	0.5	–	Sand cast Chill cast	155 190	4 –	Sand-castings, gravity and pressure die-castings. Good foundry characteristics. An inexpensive general-purpose alloy, where mechanical properties are of secondary importance.
LM 5M	–	–	4.5	0.5	–	Sand cast Chill cast	170 200	6 10	Sand-castings and gravity die-castings. Suitable for moderately stressed parts. Good resistance to marine corrosion. Takes a good polish.
LM 6M	12	–	–	–	–	Unmodified Modified	125 200	5 15	Sand, gravity- and pressure die-castings. Excellent foundry characteristics. Large castings for general and marine work. Radiators, sumps, gear-boxes, etc. One of the most widely used aluminium alloys.

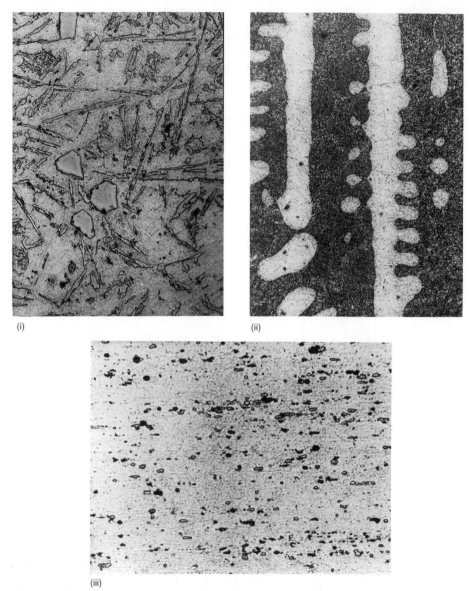

(i)

(ii)

(iii)

Fig. 17.3 *Microstructures of some aluminium alloys. (i) 12 per cent silicon in aluminium –*
unmodified, as cast. Since this alloy contains more than the eutectic amount (11.6 per cent) of
silicon (see Fig. 17.2), primary silicon (angular crystals) are present. The eutectic is coarse and
brittle and consists of 'needles' of silicon in a matrix of α solid solution because the layers of α in
the eutectic have fused together to form a continuous mass (the amount of silicon being only
11.6 per cent of the eutectic so that the layers of α would be roughly ten times the thickness of
those of silicon). (ii) The same alloy as (i) but modified by the addition of 0.01 per cent sodium.
This has the effect of displacing the eutectic point to 14 per cent silicon so that the structure now
consists of primary crystals of α (light) in a background of extremely fine-grained eutectic
(dark). The alloy is now stronger and tougher. (iii) A duralumin-type alloy in the 'as extruded'
condition (unetched). The particles consist mainly of CuAl₂ (see Fig. 17.4) elongated in the
direction of extrusion. Most of this CuAl₂ would be absorbed during subsequent solution
treatment.

considerable increase in strength and hardness of the material occurred. This phenomenon, subsequently called 'age-hardening', was unexplained at the time, since no apparent change in the microstructure was detected, and it is only in recent years that a reasonably satisfactory explanation of the causes of 'age-hardening' has been evolved. However, there are many instances in which metallurgical practice has been established long before it could be explained in terms of underlying theory – the heat-treatment of steel is an example – and Wilm's discovery was soon developed in the form of the alloy 'duralumin', which found application in the structural frames of the airships of Count von Zeppelin, which bombed England during the First World War.

17.71   As with carbon steels the heat-treatment of an aluminium alloy is related to the equilibrium diagram; and, in this instance, the important part of the diagram (Fig. 17.4) is the sloping boundary line *ABCD*, the significance of which was dealt with in 9.61. Here the slope of *ABCD* indicates that, as the temperature rises, the amount of copper which dissolves in *solid* aluminium also increases. This is a fairly common phenomenon in the case of ordinary liquid solutions; for example, the amount of salt which will dissolve in water increases as the temperature of the solution increases. Similarly, if the hot, 'saturated' salt solution is allowed to cool, crystals of solid salt begin to separate – or 'precipitate', as we say – from the solution, so that 'equilibrium' is maintained. In the case of *solid* solutions, these processes of solution and precipitation take place more slowly, because atoms find greater difficulty in moving in solid solutions than they do in liquid solutions, where all particles can move more freely. Consequently, once a solid solution has been formed at some high temperature by heat-treatment, its structure can usually be trapped by quenching, that is, cooling it rapidly so that precipitation of any equilibrium constituent has no opportunity to take place.

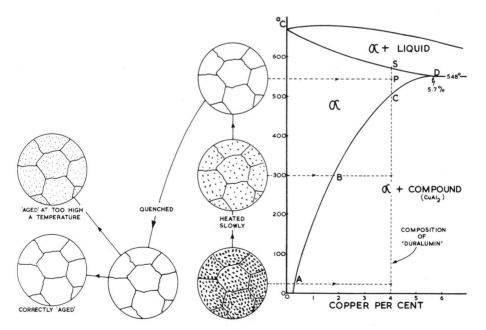

Fig. 17.4   *Structural changes which take place during the heat-treatment of a duralumin-type of alloy.*

17.72 We will assume that we have an aluminium-copper alloy containing 4 per cent copper, since this is the basic composition of some of the 'duralumin' alloys in use. Suppose the alloy has been extruded in the form of rod, and has been allowed to cool slowly to room temperature. This slow cooling will have allowed plenty of opportunity for the copper to precipitate from solid solution – not as particles of pure copper, but as small crystals of an aluminium-copper compound of chemical formula $CuAl_2$. Since this intermetallic compound is hard and brittle, it will render the alloy brittle. Moreover, since only about 0.2 per cent of copper (point A in Fig. 17.4) has remained behind in solution in the solid solution α, this solid solution will lack strength. Hence the mechanical properties of the alloy as a whole will be unsatisfactory.

Now suppose the alloy is slowly heated. As the temperature rises, the particles of compound $CuAl_2$ will gradually be absorbed into the surrounding α solid solution (just as salt would be dissolved by water). Thus at, say, 300 °C, much of the $CuAl_2$ will have been dissolved, and an amount of copper equivalent to B will now be in solution in the α. The amount of copper which dissolves increases with the temperature, and solution will be complete at C, as the last tiny particles of $CuAl_2$ are absorbed. In industrial practice, a temperature corresponding to P would be used, in order to make sure that all of the compound had been dissolved, though care would be taken not to exceed S, as at this point the alloy would begin to melt.

17.73 This heating process is known as *solution-treatment*, because its object is to cause the particles of $CuAl_2$ to be dissolved by the aluminium solid solution (α). Having held the alloy at the solution-treatment temperature long enough for the $CuAl_2$ to be absorbed by the α solid solution, it is then quenched in water, in order to 'trap' this structure, and so preserve it at room temperature. In the quenched condition, the alloy is stronger, because now the whole of the 4 per cent copper is dissolved in the α; but it is also more ductile, because the brittle crystals of $CuAl_2$ are now absent. If the quenched alloy is allowed to remain at room temperature for a few days, its strength and hardness gradually increase (with a corresponding fall in ductility). This phenomenon, which was known as *age-hardening*, is due to the fact that the quenched alloy is no longer in 'equilibrium', and tries its best to revert to its original α + $CuAl_2$ structure by attempting to precipitate particles of $CuAl_2$. In fact, the copper atoms do not succeed in moving very far within the aluminium lattice structure. Nevertheless the change is sufficient to cause a bigger resistance to movement on slip planes within the alloy, i.e. the strength has been increased. Although the copper atoms have moved closer together they still occupy positions which are part of the aluminium lattice structure. At this stage they are said to exist as a *coherent precipitate*.

17.74 This internal change within the alloy can be accelerated and made to proceed further by a 'tempering' process, formerly referred to as '*artificial age-hardening*'. A typical treatment may consist of heating the quenched alloy for several hours at, say, 160 °C, when both strength and hardness will be found to have increased considerably (curve B in Fig. 17.5). The term '*precipitation-hardening*' is now used to describe the increase in hardness produced, both by this type of treatment and by that which occurs at ordinary temperatures, as mentioned earlier. Care must be taken to avoid using too high a temperature during this precipitation treatment, or visible particles of $CuAl_2$ may form in the structure (Fig. 17.4) – that is, they now

Table 17.3 *Wrought aluminium alloys – heat-treated*

BS specifications	Composition (%)					Heat-treatment	Typical mechanical properties		Uses
	Cu	Si	Mg	Mn	Others		Tensile strength (N mm⁻²)	Elongation (%)	
—	—	0.5	0.7	—	—	Solution-treated at 520 °C; quenched and precipitation-hardened at 170 °C for 10 hours.	250	10	Glazing bars and window sections; windscreen and sliding-roof sections for automobiles. Good corrosion-resistance and surface finish.
1470: 1471: 1472: 1473: 1474: } 6082-TF	—	1.0	1.0	0.7	—	Solution-treated at 510 °C; quenched and precipitation-hardened at 175 °C for 10 hours.	310	8	Structural members for road-, rail-, and sea-transport vehicles; ladders and scaffold tubes; overhead lines (high electrical conductivity); architectural work.
2L84*	1.5	1.0	0.8	—	—	Solution-treated at 525 °C; quenched and precipitation-hardened at 170 °C for 10 hours.	390	10	Structural members for aircraft and road vehicles; tubular furniture.
2L77*	4.1	0.5	0.8	0.7	Ti } 0.3 Cr } opt.	Solution-treated at 480 °C; quenched and 'aged, at room temperature for 4 days.	400	10	Stressed parts in aircraft and other structures; general purposes. The original 'duralumin'.
1470: 1471: 1472: 1473: 1474: 1475: } 2014A-TF	4.3	0.7	0.5	0.7	Ti } 0.3 Cr } opt.	Solution-treated at 510 °C; quenched and precipitation-hardened at 170 °C for 10 hours.	510	10	Highly stressed components in aircraft stressed-skin construction; engine parts, such as connecting rods.
—	0.3	0.6	1.0	Either 0.5 Mn or 0.25 Cr	Zn – 7.0	Solution-treated at 520 °C; quenched and precipitation-hardened at 170 °C for 10 hours.	310	13	Plates, bars, and sections for shipbuilding; body panels for cars and rail vehicles; containers.
2L88* 2L95	1.6	0.5	2.5	0.3	—	Solution-treated at 465 °C; quenched and precipitation-hardened at 120 °C for 24 hours.	650	11	Highly stressed aircraft structures, such as booms; other military equipment requiring high strength/weight ratio. The strongest aluminium alloy produced commercially.
—	1.3	—	0.8	—	Li – 2.5 Zr – 1.2	—	570	6	Military aircraft – light-weight alloy (relative density 2.53)

* BS Aerospace Series, Section L (aluminium and the light alloys)

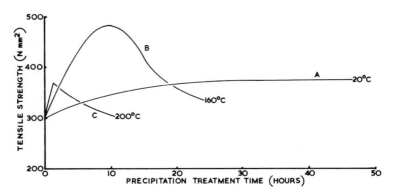

Fig. 17.5   *The effects of time and temperature of precipitation-treatment on the strength of duralumin.*

appear as a normal *non*-coherent precipitate. When this happens the mechanical properties will already have begun to fall again (curve *C* in Fig. 17.5). This is usually referred to as *over-ageing*, or sometimes, as *reversion*.

17.75   Because of the need for accurate temperature control during the heat-treatment of these alloys, solution-treatment is often carried out in salt baths, whilst precipitation-treatment generally takes place in air-circulating furnaces.

17.76   The aluminium-copper alloys are by no means the only ones which can be precipitation-hardened. Aluminium alloys containing small amounts of magnesium and silicon (forming Mg_2Si) can also be so treated. Recently aluminium alloys containing up to 2.5 per cent lithium have been developed. Here precipitation hardening depends upon the presence of the compound Al_3Li. Since lithium has a low relative density (0.51) reductions of up to 10 per cent in the densities of these alloys is achieved making them important in the aircraft industries. Since lithium is a very reactive element there are of course difficulties encountered in melting and casting these alloys.

Numerous magnesium-base alloys (18.40), some copper-base alloys (16.91 and 16.92) as well as maraging steels (13.27) can also be precipitation-hardened. Compositions, treatments and properties of some representative heat-treatable wrought aluminium alloys are given in Table 17.3.

Cast alloys which are heat-treated

17.80   Some of these alloys are of the 4 per cent copper type, as described above, but possibly the best known of them contains an additional 2 per cent nickel and 1.5 per cent magnesium. This is Y-alloy, 'Y' being the series letter used to identify it during its experimental development at the National Physical Laboratory, during the First World War. Whilst Germany was concentrating on the production of wrought duralumin for structural members of its Zeppelins, in Britain, research was being aimed at the production of a good heat-treatable *casting* alloy, for use in the engines of our fighter planes. The airframes of these wonderful machines were constructed largely of wood and 'doped' canvas, but a light alloy was needed from which high-duty pistons and cylinder-heads could be constructed for use at high temperatures. Y-alloy was the result of this research. Before I

Table 17.4 *Cast aluminium alloys – heat-treated.*

BS specifications	Composition (%)					Heat-treatment	Typical mechanical properties		Uses
	Cu	Si	Mg	Mn	Others		Tensile strength (N mm^{-2})	Elongation (%)	
1490/LM4 – TF	3.5	8	–	–	–	Solution-treated at 520°C for 6 hours; quenched in hot water or oil; precipitation-hardened at 170°C for 12 hours.	320	1	General purposes (sand-, gravity-, and pressure die-casting).
1490/LM9 – TF	–	11.5	0.4	0.5	–	Solution-treated at 530°C for 2–4 hours; quenched in warm water; precipitation-hardened at 160°C for 16 hours.	290	2	Suitable or intricate castings, due to fluidity imparted by silicon. Good corrosion-resistance.
4L53*	–	–	10	–	–	Solution-treated at 425°C for 8 hours; cooled to 390°C, and then quenched in oil at 160°C, or in boiling water.	320	18	Good strength and ductility, coupled with good corrosion-resistance. Used in marine work (seaplanes).
4L35*	4	–	1.5	–	Ni – 2.0	Solution-treated at 510°C; precipitation-hardened in boiling water for 2 hours, or aged at room temperature for 5 days.	280	–	Pistons and cylinder-heads for liquid- and air-cooled engines; general purposes. The original 'Y' alloy. Now obsolescent.
3L51*	1.4	2.5	0.2	–	Fe – 1.0 Ni – 1.3 Ti – 0.15	No solution-treatment required; precipitation-hardened at 165°C for 8–16 hours.	200	3	A good general-purpose alloy for sand-casting and gravity die-casting. High rigidity and moderate resistance to shock.
1490/LM29 – TE	1	23	1	–	Ni – 1.0	Chill cast and precipitation treated.	190	1	A gravity or pressure die-casting alloy – cylinder blocks and pistons in the automotive industries. A *hyper*-eutectic alloy in which the primary silicon is 'refined' by small additions of red phosphorus. Good wear resistance (hardness – 140 Brinell). Specification lays down microscopical examination to control crystal structure, i.e. size and distribution of primary silicon.

* BS Aerospace Series, Section L (aluminium and the light alloys)

wax too lyrical about those 'wonderful flying machines' of the early twentieth century perhaps I should remember that Leonardo da Vinci designed a craft which might well have flown if a suitable power unit had been available to him. Y-alloy (only recently obsolescent) and other heat-treatable casting alloys are enumerated in Table 17.4.

17.81 Recently a series of beryllium-aluminium alloys has been introduced containing in the region of 60% Be. Their main attribute is a low relative density of 2.16 combined with a tensile strengh of 310 N mm^{-2} (giving a specific strength of approximately 150 N mm^{-2}) with a % elongation of 3.5. These alloys are used as investment castings for such components as secondary structural members in the US Army helicopters and guidance components for advanced missiles and satellites. Naturally due to the high cost of beryllium – and its processing – these are expensive alloys.

CHAPTER 18

Other non-ferrous metals and alloys

18.10 When Columbus discovered the New World not more than half a dozen metals were known to Man. Now he has isolated all seventy of them and also 'constructed' a number of new ones, such as plutonium and neptunium, which previously did not exist in Nature. Some of these metals may never be of use to the engineer either because they have no desirable physical properties or because they 'corrode' far too readily – some instantaneously on exposure to the atmosphere. Nevertheless who would have thought, even as late as the 1950s, say, that by the end of the twentieth century we would have used samarium and neodymium (in magnets) hafnium (in nuclear plant), yttrium and lanthanum (in high-temperature alloys) and erbium (in cancer therapy generators)?

Nickel and its alloys

18.20 Although the Ancient Chinese may have used alloys similar in composition to our nickel silvers (16.81), nickel itself was not discovered in Europe until about 1750. At that time, the copper smelters of Saxony were having trouble with some of the copper ores they were using, for, although these ores appeared to be of the normal type, they produced a metal most unlike copper. This metal was given the name of 'kupfer-nickel', which, somewhat liberally translated from the Old Saxon, meant 'copper possessed of the Devil'. Later the material was found to be an alloy containing a new metal which was allowed to retain the title of 'nickel' – or 'Old Nick's metal'.

18.21 Nickel is a 'white' metal, with a faintly greyish tint. Most of the uses of commercially pure nickel depend upon the fact that it has a good resistance to corrosion, not only by the atmosphere, but by many other reagents. Consequently, much nickel is used in the electroplating industries, not only as a finishing coat, but also as a 'foundation layer' for good-quality chromium-plating.

Since nickel can be hot- and cold-worked successfully, and joined by most orthodox methods, it is used in the manufacture of chemical and food-processing plant, whilst nickel-clad steel ('Niclad') is used in the chemical and petroleum industries. A greater quantity of nickel however is used in the manufacture of alloy steels (13.21, 13.23, 13.51) whilst significant amounts are used in other alloys containing nickel, in particular the cupro-nickels (16.80), nickel silvers (16.81) and cast irons (15.81).

18.22   Electrical resistance alloys for use at high temperatures   These are generally nickel–chromium or nickel–chromium–iron alloys, the main features of which are:

- their ability to resist oxidation at high temperatures,
- high melting-ranges, and
- high electrical resistivity.

These properties make nickel–chromium alloys admirably suitable for the manufacture of resistance-wires and heater-elements of many kinds, working at temperatures up to bright red heat. Representative alloys (the Henry Wiggin 'Brightray' series) are included in Table 18.1.

Table 18.1   *High-temperature resistance alloys.*

Composition %			Resistivity $(10^{-8}\,\Omega\,m)$	Maximum working temperature (°C)	Uses
Ni	Cr	Fe			
80	20	–	103	1150	Heaters for electric furnaces, cookers, kettles, immersion-heaters, hair-dryers, toasters.
65	15	20	106	950	Similar to above, but for goods of lower quality; also for soldering-irons, tubular heaters, towel-rails, laundry-irons, and where operating temperatures are lower.
34	4	62	91	700	Cheaper-quality heaters working at low temperatures, but mainly as a resistance-wire for motor starter-resistances, etc.

18.23   Corrosion-resistant   alloys   These   alloys,   for   use   at   ordinary temperatures, all contain nickel, along with varying amounts of molybdenum and iron, and sometimes chromium and copper. Naturally, such alloys are relatively expensive, but their resistance to corrosion is extremely high. Consequently, they are used mainly in the chemical industries, to resist attack by strong mineral acids and acid chloride solutions – conditions which, in terms of corrosion, are about the most severe likely to be encountered industrially. A number of these alloys are described in Table 18.2.

Table 18.2   *Corrosion-resistant nickel-base alloys.*

Trade names	% Composition				Uses
	Ni	Mo	Fe	Others	
Corronel B	66	28	6	–	Resists attack by mineral acids and acid chloride solutions. Produced as tubes and other wrought sections for use in the chemical and petroleum industries, for constructing reaction-vessels, pumps, filter parts, valves, etc.
Ni-O-Nel	40	3	35	Cr-20 Cu-2	A 'Wiggin' alloy with characteristics similar to those of austenitic stainless steel, but more resistant to general attack, particularly in chloride solutions.
Hastelloy 'A'	58	20	22	–	Transporting and storing hydrochloric acid and phosphoric acid, and other non-oxidizing acids.
Hastelloy 'D'	85	–	–	Si-10 Cu-3 Al-1	A casting alloy – strong, tough, and hard, but difficult to machine (finished by grinding). Resists corrosion by hot concentrated sulphuric acid.

18.24 High-temperature corrosion-resistant alloys A well-known high-temperature alloy introduced many years ago is 'Inconel'. It contains 80 per cent nickel, 14 per cent chromium, and 6 per cent iron, and is used for many purposes, including food-processing plant, hot-gas exhaust manifolds, and heating-elements for cookers. It is quite tough at high temperatures, because of the very low grain-growth imparted by nickel (13.21), and it does not oxidize appreciably, because of the protective film of chromium oxide which forms on the surface (13.51).

The properties of materials like Inconel were extended in the 'Nimonic' series of alloys introduced by Messrs Henry Wiggin – alloys which played a leading part in the development of the jet engine. These 'Nimonics' are basically nickel–chromium alloys which have been strengthened – or 'stiffened' – for use at high temperatures, by adding small amounts of titanium, aluminium, cobalt, and molybdenum in suitable combinations. These elements form phases within the structure, and have the effect of raising the limiting creep stress (4.51) at high temperatures. A few of the better-known Nimonic alloys are given in Table 18.3.

18.25 Low-expansion alloys Most materials expand when they are heated and contract again as they cool. Some iron–nickel alloys, however, have extremely low coefficients of thermal expansion, making them useful in numerous types of precision equipment operating under conditions of varying temperature.

The coefficient of expansion is at a minimum for an iron-nickel alloy containing 36 per cent nickel. This alloy, originally known as 'Invar', was often used for the pendulums of clocks requiring great accuracy, as in astronomical observatories. As the amount of nickel increases above 36 per cent, the coefficient of expansion increases, so it is possible to produce alloys having a useful range of coefficients of expansion. Thus, alloys containing between 40 and 50 per cent nickel have coefficients of expansion similar to those of many types of glass, so that efficient metal-glass seals can be produced in electric lamps, TV tubes, radio valves, and the like.

Many domestic and industrial thermostats depend for their operation on the uneven expansion of two different layers of metal in a bimetallic strip. One layer is usually of brass, which has a considerable coefficient of expansion, whilst the other is of one of these iron-nickel alloys, with a very low coefficient of expansion. A change in temperature will cause the brass layer either to expand or contract whilst the nickel-iron layer will not. This causes the bimetal strip to 'warp' and so operate an electrical make/break contact. Many readers will be familiar with small thermostats of this type used in home tropical aquaria. Some low-expansion alloys are detailed in Table 18.4.

Titanium and its alloys

18.30 Titanium is probably the most important and most widely used of those metals whose technology was developed in the latter half of the twentieth century. As well as being strong and corrosion-resistant it is also a very light metal and since, at 4.5, its relative density is only just over half that of steel, this gives it an excellent *specific strength* (or 'strength/weight ratio' as it is commonly called). It is as corrosion resistant as 18/8 stainless steel but will also withstand the extreme corrosiveness of salt water. It has a high melting-point at 1668 °C.

Table 18.3 Some 'Nimonic' high-temperature alloys.

| Nimonic alloy | BS specification (Aerospace series) | Approximate composition % | | | | | | | Tensile strength (N mm^{-2}) at: | | | Uses |
		C	Cr	Ti	Al	Co	Mo	Ni	600°C	800°C	1000°C	
75	HR5	0.1	19.5	0.4	–	–	–		590	250	90	Gas-turbine flame tubes and furnace parts.
80A	2HR1	0.07	19.5	2.2	1.4	–	–	BALANCE	1080	540	75	Gas-turbine stator-blades, after-burners, and other stressed parts working at high temperatures.
90	2HR2	–	19.5	2.5	1.5	18.0	–		1080	850	170	Rotor-blades in gas turbines.
115	HR4	0.16	15.0	4.0	5.0	14.2	4.0		1080	1010	430	Excellent creep-resistant properties at high temperatures.

Table 18.4 *Iron-nickel low-expansion alloys.*

Trade name	Composition %			Coefficient of expansion (at 20 °C) $\times 10^{-6}$	Uses
	Ni	Fe	Co		
Nilo* 36 (Invar, Nivar)	36	64	–	0.9	Pendulum-rods, standard lengths, measuring-tapes, delicate precision sliding-mechanisms, thermostats for low-temperature operation.
Nilo 40	40	60	–	6.0	Thermostats for electric and gas cookers, heater-elements.
Nilo 42	42	58	–	6.2	Thermostats, also the core of copper-clad wire for glass seals in electric lamps, radio valves and TV tubes.
Nilo 50	50	50	–	9.7	Thermostats, also for sealing with soft glasses used in radio and electronic equipment.
Nilo K	29	54	17	5.7	Glass/metal seals in medium-hard glasses used in X-ray tubes and various electronic equipment.

* The trade name for these alloys, coined by Messrs Henry Wiggin.

Although discovered in Cornwall by an English priest, W. Gregor, as long ago as 1791, the industrial production of titanium did not begin until the late 1940s. This was not due to any scarcity of its ores – it is in fact the tenth element in order of abundance in the Earth's crust and fifty times more abundant than copper – but principally because molten titanium reacts chemically with most other substances, making it difficult, and hence extremely expensive, to extract, melt, cast and shape commercially. Nevertheless although chemically very reactive, once it has been successfully shaped, titanium has an excellent corrosion resistance because, like aluminium, its surface becomes coated with a dense impervious film of oxide which effectively seals it from further atmospheric attack.

18.31 Titanium is a white metal with a fracture surface like that of steel. In the pure state, it has a maximum tensile strength of no more than 400 N mm^{-2}, but, when alloyed with small amounts of other metals such as aluminium, tin, and molybdenum, strengths of 1400 N mm^{-2} or more can be obtained, and, most important, maintained at much higher temperatures than is possible with aluminium alloys. In view of the properties mentioned above – low relative density, high specific strength, good corrosion resistance and good strength at high temperatures – it is obvious that titanium will be used in increasing amounts in both aircraft and spacecraft. Titanium alloys have been used for some time for parts in jet engines, and in some modern aircraft about a quarter of the weight of the engine is taken up by titanium alloys. In structural members of aircraft, too, these alloys are finding increasing use because of their high specific strength. For example, the four engines of Concorde contain some 16 tonnes of titanium alloys.

18.32 Titanium is a polymorphic element. The α-phase (CPH in structure) is stable below 882 °C whilst the β-phase (BCC) is stable above 882 °C. As in the case of iron this polymorphic change is affected by alloying. Transformation is retarded (12.24) and if an alloy is quenched from the β-range in the region of 800 to 1050 °C (according to composition) a martensitic-type structure is formed. This is less hard than the martensite of steels but can be further hardened by an 'ageing' treatment at 500 °C for 24 hours.

18.33 Titanium alloys are used in both airframe and engine components of modern supersonic aircraft because of the combination of specific strength and corrosion resistance they offer. On a smaller scale expensive precision cameras employ titanium-alloy components requiring low inertia such as shutter blades and blinds. The high corrosion resistance of the metal has led to its use as surgical implants as well as in chemical industries.

18.34 In Japan titanium has beome established as a roofing material particularly for 'prestige' buildings. Since weight-for-weight titanium is some ten times more expensive than competitors like copper and stainless steel this may seem surprising. However, since titanium is strong, light-gauge material can be used thus reducing the cost per unit area covered. More important still this light-gauge material coupled with the low relative density of the metal means that much less massive support members are necessary in roof construction. Titanium has a high resistance to corrosion and is able to cope with the very corrosive marine atmospheres prevalent in most of Japan. This reduces maintenance costs relative to other metals so that when all of these points are considered it is claimed that titanium can compete, particularly when the architecturally pleasing result is taken into account.

Some titanium alloys are listed in Table 18.5.

Magnesium-base alloys

18.40 Magnesium is a fairly common metal in the Earth's crust and is a principal constituent of dolomite limestone. Of the metallurgically useful metals only iron and aluminium occur more abundantly. However, magnesium is relatively expensive to produce, because it is difficult to extract from its mineral ores, electrolysis being used in a process similar to that employed for extracting aluminium. Because of its great affinity for oxygen, magnesium burns with an intensely hot flame, and was used as the main constituent of incendiary bombs during the Second World War; for this reason, it is also more difficult to deal with in the foundry than are other light alloys. To prevent its taking fire, it is melted under a layer of molten flux, and, during the casting process, 'flowers of sulphur' are shaken on to the stream of molten metal as it leaves the crucible. The sulphur burns in preference to magnesium.

18.41 Apart from lithium (17.76) magnesium has the lowest relative density (at 1.7) of the metals used in general engineering. Although pure magnesium is a relatively weak metal (tensile strength approximately 180 N mm^{-2} in the annealed state), alloys containing suitable amounts of aluminium, zinc, and thorium can be considerably strengthened by precipitation-hardening treatments. Consequently, in view of these properties, magnesium-base alloys are used where weight is a limiting factor. Castings and forgings in the aircraft industry account for much of the magnesium alloys produced, e.g. landing-wheels, petrol-tanks, oil-tanks,

Table 18.5 *Some titanium-base alloys.*

			Typical mechanical properties			
Relevant BS specification	*Composition (%) (Balance Ti)*	*Heat-treatment*	*0.2% proof stress* ($N\,mm^{-2}$)	*Tensile strength* ($N\,mm^{-2}$)	*Elongation (%)*	*Typical uses*
2Ta6	Commercially pure Ti (Fe-0.2 max.)	Annealed 650–750 °C	460	650	15	Chemical plant where resistance to acids and chloride solutions is required. Fire/water systems, sea-water lift pipes, in off-shore oil and gas installations.
2Ta10 sheet & strip	Al-6 V-4	Annealed 700–900 °C	900	1120	8	*Air frame:* engine nacelles, fuselage skinning, wings. *Jet engines*: blades and discs. *Off-shore oil and gas installations*: stress joints, drilling risers in pipelines. High-pressure heat exchangers.
(Timetal LCB)	Fe-4.5 Mo-6.8 Al-1.5	Solution-treated 700–900 °C 'Aged' at 510–540 °C for 2 hours	1350	1420	10	Vehicle springs, valve springs, etc.
Ta48	Al-4 Mo-4 Sn-2 Si-0.5	Solution-treat at 900 °C – air cool. Heat at 500 °C for 24 hours	920	1125	9	*Air frame*: flaps and slat tracks, wing brackets, engine pylons. *Jet engines*: blades and discs, fan casings.

crank-cases, and air-screws, as well as other engine parts in both piston and jet engines. The automobile industry of course makes use of magnesium-base alloys. The one-time popular VW 'Beetle' was a pioneer in this respect.

18.42 In addition to the alloying elements mentioned above, small amounts of manganese, zirconium, and 'rare earth' metals are added to some magnesium alloys. Manganese helps to improve corrosion-resistance, whilst zirconium acts as a grain-refiner. The 'rare earths' and thorium give a further increase in strength, particularly at high temperatures. These alloys include both cast and wrought materials, examples of which are given in Tables 18.6 and 18.7.

Table 18.6   *Some magnesium-base casting alloys.*

| | | | Typical mechanical properties | | |
BS specification 2070	Composition (%) (Balance: Mg)	Condition	0.2% proof stress (N mm^{-2})	Tensile strength (N mm^{-2})	Elongation (%)
MAG. 1	Al-8 Zn-0.7 Mn-0.3	As die-cast Solution-treated	85 80	185 230	4 10
MAG. 5	Zn-4.2 Rare earths-1.2 Zr-0.7	Die-cast and precipitation treated	135	215	4
MAG. 8	Th-3.2 Zn-2.1 Zr-0.7	Die-cast and precipitation treated	85	185	5
DTD 5035A	Rare earths-2.5 Ag-2.5 Zr-0.6	Die-cast, solution treated and precipitation-hardened	185	240	2

Table 18.7   *Some wrought magnesium-base alloys.*

| | | | Typical mechanical properties | | |
BS 3370/3 MAG:	Composition (%) (Balance: Mg)	Condition	0.2% proof stress (N mm^{-2})	Tensile strength (N mm^{-2})	Elongation (%)
101	Mn-1.5	As manufactured	70	200	5
121	Al-6 Zn-1 Mn-0.3	As manufactured	180	270	8
151	Zn-3.2 Zr-0.6	As manufactured	180	265	8
161	Zn-5.5 Zr-0.6	Extruded and precipitation-treated	230	315	8

Zinc-base die-casting alloys

18.50   The growth of the die-casting industry was helped to a great extent by the development of modern zinc-base alloys. These are rigid and reasonably strong materials which have the advantage of a low melting-point, making it possible to cast them into relatively inexpensive dies. (Alloys of high melting-point will generally require dies in rather expensive heat-resisting alloy steels (Table 13.7.))

18.51   A very wide range of components, both for the engineering industries and for domestic appliances, is produced in a number of different zinc-base alloys sold under the trade name of 'Mazak'. Automobile fittings such as door-handles and windscreen-wiper bodies account for possibly the largest consumption, but large quantities of these alloys are also used in electrical equipment, washing-machines, radios, alarm-clocks, and other domestic equipment.

18.52   During the development of these alloys, difficulty was experienced owing to the swelling of the casting during subsequent use, accompanied by a gradual increase in brittleness. These faults were found to be due to intercrystalline corrosion, caused by the presence of small quantities of impurities such as cadmium, tin, and lead. Consequently, very high-grade zinc of 'four nines' quality (i.e. 99.99 per cent pure) is used for the production of these alloys.

Good-quality zinc-base alloy castings undergo a slight shrinkage, which is normally complete in about five weeks (Table 18.8). When close tolerances are necessary, a 'stabilizing' anneal at 150 °C for about three hours should be given before machining. This speeds up any volume change which is likely to occur.

Table 18.8   *Zinc-base die-casting alloys.*

BS specification	Composition %			Shrinkage after 5 weeks normal 'ageing' (mm/mm)	Shrinkage after 5 weeks following stabilizing (mm/mm)
	Al	Cu	Zn		
1004: Alloy A	4	–	BALANCE	0.00032	0.00020
1004: Alloy B	4	1	BALANCE	0.00069	0.00022

18.53   In recent decades these relatively low-strength zinc-base alloys (Table 18.8) have been replaced for many purposes by plastics mouldings. This, accompanied by a move away from expensive chromium-plated finishes, led to a decline in the use of these 4 per cent aluminium alloys. However a new series of zinc-base alloys has been developed containing rather higher quantities of aluminium. Like those alloys covered by BS 1004, the new series are die-casting alloys but have increased strength and hardness. As might be expected these alloys are sensitive to the presence of impurities so that these are kept to the same low limits as in those alloys covered by BS 1004. The new alloys (Table 18.9) have found use in the automobile industry, e.g. ZA27 replacing cast iron for engine mountings. The use of die-casting allows the component to be produced to closer tolerances than does sand-casting. Consequently the amount of subsequent machining is reduced.

Table 18.9   *High-strength zinc-base die-casting alloys.*

Alloy designation	Composition %				Condition	Typical mechanical properties		
	Al	Cu	Mg	Zn		Tensile strength ($N\,mm^{-2}$)	Elongation (%)	Hardness (Brinell)
ZA8	8.4	1.0	0.02	BALANCE	sand-cast	262	1.5	85
					pressure die-cast	375	8	103
ZA12	11.0	0.85	0.02		sand-cast	296	1.5	100
					pressure die-cast	403	5.5	100
ZA27	26.5	2.3	0.015		sand-cast	420	4.5	115
					pressure die-cast	424	1	119

Bearing metals

18.60   The most important properties of a bearing metal are that it should be hard and wear-resistant, and have a low coefficient of friction. At the same time, however, it must be tough, shock-resistant, and sufficiently ductile to allow for 'running in' processes made necessary by slight misalignments. Such a contrasting set of properties is almost impossible to obtain in any single metallic phase. Thus, whilst pure metals and solid solutions are soft, tough, and ductile, they invariably have a high coefficient of friction, and consequently a poor resistance to wear. Conversely, intermetallic compounds are hard and wear-resistant, but are also brittle, so they have a negligible resistance to mechanical shock. For these reasons, bearing metals are generally compounded so as to give a suitable blend of phases, and generally contain small particles of a hard compound embedded in the tough, ductile background of a solid solution. During service, the latter tends to wear away slightly, thus providing channels through which lubricants can flow, whilst the particles of intermetallic compound are left standing 'proud', so that the load is carried with a minimum of frictional losses.

18.61   White bearing metals   are either tin-base or lead-base. The former, which represent the better-quality high-duty white metals, are known as 'Babbitt' metals, after Isaac Babbitt, their originator. All white bearing metals contain between 3.5 and 15 per cent antimony, and much of this combines chemically with some of the tin, giving rise to an intermetallic compound, SbSn. This forms cubic crystals ('cuboids'), which are easily identified in the microstructure (Fig. 18.1). These cuboids are hard, and have low-friction properties; consequently, they constitute the necessary bearing surface in white metals. The background (or 'matrix') of the alloy is a tough, ductile solid solution, consisting of tin with a little antimony dissolved in it.

In the interests of economy, some of the tin is generally replaced by lead. This forms a eutectic structure with the tin–antimony solid solution. The lead-rich white metals are intended for lower duty, since they can withstand only limited pressures. Some white bearing metals are detailed in Table 18.10.

Fig. 18.1 *A 'white' bearing metal (40 per cent tin; 10 per cent antimony; 4 per cent copper; balance, lead). A copper-tin compound, Cu_6Sn_5, crystallizes out first as needle-shaped crystals (light) which form an interlocking network. This network prevents the SbSn cuboids (light), which have a lower relative density, from floating to the surface when they, in turn, begin to crystallize out. Finally the matrix solidifies as a eutectic of tin-antimony and tin-lead solid solutions.*

Table 18.10   *White bearing metals.*

Type	BS specification	Composition (%)					Characteristics
		Sb	Sn	Cu	As	Pb	
Tin-base Babbitt metals	3332/1	7	90	3	–	–	These are generally heavy-duty bearing metals.
	3332/3	10	81	5	–	4	
	3332/6	10	60	3	–	27	
Lead-base bearing alloys	3332/7	13	12	0.75	0.2	Bal.	Thse alloys are generally lower-strength, lower-duty materials.
	3332/8	15	5	0.5	0.3	Bal.	
	–	10	–	–	0.15	Bal.	

18.62   Aluminium-tin alloys containing 20 per cent tin are now used as main and big-end bearings in automobile design. Aluminium and tin form a eutectic containing only 0.5 per cent aluminium so the final structure of these bearings consists of an aluminium network containing small areas of soft eutectic tin which wears and so assists lubricant flow. Because of the long freezing range of this alloy segregation is a danger so that the cast material is usually cold-rolled and annealed to break up the eutectic leaving small islands of tin in an aluminium matrix. Bearing shells of this type are usually carried on a steel backing strip.

18.63   Copper-base bearing metals   include plain tin bronzes (10–15 per cent tin) and phosphor bronzes (10–13 per cent tin, 0.3–1.0 per cent phosphorus) (16.61). Both of these alloys follow the above-mentioned structural pattern of bearing metals. Some of the tin and copper combine to form particles of a very hard intermetallic compound, $Cu_{31}Sn_8$, whilst the remainder of the tin dissolves in the copper to form a tough solid-solution matrix. These alloys are very widely used for bearings when heavy loads are to be carried.

18.64   For many small bearings in standard sizes, sintered bronzes are often used. These are usually of the self-lubricating type, and are made by mixing copper powder and tin powder in the proportions of a 90–10 bronze. Sometimes some graphite is added. The mixture is then 'compacted' at high pressure in a suitably shaped die, and is then sintered at a temperature which causes the tin to melt and so alloy with the copper, forming a continuous structure – but without wholesale melting of the copper taking place. The sintered bronze retains its porosity, and this is made use of in storing the lubricant. The bearing is immersed in lubricating oil, which is then 'depressurized' by vacuum treatment, so that, when the vacuum seal is broken, oil will be forced into the pores. In many cases, sufficient oil is absorbed to last for the lifetime of the machine. Self-lubricating sintered bearings are used widely in the automobile industries and in other applications where long service with a minimum of maintenance is required. Consequently, many are used in domestic equipment such as vacuum cleaners, washing-machines, extractor fans and audio equipment.

18.65   Leaded-bronzes (16.65) are used in the manufacture of main bearings in aero-engines, and for automobile and diesel crankshaft bearings. They have a high wear-resistance and a good thermal conductivity, which helps in cooling them during operation. Brasses are sometimes used as low-cost bearing materials. They are generally of a low-quality 60–40 type, containing up to 1.0 per cent each of aluminium, iron, and manganese.

18.66   Plastics bearing materials are also used, particularly where oil lubrication is impossible or undesirable. The best-known substances are nylon (19.34 and Table 19.2) and polytetrafluoroethylene (19.32 and Table 19.2), both of which have low coefficients of friction. Polytetra-fluoroethylene (PTFE, or 'Teflon') is very good in this respect, and in fact feels greasy to the touch. It is also used to impregnate some sintered-bronze bearings (Fig. 18.2).

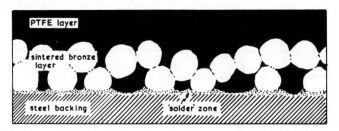

Fig. 18.2   *The structure of a 'filled' PTFE bearing (Mag. × 50). This sintered bronze (7.44) is applied to the steel backing support as a mixture of copper and tin powders. During the sintering process some of the tin 'solders' the bronze to the backing. Since it is porous it acts as a mechanical key to the PTFE-rich layer. The latter contains some particles of lead which help to form a suitable bearing surface.*

Other metals

18.70   I began this chapter by referring to a few of the more uncommon metals which have lately found use in engineering practice and it may be appropriate to close with some details of them in the form of Table 18.11 where these metals are listed in order of atomic number.

Table 18.11   Some of the less well-known metals

Metal	Symbol	Relative density	Melting point (°C)	Characteristics	Uses
Lithium	Li	0.53	180.5	A soft silvery white metal. Lowest relative density of the metals.	Because of low density used in some aluminium alloys for aircraft (17.76). Also in high-energy batteries.
Beryllium	Be	1.85	1284	Steely grey colour. Scarce and difficult to process – hence expensive. Low relative density – good specific strength.	Limited use in high-speed aircraft and rockets. As a moderator in some nuclear reactors. Mainly as an alloying element (16.91).
Gallium	Ga	5.91	29.8	Bluish-white metal.	High-temperature thermometers. Low m.pt. alloys. Semiconductors.
Germanium	Ge	5.36	937	A 'metalloid' – silvery lustre and very brittle. Chemically similar to C and Si.	Used in semiconductors but has been largely replaced by silicon.
Yttrium	Y	5.51	1500	Soft, malleable silvery metal.	As the oxide alloyed with zircona in some ceramics.
Zirconium	Zr	6.5	1860	Soft, malleable ductile metal. Chemically very reactive but good corrosion resistance because of tenacious oxide skin.	Lighter flints. In chemical industries as agitator pump and valve parts. Also in steels, magnesium (18.40) and zinc (18.50) alloys.
Niobium	Nb	8.57	2468	Originally 'Columbium' in USA. White, ductile metal. Scarce. Very corrosion resistant.	Fuel canning in nuclear industry. Used mainly in steels (13.53). Use limited by high cost.
Rhodium	Rh	12.44	1960	Rare white metal of the 'platinum group'.	Alloy element to harden platinum. Thermocouples.
Indium	In	7.31	114.8	Soft, malleable grey-white metal. Corrosion resistant.	Corrosion-protection coatings. A semiconductor. Low melting-point alloys.
Barium	Ba	3.5	710	Soft white metal. Very reactive – takes fire on exposure to moist air.	High absorption of X-rays. Hence compounds are used as screen for X-rays, and barium sulphate as 'barium meal' for diagnosis by X-rays.
Lanthanum	La	6.16	920	Malleable silvery white metal. Most abundant of the 'rare earths'.	Used in some optical glasses.
Cerium	Ce	6.77	804	Steel-grey ductile metal. High affinity for oxygen – hence powerful reducing agent.	A 'getter' (removal of last traces of oxygen from a vacuum). Lighter flints.
Neodymium	Nd	7.01	1024	Toxic silvery white 'rare earth' metal.	Modern permanent magnets. Also magnesium alloys (18.50).
Samarium	Sm	7.54	1052	A silvery white 'rare earth' metal.	Modern permanent magnets. Also as neutron absorber in nuclear industry and in laser crystals.
Dysprosium	Dy	8.56	1500	A 'rare earth' metal.	Neutron absorber in nuclear control rods. Also in laser crystals.
Thulium	Tm	9.32	1525	Malleable silver-grey metal.	Radioisotope, Tm_{170} is used as electron source in portable X-ray equipment.
Hafnium	Hf	11.4	2220	Bright metal with good corrosion resistance. Absorbs neutrons effectively.	Neutron absorption control rods in nuclear power. Cost limits use in non-nuclear fields, e.g. lamp filaments, cathodes, etc.
Tantalum	Ta	16.6	3010	Steel-blue colour. Corrosion resistant because of dense tenacious oxide film. Very malleable and ductile.	Chemical plant (very high corrosion resistance). Plugging material for repair of glass-lined tanks. Gauze for surgical implants. Cutting-tool carbide.
Osmium	Os	22.5	3045	Very hard platinum-group metal. The 'heaviest' (densest) of all metals.	Pen-nib tips. Instrument bearings (alloyed with iridium as 'Osmiridium').
Iridium	Ir	22.4	2443	A platinum-group metal. Very hard.	See osmium. Its radioactive isotope Ir_{192} is used as a γ-ray source.
Platinum	Pt	21.45	1769	Bright silvery white metal. High resistance to corrosion. High electrical resistance.	Mainly as a catalyst in processing petroleum and other materials. Gem setting in jewellery trade. Windings for high-temperature furnaces. Laboratory equipment.
Gold	Au	19.32	1063	Earliest metal known to Man. Very malleable – can be beaten to film 1.4×10^{-7} mm thick.	Mainly as jewellery and a system of exchange. Specialized uses in space craft (18.70).
Uranium	U	18.7	1130	Lustrous silver grey metal. Tarnishes in air. A naturally radioactive element.	Although discovered in 1789 its first serious use was in 'atom bomb' and subsequently nuclear power.

Of these metals gold is by no means 'new'. It is in fact the earliest metal to be used by Man since it was there for the taking, 'native' – or uncombined – in the beds of mountain streams and rivers. Gold is too soft to be of much use in engineering and is used principally in jewellery and as a system of exchange (for which reason it fills the vaults of Fort Knox). Being the most malleable of workable metals it can be beaten down to a film only 1.4×10^{-7} mm thick and it was presumably with such 'gold leaf' that Solomon covered his temple some 3000 years ago. Extremely thin films of gold leaf are transparent and will transmit green light. It was such film that Rutherford bombarded with α-particles in his experiments on the architecture of the atom in the early days of the twentieth century. By the time I became an 'undergrad' this had been translated into many pages of differential equations!

There have been a few engineering uses of gold. Some 1312 oz. of the metal were used in the construction of the *Columbia* space shuttle in the form of brazing solders, fuel cell components, electrical contacts and reflective insulation. Nevertheless on a more sober note we are now making use of more than two-thirds of the seventy metallic elements in contrast to some half-a-dozen available to metal workers in Columbus' time.

CHAPTER 19

Plastics materials and rubbers

19.10   Nylon, one of the important man-made fibres, was developed in the USA in the years just before the Second World War but made its glamorous debut to the public in the form of ladies' stockings in the spring of 1940. Both nylon and polythene, mentioned earlier in this book (1.23), are typical of the materials generally referred to as 'plastics', or more properly as 'plastics materials' – but *never* as *plastic* materials because of course many substances, including metals, undergo *plastic* deformation when the applied stress is great enough.

Though both polythene and nylon are relatively new materials, plastics development began much earlier. The technology of rubber, for example, began in 1820, when a British inventor, Thomas Hancock, developed a method for shaping raw rubber; whilst some twenty years later, in the USA, Charles Goodyear established the 'vulcanization' process by which raw rubber was made to 'set' and produce a tough, durable material, later to be so important in the growth of the motor-car industry.

Shortly afterwards, the plastics cellulose nitrate and celluloid were developed from ordinary cellulose fibre, and when, in the early years of the twentieth century, Dr Leo Bakeland, a Belgian chemist, introduced the material which was ultimately named after him – 'bakelite' – the plastics industry could be said to have 'arrived'. It is interesting to note that since this book was first published the *volume* of plastics materials produced annually has exceeded the volume of steel.

At present, the use of plastics continues to increase rapidly, as new materials are developed. A popular make of British car contains some 20 kg of plastics components. These include the radiator grille, a material being chosen which is proof against both corrosion and flying stones. A spokesman of the company said recently, 'There is no doubt that well before the end of the century, mass-produced plastic [car] bodies will be commonplace'. Those of us who have presided over the rusting away of a succession of expensive motor cars hope this promise can be fulfilled even though in recent years severe competition in the automobile industry has forced manufacturers to consider very seriously the need to make their products less prone to rust away.

Currently polythene and PVC between them account for more than half of the plastics produced in the European Union. They are followed by polypropylene, polystyrene and polyethylene terephthalate and then the more expensive ABS, polyurethanes and nylons.

19.11   Types of plastics materials   Bakelite differs from either polythene or nylon in one important respect. Whereas the latter substances will soften repeatedly whenever they are heated to a high enough temperature, bakelite does not. Once moulded to shape it remains hard and rigid and reheating has no effect – unless the temperature used is so high as to cause it to decompose and char.

Plastics materials can therefore be classified into three groups:

1. Thermoplastic materials – substances which lose their rigidity whenever they are heated, so they can be moulded repeatedly.
2. Thermosetting materials, which undergo a definite chemical change during the moulding process, causing them to become permanently rigid, and incapable of being softened again. Cold-setting plastics which become permanently hard due to a chemical reaction which occurs at ordinary temperatures are similar in basic principles. Materials used in conjunction with glass-fibre for the repair of the bodywork of decrepit motor cars fall into this class.
3. Elastomers which are also virtually thermosetting materials but which are characterized by very high elasticity but a very low *modulus* of elasticity.

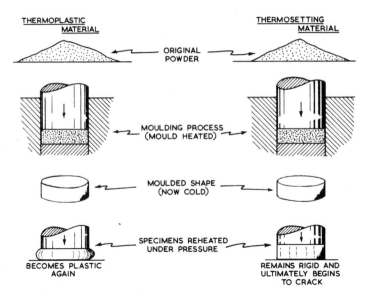

Fig. 19.1   *The behaviour of thermoplastic and thermosetting materials when reheated under pressure.*

19.12   Raw materials used in the manufacture of plastics traditionally come from three main sources:

1. animal and vegetable by-products such as casein (from cow's milk) and cellulose (mainly from cotton fibres too short for spinning) and from wood pulp (cellulosics);
2. coal by-products obtained during the destructive distillation of coal to produce coal gas (PVC, nylon, polyesters, phenolics, urea and melamine);
3. petroleum by-products obtained during the refining and 'cracking' of crude oil (polythene, PVC, polystyrene, 'Perspex').

During the latter half of the twentieth century the replacement of coal gas by natural gas, amongst other reasons, has led to a switch from coal tar to petroleum by-products on an increasing scale for the manufacture of plastics materials. New chemical processes have been developed to make this possible so that petroleum by-products are now responsible for the bulk of plastics manufacture.

19.13   General properties of plastics materials   When visiting the wild and more remote parts of Europe's coast-line one finds, cast up by the tide, along with the seaweed and driftwood, a motley collection of plastics bottles which once held detergent liquids. Because of its low relative density and comparative indestructibility, this plastics junk will presumably congregate in increasing quantity, as yet another example of Man's careless pollution of his environment.

This fact at least illustrates some of the more important properties of plastics generally.

1. They are resistant both to atmospheric corrosion and to corrosion by many chemical reagents.
2. They have a fairly low relative density – a few will just float in water, but the majority are somewhat more dense.
3. Many are reasonbly tough and strong, but the strength is less than that of metals. However, since the relative density of plastics is low, this means that many have an excellent strength/weight ratio – or *specific strength* as it is now termed.
4. Most of the thermoplastic materials begin to soften at quite low temperatures, and few are useful for service at temperatures much above 100 °C. Strength falls rapidly as the temperature rises.
5. Most plastics have a pleasing appearance, and can be coloured if necessary. Some are transparent and completely colourless.

The composition and molecular nature of plastics materials

19.20   All materials included under our present meaning of the term 'plastics' are 'organic' compounds, based on the element carbon. The use of the term 'organic' in chemical nomenclature has a similar connotation to its current use in farming practice and 'the environment' in that it implies that only living materials and substances derived therefrom are involved. Organic chemical compounds are associated mainly with living matter and those which occur naturally are of animal or vegetable origin. Cellulose fibre can be regarded as the material which constitutes the skeletons of plants, from blades of grass to conifers some hundred metres tall; whilst petroleum and coal, from the by-products of which most of our plastics are derived, were produced by the decay and fossilization of vegetable matter, millions of years ago. In addition to the element carbon, most plastics contain hydrogen, whilst many contain oxygen. A smaller number contain other elements, the chief of which are nitrogen, chlorine, and fluorine.

All organic substances exist in the form of molecules, within which individual atoms are bound together by very strong covalent bonds (1.23). The molecules in turn are attractd to each other by much weaker forces known collectively as van der Waals forces (1.30). Plastics materials consist of a mass of very large molecules, each molecule containing several thousands of atoms tightly bound one to the other by covalent bonds. Moreover, since the molecules are very large the *sum* total of the relatively weaker van der Waals forces acting between them will be considerable. The actual shape of the molecule also affects the properties of a plastics material. In thermoplastics materials, the carbon atoms are attached to each other by covalent bonds in the form of a long chain – thus in polythene a chain of about 1200 carbon atoms in length is formed, the hydrogen atoms being attached to individual carbon atoms (1.23).

Since these molecules are so large much greater van der Waals forces operate between them, particularly when they lie parallel and very close to each other. Furthermore, it is easy to imagine that a considerable amount of entanglement will exist among these long chain-like molecules:

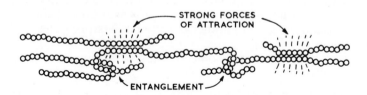

Strong binding forces operating between the molecules, as suggested above, will give rise to the formation of a solid possessing both strength and rigidity. When such a solid is heated vibration within the molecules increases in amplitude so that the distances between aligned molecules become greater, and so the forces of attraction between them will decrease.[1] Then the material will be weaker and less rigid, so that it can be moulded more easily. Such a substance is said to be *thermoplastic*.

19.21   In some plastics, a chemical change is initiated during the moulding process, and 'cross-links', in the form of strong covalent bonds are formed between adjoining molecules:

The powerful forces associated with these covalent bonds are greater than the van der Waals forces operating between the chain molecules in a thermoplastic substance. Consequently the molecules are unable to slide over each other and a strong rigid three-dimensional network is formed. Thus, *thermosetting* has taken place.

19.22   Some terms used in plastics technology   The household plastics material we call 'polythene' is more correctly known as 'polyethylene'. It is made from the gas ethylene,[2] a by-product of the petroleum industry. In the ethylene molecule, a weak 'double-bond' (indicated thus: =) exists between two carbon atoms (Fig. 19.2(i)). Such a double bond is not a sign of extra strength but of greater instability and suitable chemical treatment causes this bond to break. When this occurs simultaneously amongst many molecules of the gas, the resultant units (Fig. 19.2(ii)) are able to link up, forming long chain molecules of polyethylene (Fig. 19.2(iii)).

---

[1]   In a similar way the force acting between two magnets becomes smaller the further they are moved apart.
[2]   In the modern nomenclature of professional chemists 'ethylene' has become 'ethene' and so 'polyethylene' becomes 'polythene'. Nevertheless in this book the widely used industrial terms are retained as suggested by BSI.

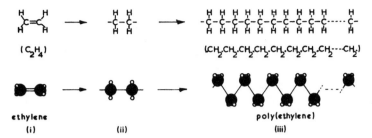

Fig. 19.2   *The polymerization of ethylene to poly-ethylene ('polythene'). The lower part of the illustration shows the arrangement of the carbon and hydrogen atoms.*

This type of chemical process is known as *polymerization*, and the product is called a *polymer*. Since many of the chain molecules so produced are extremely long, the term *super-polymer* is often applied. The simple substance from which the polymer is derived is called a *monomer*; so in this case the gas ethylene is the monomer of polyethylene – or polythene as it is popularly known.

Sometimes the term *mer* is used in this context. A mer is a single unit which, though it may not exist by itself, occurs as the simplest repetitive unit in the chain molecule. Thus in polythene the mer is the unit:

$$\left(\begin{array}{c} H \\ | \\ -C- \\ | \\ H \end{array} \quad \text{or } .CH_2. \right)$$

Frequently a polymer molecule is built up from different monomers, arranged alternately in the chain. Such a substance is called a *co-polymer*. In this way, molecules of vinyl chloride and vinyl acetate can be made to polymerize (Fig. 19.3). The result is the plastics material polyvinyl chloride acetate formerly used in the manufacture of LP gramophone records.

It is in fact possible to use more than two monomers in making a co-polymer. Thus acrylonitrile can be made to co-polymerize with butadiene *and* styrene to produce acrylonitrile-butadiene-styrene or 'ABS' as it is usually known industrially.

Some super-polymers consist of chain molecules built up from very complex monomers. This often leads to the operation of considerable forces of attraction between the resulting molecules at points where they lie alongside each other; consequently, such a material lacks plasticity, even when its temperature is increased. Cellulose is a natural polymer of this type.

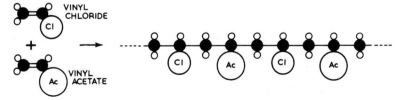

Fig. 19.3   *The formation of a co-polymer (polyvinyl chloride acetate).*

19.23 In 1846, Dr Frederick Schönbein, of the University of Basle, produced cellulose nitrate by treating ordinary cellulose with nitric acid. When heated, it proved to be slightly more plastic than ordinary cellulose. This is because the 'nitrate' side branches (Fig. 19.4(ii)) which had been attached to the chain molecules as a result of treatment with nitric acid, act as 'spacers', separating the chain molecules, and so reducing the forces of attraction between them.

In 1854, Alexander Parkes, the son of a Birmingham industrialist, began to experiment with cellulose nitrate, with the object of producing a mouldable plastic from it. He found that, by adding some camphor to cellulose nitrate, a mixture was produced which passed through a mouldable stage when hot. Not long afterwards, this substance was produced commercially, under the name of 'celluloid'. The bulky camphor molecules separate the chain molecules of cellulose nitrate by even greater distances, so that the forces of attraction between them are reduced further. Thus the plasticity of celluloid is greater than that of cellulose nitrate. In this instance, camphor is termed a *plasticizer* (Fig. 19.4(iii)). Celluloid is a very inflammable material, and this fact has always restricted its use. Before other plastic materials were freely available, much celluloid was used in the manufacture of dolls and other children's toys, often with tragic results when these articles were brought too near a fire.

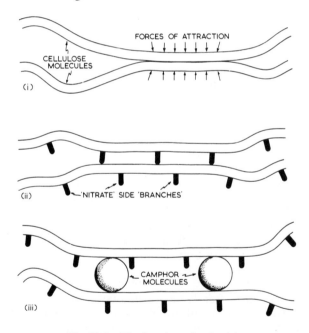

Fig. 19.4   *The function of a plasticizer.*

Some important plastics materials

19.30 It is convenient to divide these materials into three main groups – those which are thermoplastic; those which are thermosetting and those cross-linked polymers which are known as *elastomers* (natural and synthetic rubbers). In the lists which follow the chemical names are followed by customary names (where used) and by the abbreviations commonly used to designate these materials.

Thermoplastics materials

19.31 Vinyl plastics Presumably the term 'vinyl' is derived from 'vin' (as in 'vineyard') since the aromatic nature of some vinyl compounds is reminiscent of that of wine. Some of the vinyl compounds undergo polymerization of their own accord, 'monomers' linking together spontaneously at ambient temperatures to form 'polymers'. Often they begin as 'watery' liquids, but, if allowed to stand for some time, become increasingly viscous as polymerization proceeds, ultimately attaining a solid, glass-like state (Fig. 19.5). Of these plastics materials polythene (PE), polyvinylchloride (PVC), polypropylene (PP), polyethylene terephthalate (PET) and polystyrene (PS) in total account for over 80 per cent of the World's plastics market. In this section the structural formulae of the units derived from the monomers are given to show the relationship of these polymers to ethylene.

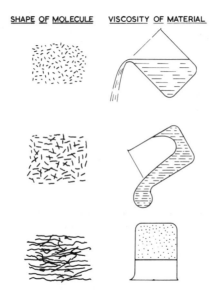

SHAPE OF MOLECULE   VISCOSITY OF MATERIAL

Fig. 19.5 *Stages in the spontaneous polymerization of some vinyl compounds.*

19.31.1 *Polyethylene (polythene) (PE)*

$$--- \!\!-\!\!\overset{\displaystyle H}{\underset{\displaystyle H}{C}}\!-\!\overset{\displaystyle H}{\underset{\displaystyle H}{C}}\!-\!--- $$

This is possibly the best known of the thermoplastics materials. It has an excellent resistance to corrosion by most common chemicals and is unaffected by foodstuffs. Polythene is tough and flexible, has a high electrical resistance, is light in weight and is easily moulded and machined. Since it is also comparatively cheap to produce it is not surprising that polythene finds such a wide range of applications. Polythene is available in several different modifications. Low-density polythene (LDPE) consists of molecules containing many side branches

(Fig. 19.6(i)) and is usually about 50 per cent crystalline (20.20). It is used mainly in the form of soft film for packaging. High-density polythene (HDPE) is approximately 90 per cent crystalline because it consists principally of linear, that is, unbranched molecules. Being harder it is used for the manufacture of bottles and similar utensils by blow-moulding (20.53.4). High *molecular weight* polythene contains molecules which in general are about twice the mass of those present in either LDPE or HDPE. Its particular feature is that it has a superior resistance to environmental stress cracking (25.83).

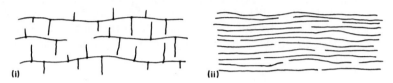

Fig. 19.6   *Different molecular structures in polythene. (i) LDPE – here side-branching of the molecules increases the distance between adjacent chains. This reduces van der Waals forces so that LDPE is softer and weaker than HDPE (ii) where there is no side-branching. In (ii) the unbranched molecules lie closer together so that crystallinity (20.20) can also be greater.*

19.31.2   *Poly(vinyl chloride) (PVC)*

$$----\underset{\underset{\text{H}}{|}}{\overset{\overset{\text{H}}{|}}{\text{C}}}-\underset{\underset{\text{Cl}}{|}}{\overset{\overset{\text{H}}{|}}{\text{C}}}----$$

The gas vinyl chloride was discovered more than a century ago. It was found that, on heating, it changed to a hard, white solid, later identified as polyvinyl chloride. Having rather a high softening temperature, PVC was difficult to mould, and it was not until the late nineteen-twenties that it was discovered that it could be plasticized. By adjusting the proportion of liquid plasticizer used, a thermoplastic material can be produced which varies in properties from a hard rigid substance to a soft rubbery one. Consequently, during the Second World War, and since then, PVC has been used in many instances to replace rubber. Protective gloves, raincoats and garden hose are examples of such use. PVC can be shaped by injection moulding, extrusion, and the normal thermoplastic processes. It can also be compression-moulded without a plasticizer, to give a tough, rigid material such as is necessary in miners' helmets. The effects on strength and ductility of PVC relative to the degree of plasticization are indicated in Table 19.1.

Table 19.1   *Mechanical properties (average) of PVC related to the degree of plasticization.*

Degree of plasticization	Tensile strength $(N\,mm^{-2})$	Elongation (%)
None	55	20
Low	35	200
High	17	400

19.31.3　*Poly (vinyl acetate) (PVA)*

```
        H  H
        |  |
----—C-C—---   H
        |  |        |
        H  C-O-O-C-H
                   |
                   H
```

This polymer softens at too low a temperature for it to be of much use as a mouldable plastic, but it is useful as an adhesive, since it will stick effectively to almost any surface. It is supplied as an emulsion with water; when used in the manner of glue, the water dries off, leaving a strong, adhesive polyvinyl acetate film for bonding the parts together. It is possibly best known, as an emulsion with water, as the basis for 'vinyl emulsion paints'. The introduction of these was certainly a boon to those of us who, in our earlier days, battled with 'glue-size' based 'distemper' for our home decoration.

19.31.4　*Poly(vinyl acetate/chloride) co-polymers* have already been mentioned (19.22). In these, the very high softening temperature of PVC is reduced by combining it with vinyl acetate to form a co-polymer of lower softening temperature. The LP gramophone records made from this have been superseded by polycarbonate compact discs.

19.31.5　*Poly(ethylene-vinyl acetate)* (EVA) is a co-polymer of ethylene and vinyl acetate in which the proportions of vinyl acetate are varied between 3 and 15 per cent to produce materials with different ranges of properties. The principal features of EVA are high flexibility, toughness, clarity and a high resistance to stress cracking (25.83), together with resistance to damage from ultra-violet radiation and ozone.

The rubbery nature and glossy finish of some forms of EVA make it useful for meat packaging and for cling-wrap purposes. Other forms are used for the moulding of automotive parts, ice-cube trays, road-marker cones, medical equipment, turntable mats and garden hose, whilst a harder cross-linked form is used for shoe soles. EVA is also the basis for a number of hot-melt adhesives.

19.31.6　*Poly(propylene) (PP)*

```
        H  H
        |  |
----—C-C—---
        |   \
        H  H-C-H
              |
              H
```

is similar in structure and properties to polyethylene but it has a higher temperature tolerance. Whilst PE is produced from the gas ethylene, PP is polymerized from the gas propylene. PP is stronger than PE and is used for a wide variety of mouldings where greater strength and rigidity than can be obtained with PE are required. The higher melting point of PP as compared with HDPE make it more suitable for fibre manufacture, whilst large amounts are also produced as clear film for wrapping cigarettes and crisps.

19.31.7　*Poly(propylene-ethylene) co-polymers*　A disadvantage of PP is that it becomes brittle at about 0 °C as the amorphous regions of the polymer become glassy (20.22). To reduce this tendency PP is co-polymerized with PE. If 5–15 per cent PE is used toughness is increased without loss of rigidity but increasing the amount of PE to 60 per cent produces a rubbery co-polymer. Such elastomers are resistant to atmospheric attack.

19.31.8 *Poly(styrene) (PS)*

The 'ring' of carbon atoms shown in the above formula is really the nucleus of a benzene molecule, C_6H_6, and the significance of the 'circle' in its centre is to indicate that all of the valency electrons joining these six carbon atoms together are shared equally by the six carbon atoms. In structural formulae this 'benzene ring' is usually depicted so:

PS was developed in Germany before the Second World War and is made by the polymerization of styrene, a chemical substance known since 1830. PS has become one of the most important of modern thermoplastics materials. It is a glassy, transparent material similar in appearance to PMMA (19.37) and has a high electrical resistance. In this homo-polymeric state it is brittle and for this reason toughened versions of PS are used for most engineering and many other applications. This *high-impact polystyrene* (HIPS) is produced by the addition of some polybutadiene – a rubber – which is present in the structure as tiny spheres. PS hollow-ware is fairly easy to identify as such since it emits a resonant 'metallic' note when tapped smartly. PS can be cast, moulded and extruded and shaped easily by the ordinary hand methods of sawing, filing and drilling. Mouldings in HIPS have a lower tensile strength than those in ordinary PS but a much higher impact value particularly at low temperatures for which reason they are extremely useful as refrigerator parts. Nevertheless PS is probably better known in its 'expanded' or foamed form (20.45).

19.31.9 *Acrylonitrile-butadiene-styrene (ABS)*   The 'rubber toughening' of styrene has been described above and the same principle is carried further in these ABS plastics materials which are co-polymers of acrylonitrile:

butadiene:

and styrene. In these materials some of the butadiene polymerizes as globules as in HIPS but the remainder co-polymerizes with acrylonitrile and styrene producing a structure of outstanding resistance to fracture by impact combined with a high tensile strength and abrasion resistance. So much so that luggage made from ABS resists all but the most serious attacks by airport baggage-handling systems! ABS is resistant to acids and alkalis as well as to some organic solvents and is available in the form of moulding powders for injection-moulding and extrusion; sheet is also available for vacuum-forming processes.

Large amounts of ABS are used in the automobile industry where glass-fibre reinforced plastics are increasingly used for panel work. An added bonus for ABS and PP was that they could be chromium plated when used in radiator grilles and similar components. In recent years the fashion has turned away from 'chrome' in favour of more sombre hues so that now everything from car 'bumpers' to hi-fi and TV cabinets and cameras are 'black as the ace of spades'. Of course carbon black is much cheaper than increasingly expensive chromium. However I suppose I must suppress this unwarranted cynicism!

19.32 Fluorocarbons The most important of these materials is *poly(tetrafluoroethylene)* (PTFE). It can be regarded as polythene in which all of the hydrogen atoms have been replaced by atoms of the extremely reactive gas fluorine:

$$\cdots-\overset{\displaystyle \overset{F}{|}}{\underset{\displaystyle \underset{F}{|}}{C}}-\overset{\displaystyle \overset{F}{|}}{\underset{\displaystyle \underset{F}{|}}{C}}-\cdots$$

Since fluorine is so chemically reactive its compounds, once formed, are difficult to decompose. Thus PTFE will resist attack by all solvents and corrosive chemicals; for example hot concentrated sulphuric acid does not affect it, whilst *aqua regia* (a mixture of concentrated nitric and hydrochloric acids which will dissolve gold and platinum) leaves PTFE unharmed. It will also withstand a wider range of temperatures than most plastics materials and can be used for continuous working between -260 and $+250\,°C$ ($+300\,°C$ for intermittent use). It is an excellent electrical insulator but above all it has the lowest coefficient of friction (μ) of any solid material, a feature which also makes it difficult to coat on to other surfaces.

The high cost of fluorine – and hence of PTFE – at present restricts the use of this material except where its particular properties of low μ value (bearings and non-stick coatings for frying pans) and resistance to chemical attack (chemical plant) are utilized. Because of its very low μ, bearing surfaces coated with PTFE (18.66) can be used without lubrication at temperatures approaching $250\,°C$. Similarly chemical plant carrying corrosive liquids can be used at that temperature.

19.33 Cellulose-base plastics (cellulose esters) are derived from natural cellulose, one of the world's most plentiful raw materials. It occurs in many forms of plant life but much of the raw cellulose used in the plastics industry is as cotton linters (cotton fibres too short for spinning to yarn). These 'cellulosics' were amongst the first thermoplastics materials to be developed. In the early days of the cinema the film base was of cellulose nitrate (19.23). This was so inflammable that the unfortunate projectionist was often incarcered, along with his film and a powerful carbon-arc 'lamp', in a fire-proof cabin in order to safeguard the audience. This very dangerous 'nitrate' film has since been replaced by *cellulose acetate* (CA) for this and other purposes since CA is virtually non-flammable.

Although only moderately strong and tough CA is fairly cheap and is used for moulding a wide range of articles from pens and pencils to toys and toothbrush handles.

Cellulose aceto-butyrate (CAB) is tougher and more resistant to moisture than is ordinary CA and at the same time retains the other useful properties of CA. Because of its greater resistance to moisture it is useful for the manufacture of handles for brushes and cutlery.

Cellulose aceto-propionate (CAP) is slightly tougher and more ductile than the other cellulosics but is used for similar purposes.

19.34   Polyamides (PA) include a number of compounds better known by the collective name of *nylon*. The more common members of this series of polymers are designated nylon 6; nylon 6.6; nylon 6.10, nylon 11 and nylon 12, of which the most popular forms are 6 and 6.6. In each case the first digit indicates the number of carbon atoms in the repeating unit or mer (19.22) in the polymer chain. Although nylon 6.6 will always be associated with ladies' hosiery and underwear it is in fact one of the more important engineering thermoplastics. It is a strong, tough, hard-wearing material with a low coefficient of sliding friction. Its softening temperature is relatively high so that it can be safely heated in boiling water. However nylons do absorb considerable amounts of moisture which reduces strength whilst increasing toughness.

Nylons have very good resistance to most organic solvents, oils and fuels and are inert to inorganic reagents with the exception of mineral acids and chloride solutions containing Cl^- ions.

The high strength of nylon fibre claimed my personal interest and attention about half way through my rock-climbing career when nylon climbing ropes were introduced. I soon realized that a nylon 'line' with an equivalent 'potential life-saving capacity' to that of the old hemp rope was but a fraction of the weight of the latter – a factor to be considered when a long trek in high mountainous country was involved. The conservative 'hemp rope' school warned me that friction against rock could cause nylon to melt! However it was never my intention to find myself dangling at the end of a long rope swinging pendulum-wise across a cliff face whilst wondering what the heck to do next.

In addition to the use of nylon 6.6 in the fibre and textile fields nylons have many applications in engineering industries particularly as wearing and moving parts where high impact strength is necessary. Small gears operating at low noise levels in music-reproduction equipment and electric clocks are examples. These polymers are injection moulded or extruded as rod, sheet or tube from which components can be machined. Some forms of nylon 6 are cast. The high-pressure lubricant molybdenum disulphide, MoS_2, is sometimes added to nylon 6 destined for use as a bearing material.

19.35   Polyesters   These are sometimes of the 'setting' variety (19.42) but one important member of the group is thermoplastic. This is polyethylene terephthalate (PET) – better known as 'Terylene'. Like nylon this was originally produced as a 'man-made' fibre developed in Britain during the Second World War and adopted by the textile industry. Since then uses of

PET have developed further, first in the injection moulding of electrical components and then to the manufacture of bottles for a wide range of beers and soft drinks. When these drinks contain dissolved gases such as CO_2 under pressure diffusion of the gas through the walls of the PET bottle can take place. Such diffusion is more rapid through an amorphous (glassy) structure than through a crystalline structure hence the forming process is controlled to produce a highly crystalline structure. Thus the process of 'stretch blow moulding' was developed.

In this process extruded PET tube is heated and then simultaneously stretched and blown into the mould, the temperature and speed being so controlled that just below the glass transition temperature (20.22) very small crystals are formed from the amorphous structure. Because the crystals are very small the bottle retains its transparent appearance and is also strong. The stretching of the PET in two directions during blow moulding promotes the crystallization process. Nevertheless to make such bottles completely gas tight they can be given an outer coating or poly(vinylidene chloride) (PVDC). This is important to prevent the ingress of oxygen which would spoil the contents of beer bottles.

19.35.1 *Polycarbonates (PC)* are structurally linear-chain thermoplastic polyesters. They have excellent mechanical properties – particularly high strength and high impact toughness. Since they are also transparent they are useful for the manufacture of vandal-proof light globes and for babies' feeding bottles. Their temperature tolerance is good so that they are used in high-temperature lenses, coffee pots and the like. A combination of suitable optical properties along with mechanical toughness, a high heat-distortion temperature, ease of processing and good solvent resistance has led to the use of PC for the manufacture of CDs and similar data-storage products.

19.36 Polyacetals 'Acetal resin' in its simplest (homo-polymer) structure is a highly crystalline form of polymerized formaldehyde or poly(oxymethylene) (POM).

$$
\cdots\!\!-\!\!O\!-\!\!\underset{\underset{H}{|}}{\overset{\overset{H}{|}}{C}}\!-\!O\!-\!\!\underset{\underset{H}{|}}{\overset{\overset{H}{|}}{C}}\!\!-\!\!\cdots
$$

It has a high yield strength both at ambient and at high temperatures, associated with a high modulus of elasticity. Resistance to creep is greater than with most other plastics. Impact toughness is high and is only slightly less at $-40\,^{\circ}\text{C}$ than at ambient temperature but it is notch sensitive so that sharp corners in desgin are best avoided. It is a hard material with good dimensional stability whilst abrasion resistance is good and its coefficient of friction low. Moisture absorption is very low, chemical resistance high and electrical insulation properties good. Heat distortion temperature is very high – near to its melting point.

POM can be shaped by injection moulding or extrusion, though wrought forms cannot be shaped by heating and stretching. However if necessary it can be re-ground and used again. Its machining properties are excellent and are similar to those of α/β brasses. BUT – it is expensive.

Co-polymer polyacetals contain extra $-\!\!\underset{\underset{H}{|}}{\overset{\overset{H}{|}}{C}}\!\!-$ units in the polymer chain.

They are generally less strong than the homo-polymer but tend to retain strength at elevated temperatures over longer periods. Polyacetals are particularly useful in applications requiring a high degree of dimensional accuracy especially under conditions of varying humidity. Their combination of properties enables them to replace metals in many cases, e.g. die-castings.

19.37   Acrylics   are a group of vinyl plastics of which the most important is polymethyl methacrylate (PMMA). It is a clear, glass-like, plastics material better known as 'Perspex' (in Britain) or 'Plexiglas' (in the USA) and was developed during the Second World War, for use in aircraft. Not only is it much tougher than glass, but it can easily be moulded. It is produced by the polymerization of methyl methacrylate.

Since it will transmit more than 90 per cent of daylight, and is much lighter and tougher than glass, it can be used in the form of corrugated sheets, interchangeable with those of galvanized iron, for use in industrial buildings. Lenses can be made from PMMA by moulding from powder. This is an inexpensive method of production as compared with the grinding and polishing of a glass blank. Unfortunately PMMA lenses scratch very easily.

19.38   High-temperature thermoplastics   The rapid development of plastics materials following the Second World War led to a demand for thermoplastics with higher temperature tolerance. Some of these materials are of complex structure and are produced by involved processes so these matters will not be discussed here. One important point however which needs mention is that the linear chain molecules are built up from bulky ring-type units. Van der Waals forces, acting between these large units in adjacent chain molecules, are therefore relatively high and this results in a high softening temperature, high thermal stability and exceptional heat resistance.

19.38.1   *Polyimides*   are, structurally, similar to the poly*a*mides (nylons) and were the first of the high-temperature thermoplastics of this group to be developed in the early 1960s.

As described above the bulky, complex ring structures, which are units along the polymer chain, give rise to increased van der Waals forces between the long chain-like molecules so that a higher temperature is necessary to cause their separation – greater heat input causes the molecules to vibrate more vigorously until they separate. Thus polyimides have a very high heat resistance and are useful for high-temperature service. Even after exposure for 1000 hours in air at 300 °C about 90 per cent of the original tensile strength is retained.

Polyimides also have excellent electrical resistance properties, are solvent resistant, resistant to most chemicals (except alkalis) and are very resistant to abrasion.

19.38.2   *Polysulphones*   are in some respects similar in structure to the polycarbonates in that the linear molecular chains contain the bulky unit:

already mentioned (19.31.8) giving a structure:

These large units contribute to increased van der Waals forces and so higher softening temperatures. The polysulphones are more expensive than polycarbonates and are therefore only used when the properties of the latter are inadequate. The creep resistance of polysulphones is excellent and superior to that of polycarbonates. They have good temperature resistance and rigidity and also transparency.

19.38.3   *Poly ether ether ketone* (PEEK):

is an outstanding heat-resistant thermoplastic material which also relies for its high-temperature properties on the presence of the 'benzene ring' as a unit in the linear molecules. In addition to its high softening temperature it is resistant to oxidation and has a low flammability. It is useful as a moulding material for aggressive environments. It has superior toughness and as a coating for wire it resists cuts and fracture caused by sharp corners. Fatigue resistance is very high.

19.40   Thermosetting materials

19.41   Phenolics   are possibly the best known of the thermosetting group.

19.41.1   *Phenol formaldehyde* (PF)   During the cynical closing years of the twentieth century the Crimean War (1854–56) is remembered chiefly for the military blunders which led to the decimation of the 'Gallant Six Hundred' in the Charge of the Light Brigade. Nevertheless in contemplating this melancholy prospect – as Sir Winston Churchill might have described it – we should reflect on the heroic works of one, Florence Nightingale, whose efforts in the reorganization and discipline of field hospitals reduced the death rate of wounded soldiers from 42 per cent to a little over 2 per cent. In those days the badly injured had little hope of survival as the inevitable *septicaemia* took control. Then the introduction of a new disinfectant known as 'carbolic acid' was an important factor in this clinical revolution.

Carbolic acid – or phenol as we now know it – is a colourless, crystalline solid and a weak acid. It was used, in 1907, by Dr Leo Bakeland to develop the first synthetic plastics material 'bakelite'. This name became synonymous with plastics materials generally as the plethora of 'brown-bakelite-and-chrome' artefacts of the 1930s filled our homes, but we can't blame Dr Bakeland for that!

In the manufacture of bakelite phenol reacts with the gas formaldehyde (H.CHO). The phenol molecule is based on that of benzene (described in 19.31.8) in that one hydrogen atom has been replaced by the —OH radical:

Table 19.2 *Properties and uses of thermoplastics materials.*

Group	Compound		Relative density	Tensile strength (N mm⁻²)	Chemical resistance	Safe working temperature (°C)	Relative cost	Typical uses
Vinyls	Polythene (PE)	HD	0.95	30	Excellent	120	Low	Acid-resisting linings, babies' baths, kitchen equipment and other household ware. Piping, toys, fabric filaments.
		LD	0.92	13	Excellent	80	Low	Sheets, wrapping material, polythene bags, squeeze bottles, electrical insulation, bottle caps, tubing for ball-point pens, ink cartridges.
	Polyvinyl chloride (PVC)	Unplasticized	1.40	55	Good	70	Low	Domestic and industrial piping (rainwater, waste, etc.), light fittings, curtain rail (with metal insert), radio components, safety helmets, ducting, plating vats.
		Plasticized	1.30	35	Good	100	Low	Artificial leather cloth, gloves, belts, raincoats, curtains, packaging, cable covering, protective clothing.
	Polyvinyl chloride/acetate		1.3	25	Fairly good	70	Moderate	Gramophone-records, containers, chemical equipment, screens, protective clothing.
	Ethylene vinyl acetate (EVA)		0.94	25	Good	70	Moderate	Meat packaging, cling wraps, turntable mats, automobile parts, garden hose, car-door protectors, shoe-soles, road cones, surgical ware. Adhesives.
	Polypropylene (PP)		0.90	33	Excellent	100	Low	Packaging, pipes and fittings, cable insulation, battery boxes, refrigerator parts, sterilizable bottles and other uses where boiling water is involved, cabinets for TV and radio sets, fan blades, crates and containers, stackable chairs.
	Polystyrene (PS)	General purpose	1.05	45	Fairly good	80	Low	Moulded containers (food and cosmetics), boxes, toilet articles. *Foams*: ceiling tiles, heat insulation, packaging for fragile equipment.
		HIPS	1.02	20	Fairly good	80	Low	Radio and TV cabinets, vacuum cleaners, kitchen equipment, refrigerator parts, vending machine cups, cases for cheap cameras.
	Acrylonitrile-butadiene-styrene (ABS)		1.01	35	Very good	80	Moderate	Pipes, radio cabinets, tool handles, protective helmets, textile bobbins, pumps, battery cases, luggage, typewriter and camera cases, telephone handsets, hair driers, large amounts in automobile bodywork.
Fluorocarbons	Polytetrafluoroethylene (PTFE; 'Teflon')		2.15	25	Excellent	250	Very high	Gaskets, valve packings, inert laboratory equipment, chemical plant, piston rings, bearings, non-stick coatings (frying pans), filters, electrical insulation.

Table 19.2 – *continued*

Cellulosics	Cellulose acetate (CA)	1.30	35	Fair	70	Fairly high	Artificial leather, brush backs, combs, spectacle frames, photographic film base, mixing bowls, lamp shades, toys, laminated luggage, knobs, wire- and cable-covering.
	Cellulose aceto-butyrate (CAB)	1.18	35	Fair	70	Fairly high	Illuminated road signs, extruded pipe, containers.
	Cellulose aceto-propionate (CAP)	1.21	55	Fair	70	Fairly high	Steering wheels, packaging, toothbrushes, door knobs.
Polyamides	Nylon 66 (PA66)	1.12	Moulded-60 Filaments-350	Good	140	High	Raincoats, yarn (clothing), containers, cable-covering, gears, bearings, cams, spectacle frames, combs, bristles for brushes, climbing ropes, fishing lines, shock absorbers.
Polyesters (thermoplastic)	Polyethylene terephthalate ('Terylene', 'Dacron') (PET)	1.38	Moulded-60 Fibres-175	Moderate	85	Moderate	*Fibres*: a wide range of clothing. *Tape and film*: music and recording tapes, insulating tape, gaskets. *Moulded*: electrical plugs and sockets. *Blow mouldings*: bottles for beer and soft drinks.
	Polycarbonates (PC)	1.2	66	Good	140	High	Very tough – protective shields (police vehicles), hairdrier bodies, telephone parts, automobile tail light lenses, tool handles, machine housings, baby's bottles, vandal-proof street light covers, safety helmets, CDs (music and data storage).
Polyacetals	Homopolymer Co-polymer	1.4 1.4	70 60	Fairly good Fairly good	95 95	– 	Bearings, cams, gears, flexible shafts, office machinery, carburettor parts, pump impellers, car instrument panels, knobs, handles, water pumps, washing machine parts, seat-belt buckles.
Acrylics	Polymethyl methacrylate (PMM) ('Perspex', 'Plexiglas')	1.18	55	Fairly good	95	Moderate	Aircraft glazing, building panels, roof lighting, baths, sinks, protective shields, advertising displays, windows and windscreens, automobile tail lights, lenses, toilet articles, dentures, knobs, telephones, aquaria, double glazing, garden cloches, shower cabinets.
High-temperature thermoplastics	Polyimides	1.42	90	Good	350	High	Seals, gaskets, piston rings, jet engine compressor seals, data processing (pressure discs, bearings, friction elements).
	Polysulphones	1.24	70	Good	175	High	Hairdriers, oven, iron and fan heaters, microwave oven parts, pumps for chemical plant, transparent pipelines, hot-water dispenser cups for drinks machines.
	Polyether ether ketone (PEEK)	1.28	92	Good	150	Very high	Aggressive environments (nuclear plant), oil and geothermal wells, high-pressure steam valves, aircraft and automobile engine parts, wire covering (cut resistant), filament for weaving high-temperature filtration cloth.

During subsequent chemical reactions the hydrogen atoms at (a), (b) and (c) are eliminated, leaving points at which covalent bonds can then join units together. For the sake of simplicity we will represent the phenol molecule so:

i.e. omitting the —OH group and the hydrogen atoms which take no part in the reaction. If phenol is mixed with a *limited* amount of formaldehyde then molecules of each react to form a long-chain molecule thus:

where the unit

is released from the formaldehyde molecule (H.CHO). In this condition, the material is brittle but thermoplastic, and can be ground to a powder suitable for moulding. This powder (or 'novolak') contains other materials, which, on heating will release more formaldehyde, so that, during the final moulding process, cross-links are formed between the chain molecules:

The 'network' structure represented above can only be shown two-dimensionally on flat paper but it is in fact a three-dimensional cage-like network in which the units have 'set' *permanently*. Since a non-reversible chemical reaction has taken place phenol formaldehyde cannot be softened again by heat.

This type of plastic is hard and rigid, but tends to be rather brittle in thin sections. However, it has a good electrical resistance, so it is not surprising that the 'bakelite' industry grew in conjunction with the electrical industries from the late nineteen-twenties onwards. Various mouldings for wireless-cabinets, motor-car parts, bottle-tops, switchgear, electric plugs, door-knobs, and a host of other articles are commonly made from bakelite. In most of these articles, a wood-flour filler is used; but, if greater strength is required,

 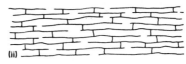

Fig. 19.7 *Thermosetting in a polymer material. (i) Illustrates the structure at the 'novolak' stage. Here the material is still thermoplastic and the chain molecules are held together only by van der Waals forces. (ii) Shows the effects of thermosetting. Here the linear molecules are now held firmly together by powerful covalent bonds ('cross-links').*

a fibrous filler (paper, rags, jute, or sisal) is employed.

19.41.2 *Urea-formaldehyde (UF)* plastics are basically similar to the phenol-formaldehyde types, since they depend upon the two-stage cross-linking reaction, but in this instance between molecules of urea and formaldehyde. The first stage of the reaction, however, results in the formation of a syrupy material. This is mixed with the filler, to give a moist, crumbly mass. After mixing with other reagents, this is allowed to dry out, producing the urea-formaldehyde moulding powder.

The fact that the urea-formaldehyde mixture passes through a syrupy stage makes it useful in other directions. For example, paper, cloth, or cardboard can be impregnated with it, and then moulded to the required shape, before being set by the application of heat and pressure. Weather-resistant plywood can also be made, using urea-formaldehyde as the bonding material, whilst the syrup can also be used as a vehicle for colouring materials in the coating of metal furniture, motor cars, washing machines, and refrigerators with a layer of enamel.

19.41.3 *Melamine-formaldehyde (MF)* plastics are of a similar type structurally to the two foregoing thermosets but are much harder and more heat resistant. Moreover, they are most resistant to water and consequently find use in the household as cups and saucers, baths and sundry kitchen utensils, particularly where greater heat-resistance is necessary.

19.42 Polyester resins of the thermoplastic type have already been mentioned (19.35). These are of the 'straight-chain' molecule form, which are incapable of linking up with other adjacent chain molecules; they are therefore thermoplastic. By using special monomers, however, polyesters can be produced with side-chains capable of forming cross-links between adjacent molecules, and these are therefore 'setting plastics', of which some are of the cold-setting type. For example, if styrene is added to the liquid polyester, it provides the necessary cross-links, and the liquid gradually sets without the application of any heat. Used with glass-fibre, to provide reinforcement, these materials are useful for building up such structures as the hulls of small boats, wheelbarrows, and car bodies. The glass fibre, impregnated with the liquid mixture, is built up a layer at a time on a suitable former. Pressure is not required, except to keep the material in position until it has set. These glass-fibre/polyester resin composites are very strong and durable.

19.42.1   *Alkyd resins*   constitute a further group of polyesters. They are thermosetting materials originally introduced as constituents of paints, enamels, and lacquers. They have a high resistance to both heat and electricity, and are dissolved by neither acids nor many organic solvents. These resins are now available in powder form, which is generally compression-moulded (20.53.1). Liquid alkyds are used in enamels and lacquers for automobiles, stoves, refrigerators, and washing machines.

19.43   Polyurethanes   constitute a group of very adaptable plastics, comprising both thermoplastic and thermosetting materials. They are generally clear and colourless. One type is used in the manufacture of bristles, filaments, and films.

In another group of polyurethanes, carbon dioxide is evolved during the chemical process necessary for establishing cross-links between the chain molecules. This carbon dioxide is trapped by the solidifying polymer, thus producing a foam. The mechanical properties of the foam can be varied by using different materials to constitute the polymer. Thus, some of the foams become hard and rigid, whilst others are soft and flexible. Rigid thermosetting polyurethane foams are used as heat-insulators, and for strengthening hollow structures, since they can be poured into a space where foaming and setting will subsequently occur. Aircraft wings can be strengthened in this manner. Flexible sponges, both for toilet purposes and for seat upholstery, are generally of polyurethane origin.

19.44   Epoxy resins are produced in a similar way to polyesters: by being mixed with a cross-linking agent, which causes them to set as a rigid network of polymer molecules. They are used for manufacturing laminates, for casting, and for the 'potting' of electrical equipment. Excellent adhesives can be derived from the epoxy resins available as syrups, and they are particularly useful for metal glueing (26.21). The reader will no doubt be familiar with these two-component adhesives available at DIY stores – and be aware of how effective the cold-setting process really is if he gets the tube caps mixed!

Solid epoxy resins are sometimes mixed with phenolic resins for moulding purposes.

19.45   Polyimides are high-temperature resistant polymers which can be either thermoplastic or thermosetting according to their formulation. They are available as film or as solid parts which retain mechanical properties at 300 °C over long periods. Even at 500 °C properties are retained for a short time.

19.46   Silicones   The polymers mentioned so far in this chapter are based on the element carbon, and are known as organic compounds. Unfortunately, they all soften, decompose, or burn at quite low temperatures. However, more than a century ago, it was realized that there was a great similarity in chemical properties between carbon and the element silicon. Since then, chemists have been trying to produce long-chain molecules based on silicon, instead of carbon, hoping that in this way materials would be discovered which lacked many of the shortcomings of carbon polymers. Common compounds of silicon include quartz, glass, and sand – all substances which are relatively inert, and which remain unchanged after exposure to very high temperatures. Consequently, the more fanciful of our science-fiction writers have speculated on the possibility of a system of organic life, based on silicon instead of carbon, existing on the hot planets Venus and Mercury. Whilst their 'bug-eyed monsters' may remain a figment of the imagination, some progress has taken place here on Earth in producing polymers based on silicon, and called organo-silicon compounds, or 'silicones'.

Some of these compounds – of which sand is the basic raw material – have properties roughly midway between those of common silicon compounds, such as glass, and the orthodox organic plastics. Silicones are based on long-chain molecules in which carbon has been replaced by silicon and oxygen. They are available as viscous oils, greases, plastics, and rubbers. All are virtually non-combustible, and their properties remain constant over a very wide temperature range. Thus, silicone lubricating oils retain their fluidity more or less unchanged at temperatures low enough to make ordinary oils congeal. Similarly, silicone rubbers remain flexible from low subzero temperatures up to temperatures high enough to decompose ordinary rubbers.

Silicones are water repellent, and are widely used for waterproofing clothes, shoes, and other articles, whilst silicone jelly is useful as a moisture-proof coating and sealing compound.

By forming still longer chain molecules, silicone plastics can be produced. These are very useful as an insulating varnish for electrical equipment designed to work at high temperatures. As moulding plastics, silicones are used in the manufacture of gaskets and seals for engineering purposes where high temperatures are involved, since they retain their plasticity and sealing efficiency under such conditions.

Silicone-resin paints provide durable finishes which clean easily, and which do not deteriorate appreciably. Such a finish applied to a motor car would normally outlast the car, and with a minimum of attention.

Elastomers

19.50 Christopher Columbus and his sailors may well have been the first Europeans to handle natural rubber since it is reported that some tribes of South American Indians played ball games well before the days of Pele. Natural rubber is in fact derived from the sap of the rubber tree *Hevea brasiliensis*. Following the work of Goodyear in vulcanizing rubber, production grew rapidly in Brazil, the home of the rubber tree and the city of Manaus, complete with an opera house and many other trappings of civilized living, mushroomed in the middle of the Amazon jungle as capital of a Rubber Empire. In 1876 however, some seeds of the rubber tree were smuggled out of Brazil by British botanists and planted in greenhouses in Kew Gardens. The young plants were sent to the Dutch East Indies (now Indonesia) and Malaya so that Britain was able to establish and control its own rubber industry. This naturally led to a severe contraction of the Brazilian rubber industry and a rather shabby Manaus is all that remains of its former glory.

19.51 In natural rubber the long chain molecules are more complex in structure than in the simpler polymers such as polythene, and consist of chains 44 000 or more carbon atoms in length. The monomer present in these rubber molecules is the carbon-hydrogen compound *isoprene*, so that natural rubber is really *polyisoprene*. When the isoprene mers are joined to form the large molecular chain (Fig. 19.8) the side group makes the

Table 19.3 Properties and uses of thermosetting materials.

Group	Compound (with abbreviations/ trade names)	Relative density	Tensile strength (N mm^{-2})	Chemical resistance	Safe working temperature (°C)	Relative cost	Typical uses
Phenolics	Phenol formaldehyde ('bakelite'; 'phenolic') (PF)	1.45	50	Very good	120	Low	Electrical equipment, radio-cabinets, vacuum-cleaners, ashtrays, buttons, cheap cameras, automobile ignition systems, ornaments, handles, instrument-panels, advertising displays, novelties and games, dies, gears, bearings (laminates), washing-machine agitators
	Urea formaldehyde ('urea') (UF)	1.48	45	Fair	80	Moderate	Adhesives, plugs and switches, buckles, buttons, bottle-tops, cups, saucers, plates, radio-cabinets, knobs, clock-cases, kitchen equipment, electric light-fittings, surface coatings, bond for foundry sand
	Melamine formaldehyde ('melamine') (MF)	1.49	50	Good	130	Moderate	Electrical equipment, handles, knobs, cups, saucers, plates, refrigerator coatings, trays, washing-machine agitators, radio-cabinets, light-fixtures, lamp-pedestals, switches, buttons, building-panels, automotive ignition-blocks, manufacture of laminates
Polyesters (setting types)	Polyester	1.3	40	Fairly good	95	Moderate	Adhesives, surface coatings, corrugated and flat translucent lighting-panels, lampshades, radio grilles, refrigerator parts; polyester laminates are used for hulls of boats, car bodies, wheelbarrows, helmets, swimming-pools, fishing-rods and archery-bows
	Alkyd resins	2.2	25	Fair	230	Moderate	Enamels and lacquers for cars, refrigerators, and stoves. Electrical equipment for cars, light-switches, electric-motor insulation, television-set parts
Polyurethanes	Polyurethane (PUR)	1.2	Mainly foams	Good	120	High	Adhesives (glass to metal), paint base, wire-coating, gears, bearings, electronic equipment, handles, knobs. Foams are used for insulation, upholstery, sponges, etc. Rigid foams are used for reinforcement of some aircraft wings
Epoxy resins (EP)	Heat-resistant type	1.15	70	Good	200	Fairly high	Adhesives (metal gluing), surface coatings, casting and 'potting' of specimens. Laminates are used for boat hulls (with fibreglass), table surfaces and laboratory furniture, drop-hammer dies. Epoxy putty is used in foundries, to repair defective castings
	General purpose	1.15	63	Good	80	Fairly high	
Polyimides (thermosetting)	Polyimide	1.43	40	Good	300	Fairly high	Bearings; compressor valves; piston rings; diamond abrasive wheel binders

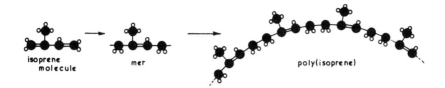

Fig. 19.8 *The 'side groups' (consisting of one carbon atom and three hydrogen atoms) cause the polyisoprene molecule to 'bend' into a coiled form. (Carbon atom – ●; hydrogen atom – ○.)*

molecule rather 'lopsided' and this causes the molecule to bend into a folded or coiled form in order to accommodate these large side groups.

19.52   Since these rubber molecules are folded and coiled they possess elasticity in a similar manner to that of a coiled spring and immediately stress is applied there is an elastic response. Nevertheless, natural rubber stretches like dough when stressed because the chain molecules slide past each other into new positions (Fig. 19.9). This plastic flow takes place slowly because weak van der Waals forces acting between molecules are overcome progressively as entangled molecules become very slowly disentangled. For this reason a piece of unvulcanized natural rubber is both *elastic and plastic* at the same time, and, once stretched, will not return to its original shape.

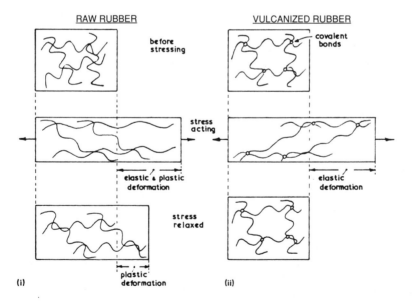

Fig. 19.9 *Due to their folded or coiled form, rubber molecules become extended in tension but return to their original shapes when the stress is removed. In raw rubber (i) a steady tensile force will also cause separate molecules to slip slowly past each other into new positions because only weak van der Waals forces attract them to each other; so when the force is relaxed some plastic deformation remains, though elastic deformation has disappeared. By vulcanizing the raw rubber (ii) the chain molecules are covalently bonded at certain points so that no permanent plastic deformation can occur and only elastic deformation is possible. This disappears when stress is relaxed.*

19.53   When up to 3 per cent sulphur is added to rubber prior to moulding, 'vulcanization' will occur if the moulding temperature is high enough. Atoms of sulphur form strong covalent bonds between the rubber molecules so that they are permanently anchored to each other. Thus, whilst the vulcanized rubber retains its elasticity due to the folded or coiled nature of its molecules, it retains its shape permanently, because of this cross-linking. If the vulcanization process is carried further – that is, if more sulphur is used – so many cross-links are produced between adjoining molecules that the whole mass becomes rigid, and loses most of its elastomer properties. The product is a black plastics material known as 'ebonite' or 'vulcanite' and was used extensively some half century ago for the manufacture of such articles as fountain-pens, before other more suitable materials were developed.

The Japanese invasion of Britain's Far East rubber plantations in 1942 led to the development of synthetic rubber substitutes. In fact synthetic polyisoprene was produced. Since those days the properties of these synthetic rubbers have considerably improved and rubbers with new properties developed, so that the industrial output of synthetic rubbers now exceeds that of natural rubber.

As with natural rubber, sulphur is used to provide the cross-linking atoms in the vulcanization process for many of these synthetic rubbers but in some cases oxygen cross-linking atoms are used, the oxygen atoms being provided by magnesium oxide or zinc oxide.

To be of value in engineering a rubber must have a 'low hysteresis', that is, it must return within very close limits to its original shape following each successive deformation cycle. It must also have a low heat build-up. That is, heat generated as a result of friction (braking in a motor vehicle) or as a result of absorbing vibrational energy, must be dissipated adequately.

19.54   Engineering elastomers   In addition to natural rubber some twenty elastomers are available for engineering uses. Some of these are expensive but with special properties. The more important elastomers in general use include:

19.54.1   *Natural rubber (NR) and polyisoprene (IR)*   The latter is manufactured from petrochemicals and has a molecular structure similar to that of natural rubber. Both are used for commercial vehicle tyres, anti-vibration and anti-shock mountings and dock fenders. They have good mechanical properties in respect of strength, high elasticity and low hysteresis in the important working range. Creep properties are also good. Unfortunately the resistance to attack by oxygen and ozone is poor though this can be overcome by the use of additives (25.82) for most applications.

19.54.2   *Styrene-butadiene rubber (SBR)*   is possibly the best known and most widely used of elastomers and was originally a replacement for natural rubber following the Japanese occupation of Malaya during the Second World War. It is a co-polymer of styrene and butadiene with physical properties slightly inferior to those of natural rubber. Although ageing properties (25.81) and abrasion resistance of SBR are slightly superior to those of natural rubber it suffers greater heat build-up. Nevertheless it is widely used for car tyres where adhesion and wear resistance are of paramount importance. Heat build-up is not a great problem here since modern car tyres are heavily reinforced and consequently have thin walls which lose heat more quickly.

19.54.3   *Butadiene rubber (BR)* – monomer, butadiene – is noted for its high resilience but poor tensile strength and poor tear resistance for which reasons it is seldom used on its own but generally blended with other elastomers. Since it has a low heat build-up it is used for vehicle tyres – blended with natural rubber for truck tyres and with SBR for motor car tyres. BR imparts good tread wear to tyres and also has good low-temperature properties. It is also used as a sealant compound in caulking.

19.54.4   *Polychloroprene rubber (CR)* – of which the monomer is of course chloroprene – is also known as *neoprene*. It has the good physical properties of natural rubber coupled with a much better resistance to oxygen, ozone and some oil products. CR retains its properties well up to 90 °C but tends to become hard and stiff below −10 °C. It finds use in a variety of products such as divers' wet suits, anti-vibration mountings, conveyor belts, tank linings, hoses and cable sheathing.

19.54.5   *Acrylonitrile-butadiene   rubbers   (NBR)* are   co-polymers of acrylonitrile and butadiene and are usually called *nitrile rubbers*. The normal grades have quite good physical properties though slightly inferior to those of natural rubber and SBR. Nitrile rubber is relatively expensive and is used principally for sealing applications (gaskets) where its excellent resistance to oil products and its good temperature resistance (95 °C) are essential. It is also used for hoses, conveyor belts and cable sheathing where similar conditions prevail.

19.54.6   *Butyl rubber (IIR)*   is a co-polymer of isobutane and small amounts of isoprene. It is cheaper than natural rubber to which its physical properties are, however, inferior. Since it has a very low permeability to air and other gases it is used mainly for tyre inner tubes and also for the liners of tubeless tyres. It has excellent resistance to ozone and to ageing generally so that it is less apt to perish. 'Butyl' – as it is generally called – is also used for air-bags and supporting equipment as well as for more general purposes such as hoses and tank linings since it is impervious to many chemicals.

19.54.7   *Ethylene-propylene rubber (EPM)* is a co-polymer of the gases ethylene and propylene in which the quantity of either monomer can vary up to 65 per cent. During polymerization any weak double-bonds are 'used up' and the resulting rubber is therefore chemically inert so that, like polythene, it has good resistance to ageing, weathering and ozone attack. Its service-temperature range of −50 to 125 °C is also good but it has a poor resistance to some solvents and a rather high creep rate. Since EPM has good electrical insulation properties it is used for cable covering.

   A further -*diene* material is often added to provide extra cross-linking sites (thus toughening the polymer and increasing its heat tolerance). It is then designated EPDM and has a number of uses in the automobile industry.

19.54.8   *Silicone rubbers (SI)*   In these materials silicon and oxygen replace carbon in the polymer chains. They are expensive but their high temperature tolerance (200 °C) makes them very useful for many purposes. Their resistance to low temperatures too is good but resistance to some oils and other organic liquids is only moderate. This can be improved by replacing some of the hydrogen atoms by fluorine, but at a great increase in cost.

19.54.9   Of some dozen other synthetic rubbers few are of engineering significance. *Polysulphide rubber* has a good resistance to oils and solvents and is used as an internal coating for oil and paint piping. *Styrene-butadiene-styrene rubber* (SBS) is thermoplastic at temperatures above that of glass transition so that it can be moulded by conventional methods. For this reason it is used as a carpet-backing material and in some adhesives.

CHAPTER 20

The structure and properties of plastics materials

20.10   In the previous chapter we considered a polymer material as consisting of an intertwined mass of giant molecules each containing several thousands of atoms. In thermoplastics materials these molecules are attracted to each other by relatively weak van der Waals forces, whilst in thermosetting substances they are joined to each other by strong, permanent covalent bonds. This means that whilst thermosetting materials are sometimes stronger then thermoplastics materials, they are inevitably more rigid and much more brittle.

The physical structure of polymers

20.20   Most solid substances – metals, salts and minerals – exist in a purely crystalline form. That is the atoms or ions of which they are composed are arranged in some regular geometric pattern. This is generally possible because the atoms (or ions) are small and very easily manoeuvrable. With polymers, however, we are dealing with very large molecules which entangle with each other and are consequently much less manoeuvrable. As a result only a limited degree of crystalline arrangement is possible with most polymers as they cool, and only restricted regions occur in which the linear molecular chains arrange themselves in an ordered pattern (Fig. 20.1). These ordered regions are called *crystallites*.

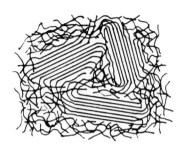

Fig. 20.1   *'Crystallites' in a solid plastics material.*

Thus in a solid state polymers consist of both crystalline and amorphous regions. Highly crystalline polymers contain up to 90 per cent crystalline regions whilst others are almost completely amorphous.

20.21   Melting points of polymers   A pure, completely crystalline solid such as a metal or an inorganic salt, melts at a single definite temperature (Fig. 2.1) called the melting-point, because *single* atoms (or ions) are easily mobile. Amorphous and partly crystalline materials however merely become progressively less rigid when heated and show no clear transition from solid to liquid. This is what we would expect since liquids (which are amorphous) become less viscous as the temperature rises – viscous liquids like treacle and tar flow more readily when they are heated. With highly crystalline polymers on the other hand a sudden change in the rate of expansion occurs on heating to a certain temperature. This is the point at which all crystalline regions have become amorphous – as happens of course when all crystalline metals melt to form amorphous liquids. In polymer materials this temperature is designated T_m, the melting point.

As will be noted in Table 20.1 the greater the degree of crystallinity of polythene the higher its melting point. Presumably this is because van der Waals forces are much greater in the crystalline regions since the molecules are closer together there.

Table 20.1   *Melting points of some crystalline polymers.*

Polymer	Melting point (T_m) (°C)
Polythene (PE)(50% crystalline)	120
Polythene (PE)(80% crystalline)	135
Polypropylene (PP)	170
Polyvinyl chloride (PVC)	212
Polyethylene terephthalate (PET)	255
Polyamide (Nylon 6:6)	265
Polytetrafluoroethylene (PTFE)	327
Polyether ether ketone (PEEK)	334

20.22   Glass transition temperature   At ambient temperatures polymers may be either soft and flexible or hard, brittle and glassy. When a soft, flexible polymer is cooled to a sufficiently low temperature it becomes hard and glassy. For example a soft rubber ball when cooled in liquid air shatters to fragments if any attempt is made to bounce it whilst it is still at that very low temperature. The temperature at which this soft-to-glassy change occurs is known as the *glass transition temperature*, T_g. For some plastics materials T_g is above ambient temperature so that they become soft and flexible when heated past T_g. Polystyrene (Table 20.2) is a polymer of this type. Thus

Table 20.2   *Glass transition temperatures of some polymers.*

Polymer	Glass transition temperature (T_g) (°C)
Polythene (PE)	−120
Natural rubber (NR)	−73
Polypropylene (PP)	−27
Polymethylmethacrylate (PMMA)	0
Polyamide (Nylon 6:6)	60
Polyethylene terephthalate (PET)	70
Polyvinyl chloride (PVC)	87
Polystyrene (PS)	100
Cellulose acetate (CA)	120
Polytetrafluoroethylene (PTFE)	126
Polyether ether ketone (PEEK)	143
Polycarbonate (PC)	149

polystyrene kitchen hollow-ware is hard and rigid at ambient temperatures and gives a characteristic 'ring' when lightly struck, but, on heating above 100 °C (T_g) it loses that rigidity.

The glass transiton temperature is associated with the thermal vibrations of atoms within polymer chains. Above T_g these vibrations reach proportions where they impart flexibility to the polymer.

20.23 Vicat softening temperature This is a value which can be determined more precisely in numerical terms than either T_m or T_g, so that it offers a useful means of comparison between plastics materials as far as their response to temperature change is concerned. It measures the temperature at which a standard indentor penetrates a specific distance into the surface of the plastics material using a standard force. During the test the temperature is raised at a specified uniform rate and the penetration is recorded by a dial gauge micrometer. The *vicat temperature* is read when the specified penetration (usually 1 mm) has been attained.

Table 20.3   *Vicat softening temperatures for some polymers.*

Polymer	Vicat softening temperature (°C)
Ethylene vinyl acetate (EVA)	51
Cellulose acetate (CA)	72
Polyvinyl chloride (PVC)	83
Low-density polythene (LDPE)	85
Polymethylmethacrylate (PMMA)	90
Polystyrene (HIPS)	94
Polystyrene (PS)	99
Acrylonitrile-butadiene-styrene (ABS)	104
High-density polythene (HDPE)	125
Polypropylene (PP)	150
Polycarbonate (PC)	165
Polyacetal	185
Polyamide (Nylon 6:6)	185

Mechanical properties and testing of plastics materials

20.30 The strength of plastics materials is generally much lower than that of most other constructional materials. Nevertheless plastics are light materials with a relative density between 0.9 and 2.0 so that when considered in terms of strength/weight ratio they compare favourably with some metals and alloys. Fig. 20.2 indicates the types of stress/strain relationship obtained for different groups of polymers.

The mechanical properties of most engineering metals and alloys vary very little within the range of ambient temperatures encountered in service. This is to be expected since no structural changes occur (other than thermal expansion) until the recrystallization temperature (6.22) is reached and this is usually well above 100 °C. With many polymers – particularly thermoplastic – materials, however, mechanical properties vary considerably with temperature in the ambient region. Both T_g and T_m will affect mechanical properties as will the gradul reduction in van der Waals forces with rise in temperature. Thus a thermoplastic polymer may have a tensile strength of, say, 70 N mm^{-2} at 0 °C, falling to 40 N mm^{-2} at 25 °C and to no more than 10 N mm^{-2} at 80 °C. As tensile strength falls with rise in temperature there is a corresponding increase in percentage elongation (Fig. 20.3).

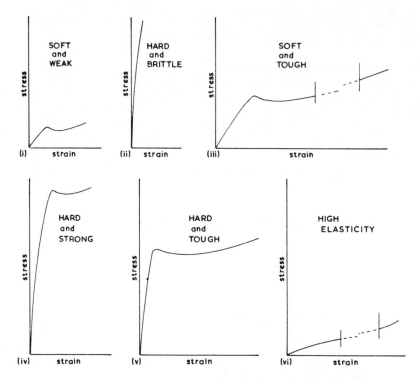

Fig. 20.2 *Types of stress-strain diagram for different polymer materials (after Carswell and Nason). (i) Low elastic modulus; low yield stress, e.g. PVA and PTFE. (ii) High elastic modulus; low elongation, e.g. PF, PMMA and PS. (iii) Low elastic modulus; low yield stress but high elongation and high stress at break, e.g. PE and plasticized PVC. (iv) High elastic modulus; high yield stress; high tensile strength and low elongation, e.g. rigid PVC and modified PS. (v) High elastic modulus; high yield stress; high tensile strength and high elongation, e.g. nylons and polycarbonates. (vi) Very low elastic modulus; low yield stress and low tensile strength but very high elastic elongation, e.g. natural rubber and other elastomers.*

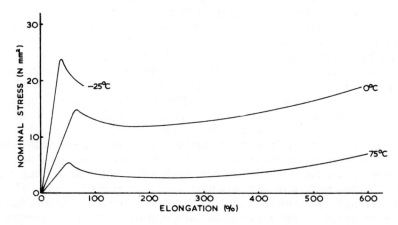

Fig. 20.3 *Tensile stress-strain curves for low-density polythene at −25 °C, 0 °C and 75 °C.*

20.31   Tensile testing of plastics materials   Clearly if mechanical tests on plastics are to have any meaning then a *fixed testing temperature* must be specified. Thus BS 2782 (Part 3) requires a testing temperature maintained at 23 ± 2 °C with an atmospheric humidity of 50 ± 5 per cent for many thermoplastics materials; moreover the test-pieces must be maintained under these conditions for 88 hours prior to the test. In some cases, for example thermosetting plastics, this conditioning is unnecessary since the permanent covalent bonds between the polymer chains render these materials much less temperature sensitive.

Plastics materials are also very time-sensitive as far as mechanical testing is concerned. This is because total deformation depends upon:

1. bond bending of the carbon-carbon covalent bonds in the polymer chain – this is manifested as the ordinary elasticity and is an *instantaneous* deformation;
2. uncoiling of the polymer chains – this gives rise to high elasticity and is *very time dependent*;
3. slipping of polymer chains past each other – this produces *irreversible* plastic flow and is also *very time dependent*.

The sum total of deformation arising from (1), (2) and (3) is generally termed *viscoelastic*, and the rate of strain has a considerable effect on the recorded mechanical properties. Generally as the straining rate increases so does the recorded yield strength and again BS 2782 (Part 3) lays down different conditions for different plastics materials as far as tensile testing is concerned. Thus the speed of separation of the test-piece grips varies between 1 and 500 mm/min for different materials and forms of material.

The geometry of test-pieces also differs from that used in metals testing. Abrupt changes in shape cause stress concentrations which are likely to precipitate failure so tensile test-pieces are generally of the form shown in Fig. 20.4.

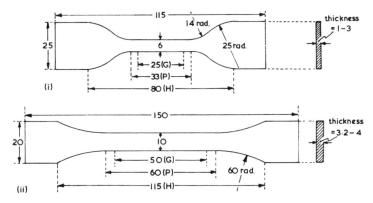

Fig. 20.4   *Two principal forms of tensile test-piece used for plastics materials. Gauge length = G; parallel section = P; distance between grips = H (measurements are in mm).*

20.31.1   *Secant modulus*   Many plastics do not obey Hooke's law; that is, elastic strain produced is not proportional to the stress applied so that it becomes impossible to derive Young's modulus of elasticity since this value applies only to materials with Hookean characteristics. As an alternative it is usual to calculate the secant modulus of the material. This is defined as the ratio of stress (nominal) to corresponding strain at some specific point

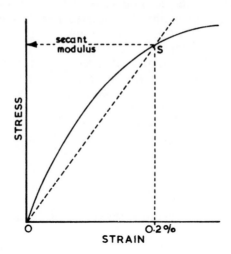

Fig. 20.5   *Derivation of the secant modulus for non-Hookean polymer materials.*

on the stress/strain curve. In Fig. 20.5 the secant modulus associated with a strain of 0.2 per cent is shown and is in fact the slope of the line *OS*.

At the commencement of the tensile test an initial force of *w* (usually about 10 per cent of the expected force necessary to produce 0.002 strain) is applied to 'take up slack' and straighten the test-piece. With this force applied the extensometer is set to zero. The force is then gradually increased (in accordance with the straining rate specified) until the necessary force, *W*, is reached to produce 0.2 per cent strain in the gauge length.

$$\text{Elastic (secant) modulus} = \frac{\text{stress}}{\text{strain}}$$

$$= \frac{W - w}{A} \div 0.002$$

$$= \frac{W - w}{0.002\,A}$$

where *A* is the initial cross-sectional area of the test-piece at the gauge length. Other values of strain between 0.1 per cent and 2.0 per cent may be used depending upon the type of material. This strain value must therefore be stated when quoting the secant modulus.

20.32   Hardness tests   Ball-indentation tests similar to those used for metals are applied to some of the harder plastics, e.g. ebonite and hard rubbers, and the hardness index is derived by the same principle:

$$H = \frac{\text{applied force}}{\text{surface area of impression}}$$

When using the Rockwell test (4.36) for these hard plastics materials scales M, L or R are appropriate. As with metals testing the hardness index is prefixed with the appropriate letter (M, L or R).

For other plastics a form of indentation hardness test is used in which the depth of penetration is measured under the action of a standard force. In the *durometer* (which measures Shore hardness) two different indentors are available. Type A (Fig. 20.6) is used for soft plastics materials and type D for hard plastics.

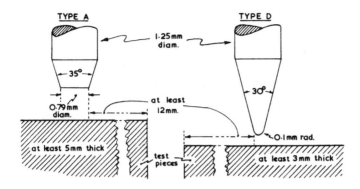

Fig. 20.6   *Indentors used in the durometer (Shore hardness).*

As is the case when hardness testing metals the thickness and width of test-pieces must be adequate as indicated in Fig. 20.6. The load is applied for 15 s whilst the conditions of temperature and humidity (23 °C and 50 per cent humidity) specified in the case of the tensile test also apply here. In the Shore hardness test the hardness index is related inversely to the depth of penetration. It was suggested earlier in the book (4.30) that values described as 'hardness indices' are in fact generally measurements of the resistance of a material to penetration rather than the resistance of its surface to abrasion. This is particularly true of plastics. Thus high-density polystyrene will resist penetration in tests of this type better than will polythene, yet the surface of polystyrene can be *scratched* much more easily.

20.33   Impact tests   Both Izod and Charpy tests are used to assess the toughness of plastics materials which fall into three groups as far as toughness is concerned (Table 20.4). In this table those materials described as 'brittle' will break even when unnotched, whilst those described as 'notch-brittle' will not break unnotched but will break when sharp-notched. The 'tough' materials will not break completely even when sharp-notched. For this reason a variety of notches are used in impact testing (Fig. 20.7) in order to suit the material under test and to make possible the production of comparable test results.

Table 20.4   *Relative impact brittleness of some common plastics materials.*

Brittle	Notch brittle	Tough
Acrylics	Polyvinyl chloride (PVC)	Low-density polythene (LDPE)
Polystyrene (PS)	Polypropylene (PP)	Acrylonitrile-butadiene-styrene (ABS) (some types)
Phenolics	High-density polythene (HDPE) Acrylonitrile-butadiene-styrene (ABS) (some types)	Polythene-propylene co-polymers

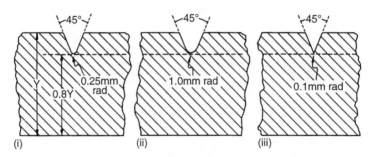

Fig. 20.7   *Test-piece notches for impact tests on plastics materials. (i) and (ii) are used in the Izod test whilst all three are available for the Charpy test.*

20.34 Creep, the gradual extension under a constant force, must be considered seriously for metals particularly if they are to work at high temperatures (4.50). Plastics materials, however, are much more prone to this form of deformation since they undergo the viscoelastic type mentioned earlier (20.31). Creep rates like other mechanical properties for polymers are very temperature dependent and even a plastics raincoat must be capable of *supporting its own weight* so that it can hang in a wardrobe possibly for long periods without slowly flowing to the floor during a period of warm weather.

The creep curves shown in Fig. 20.8 are typical of those for most thermoplastics materials. These show that initial creep is rapid but then decreases to a constant value which approaches zero for small values of stress. Naturally the greater the applied force the greater the creep rate and elongation and quite small rises of temperature will produce much greater rates of creep.

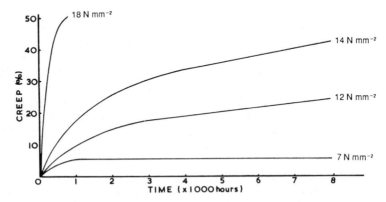

Fig. 20.8   *Creep curves for cellulose acetate at 25 °C. (At a stress of 18 N mm^{-2} rapid creep soon leads to catastrophic failure, whilst at a stress of 7 N mm^{-2} a small initial amount of creep ceases after about 1000 hours and no further measurable creep occurs.)*

20.35 Other mechanical tests are used to measure compressive properties; tear strength; shear strength and deflection in bend under an applied load. All of these tests, together with those described briefly here, are dealt with in BS 2782: Part 3.

Fillers and other additives

20.40   Various materials can be introduced into a polymer at the moulding stage by mixing the additive thoroughly with the moulding powder. *Fillers* are used principally in the interests of economy when up to 80 per cent of the filler may be incorporated. Generally fillers are finely powdered, naturally occurring, non-polymeric substances which are readily available at low cost. Phenolic (bakelite)-type resins are nearly always compounded or filled with substances such as sawdust, wood flour, short cellulose fibres from wood pulp, or cotton flock, whilst fillers like paper, rag and cotton fibre will increase the strength of bakelite. Some soft thermoplastics materials are blended with up to 80 per cent mineral solids such as crushed quartz or limestone. Here the plastics material operates in much the same way as does the cement film in concrete (Fig. 23.3). Thus some fillers serve to increase strength and hardness and to reduce creep and thermal expansion.

20.41   Anti-static agents   The act of walking across a polypropylene fibre carpet can easily produce a considerable charge of static electricity, whilst most readers will be aware that the simple act of removing a synthetic-fibre garment, particularly in a dry atmosphere, will generate 'static' – a charge more than sufficient in fact to create a spark capable of igniting a petrol-air mixture. One wonders how many unexplained fires are started in this way.

Since the valency electrons (1.20) are held captive within the covalent bonds throughout a plastics material such a material is a non-conductor of electricity. In fact most polymers are excellent insulators. It follows therefore that a plastics material can become highly charged with 'static' electricity because this static is unable to be conducted away. This may be undesirable or indeed hazardous and in such cases an anti-static agent is added to the mix. Additives to achieve this result are substances which will lower the electrical resistivity of the surface of the polymer, generally by attracting moisture to its surface since moisture is sufficiently conductive to provide a path by which free electrons (which *are* both 'static' and 'current' electricity) can 'leak' away.

Many such additions are 'waxes' in which the molecules have strong dipole moments (1.31). They therefore attract water molecules which also have strong dipole moments so that the two are held together by van der Waals forces. Such a strategy is of course not effective in very dry atmospheres (below 15 per cent humidity). Nevertheless such 'waxes' can be added to PE and PS since their moulding temperatures are low enough to avoid decomposition of such additives. Other organic compounds (amines and amides), also phosphoric acid derivatives and sulphonic acids, are sometimes used for this purpose.

20.42   Flame retardants   Many plastics materials constitute a fire risk as witnessed by a number of leisure centre disasters in past years. Plastics foam fillings of cushions can also be hazardous. To reduce such fire risks a large number of flame retardant additives are available. Their object is to interfere with those chemical reactions underlying combustion. One such substance is hexabromocyclodecane (HBCD) which is incorporated in materials like polypropylene. If combustion of the plastics material begins hydrogen bromide is released from HBCD as a vapour which immediately combines with those portions of the plastics molecule, generally – OH groups, which would otherwise burn and lead to a rapidly spreading conflagration. Thus combustion is stifled.

Some organic chlorides have a similar action in releasing hydrogen chloride which similarly neutralizes the –OH groups. Their effectiveness can often be improved by the addition of antimony oxide. This reacts with the organic chloride in the flame producing antimony trichloride which increases flame retardation.

20.43   Friction modifiers   Two apparently flat surfaces may in fact be making contact at only a few high spots and so friction between these surfaces will appear misleadingly low. Generally the closer the contact between two surfaces the greater the friction which, in the case of plastics materials, is the sum of all the van der Waals forces acting between the molecules of one surface adjacent to those of the other surface. Coefficients of friction (μ) for different plastics materials vary widely since the different groupings of atoms at molecule surfaces affect the magnitude of these van der Waals forces. μ is as low as 0.04 for polytetrafluoroethylene and as high as 3.0 for natural rubber. Consequently PTFE is widely used as a bearing material (18.66) as is nylon ($\mu = 0.15$) and ultra-high molecular weight polythene ($\mu = 0.08$ to 0.2).

The addition of graphite to a plastics material will reduce μ, as will particles of added PTFE and the high-pressure lubricant molybdenum disulphide. Such materials will reduce wear in bearings and bushings.

Some polymer materials suffer from 'blocking', that is the tendency of two very flat surfaces to stick together. Again this is due to the operation of van der Waals forces between adjacent large molecules across the narrow gap between the surfaces. It is generally overcome by increasing the roughness of the surfaces so that they are spaced further apart. 'Silica flour' is often used as an additive to achieve this.

20.44   Other additives   include *colourants* which are either insoluble, finely divided pigments or soluble dyes, usually chosen for colour stability, inertness and non-toxicity. In some instances pigments may act as a filler or serve some other purpose such as that of a stabilizer.

The addition of *asbestos* to a plastics material will increase its heat tolerance whilst *mica* improves its electrical resistance.

20.45   Foamed or 'expanded' plastics materials   Foamed rubber was developed many years ago but at present a number of polymers are available as 'foams'. 'Expanded' polystyrene is probably the best known of these. It is used as a very light-weight packing material for fragile equipment. Without it one wonders how many of the cameras and TV sets from the Far East would reach us intact. Up to 97 per cent of its volume consists of air bubbles so that it is useful as a thermal insulator and for the manufacture of DIY ceiling tiles, as well as being an excellent flotation material where, unlike cork, it does not become waterlogged.

Polyurethane foams are available in two different forms, either as flexible foams for the production of sponges and upholstery, or as rigid foams used for the internal reinforcement of some aircraft wings. Both thermoplastic and thermosetting polymers (including elastomers) can be foamed and four basic methods are available for introducing the gas pockets into the polymer:

1. By mechanically mixing air with the polymer whilst the latter is still molten.
2. By dissolving a gas such as carbon dioxide into the mix under pressure.

As hardening proceeds pressure is reduced so that gas is released from solution, rather like the bubbles of carbon dioxide which are released from mineral water – or of course, champagne – when the stopper is removed. The bubbles in this case are trapped in the hardening polymer.

3. A volatile component may be added to the hot mix. This vaporizes during hardening, forming bubbles.

4. In some processes a gas is evolved during the actual polymerization process. This can be utilized to cause foaming. Alternatively a 'blowing agent' may be added which will release gas when the mix is heated for the moulding process. This is useful for thermoplastics materials. A simple reagent of this type is sodium bicarbonate which releases carbon dioxide when heated. It behaves in much the same way when used as 'baking powder' when bubbles of carbon dioxide released during baking make a cake 'rise'.

Methods used to shape plastics materials

20.50   Most of the larger plastics manufacturers supply raw materials in the form of powders, granules, pellets, and syrups. These are purchased by the fabricators for the production of finished articles. As with metals, plastics can be purchased in the form of extrusions and sheet on which further work is to be carried out.

The principal methods of fabrication are calendering, extrusion, moulding, blow moulding and vacuum forming but raw materials available in liquid form may be cast.

20.51   Calendering   is used to manufacture thermoplastics sheet, mainly PE and PVC. The calendering unit consists of a series of heated rolls into which the raw material is fed as a heated 'dough' (Fig. 20.9). The formed sheet is cooled by a chill roll. Fabric, paper and foil are coated with a film of plastics material using a similar process.

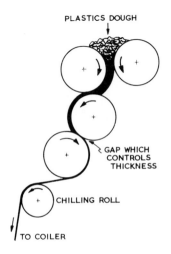

PLASTICS DOUGH

GAP WHICH
CONTROLS
THICKNESS

CHILLING ROLL

TO COILER

Fig. 20.9   *The principles of calendering.*

20.52   Extrusion   The principle of extrusion is shown in Fig. 20.10. Here the plastic material is carried forward by the screw mechanism, and, as it enters the heated zone, it becomes soft enough to be forced through the die. The die aperture is shaped according to the cross-section required in the product.

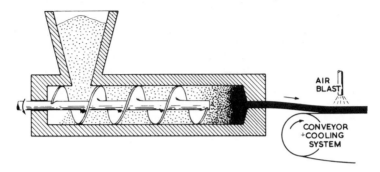

Fig. 20.10   *The extrusion of plastics. Often known as 'screw-pump' extrusion.*

Man-made fibres are extruded using a multi-hole die, or 'spinneret'. Wire can be coated with plastics by extrusion, assuming that the machine is adapted so that the wire can pass through the die in the manner of a mandrel.

20.53   Moulding   Several important hot-moulding processes are commonly used.

20.53.1   *Compression-moulding*   (Fig. 20.11) is probably the most important, and is used for both thermoplastic and thermosetting materials, though it is particularly suitable for the latter. In either case the mould must be heated, but for thermoplastic substances it has to be cooled before the work-piece can be ejected. A carefully measured amount of powder is used, and provisions are made to force out the slight excess necessary to ensure filling of the mould cavity.

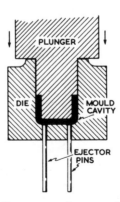

Fig. 20.11   *One system of compression-moulding.*

20.53.2   *Injection-moulding*   (Fig. 20.12) is a very rapid process, and is widely used for moulding such materials as polythene and polystyrene. The material is softened by heating it in the injection nozzle. The mould itself is

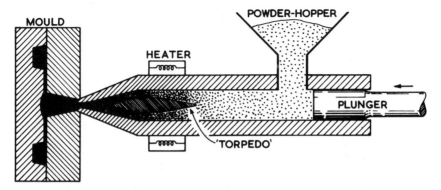

Fig. 20.12   *The principle of injection-moulding.*

cold, so that the plastic soon hardens and can be ejected. The work-piece generally consists of a 'spray' of components, connected by 'runners' which are subsequently broken off.

20.53.3   *Transfer-moulding* is used for thermosetting plastics. The material is heated, to soften it, after which it is forced into the heated mould (Fig. 20.13), where it remains until set. Rather more intricate shapes can be produced by this process than by compression-moulding.

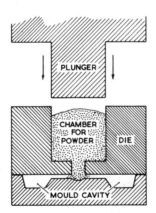

Fig. 20.13   *Transfer-moulding.*

20.53.4   *Blow-moulding* (Fig. 20.14)   is used to produce hollow articles. The plastic is first softened by heating, and is then blown by air pressure against the walls of the mould. In the diagram, extruded tube (parison) is being used. A variety of containers such as bottles and coffee jars are produced in this way.

A similar process known as *stretch-blow moulding* is used in the manufacture of PET drinks bottles (19.35). Here a preform is stretched and blown within the mould so that crystallization of the PET occurs within the amorphous structure. This process provides increased yield strength and stability in bottles which will be filled with liquids at around 90 °C.

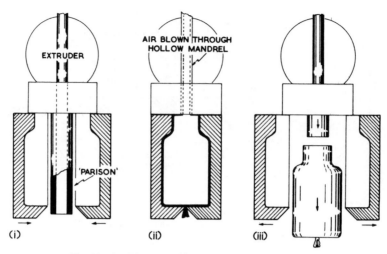

Fig. 20.14   *Blow-moulding by interrupted extrusion.*

20.53.5   *Film blowing* (Fig. 20.15)   Film can be blown continuously from heated thermoplastics delivered straight from the extruder. This material has a nominal thickness of not more than 0.25 mm. The process is used for the manufacture of PE film and 'tube' (Fig. 20.15) for parting off as bags or other forms of packaging.

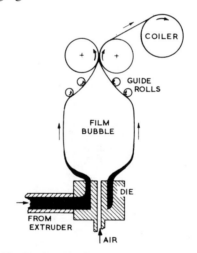

Fig. 20.15   *Film-blowing of plastics bags.*

20.53.6   *Vacuum-forming*   is also used in producing simple shapes from thermoplastic materials in sheet form (Fig. 20.16). The heated sheet is clamped at its edges, and is then stretched by the mould as it advances into position. The ultimate shape is produced by applying a vacuum, so that the work-piece is forced into shape by the external atmospheric pressure.

20.54   Casting   is limited to those plastics materials whose ingredients are in a liquid form, that is, as monomers which polymerize in the mould. The mixed liquids are poured into the open mould and are allowed to remain at atmospheric pressure until setting has taken place.

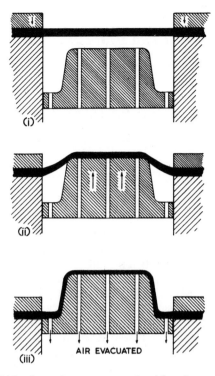

Fig. 20.16   *Stages in a vacuum-assisted forming process.*

20.54.1   *Slush-moulding*   is essentially a casting process rather than the moulding process its name implies. It is used mainly for PVC, the mix being prepared as a thin paste which is injected into the mould. The mould is then rotated and heated until the paste forms a solid 'gel' on the inner surface of the mould. The excess fluid is poured out and the *hollow* formed article removed from the mould. The process is used for the manufacture of dolls' heads and other semi-soft toys. The 'slush casting' of aluminium tea-pot spouts and of lead soldiers in former days followed a similar method.

CHAPTER 21

Ceramics

21.10 The term 'ceramics' is derived from the Greek *keramos* – 'potter's clay'. Gradually this meaning has been extended to include all products made from fired clay such as bricks, tiles, fireclay refractories, wash-hand basins and other sanitary ware, electrical porcelains and ornaments as well as pottery tableware. Many substances now classed as ceramics in fact contain no clay though most are relatively hard, brittle materials of mineral origin with high fusion temperatures. Thus hydraulic cement is usually classed as a ceramic material whilst a number of metallic oxides such as alumina, magnesia, zirconia and beryllia form the basis of high-temperature ceramic refractories. Some of the latter, in particular alumina and zirconia along with recently developed materials like silicon nitride and the 'sialons', boron nitride and boron carbide, have more sophisticated engineering uses.

21.11 Most of the ceramics materials mentioned above are either completely crystalline in structure or are a mixture of crystalline regions cemented together by an infilling of amorphous networks. Glasses on the other hand are materials which, at ambient temperatures, are still in an amorphous state, that is they are virtually still in a liquid condition.

Glass comprises a range of substances from boiled sweets to window panes, beer tankards and even some metallic materials which can be cooled quickly enough from the molten state – speeds of about one million degrees Celsius per second are necessary – to prevent them from crystallizing. At the other extreme plastics materials are either completely 'glassy' or contain both crystalline and glassy regions (Fig. 20.1) because they are composed of very large and cumbersome linear molecules which can 'wriggle' into position only with difficulty so that this process is overtaken by fall in temperature.

Silicate-based ceramics

21.20 Those ceramic materials derived from sand, clay or cement contain the elements silicon and oxygen (the two most abundant elements on planet Earth) in the form of *silicates*, which also contain one or more of the metals sodium, potassium, calcium, magnesium and iron. The most simple silicon-oxygen unit in these compounds is the group SiO_4^{-4} in which a small silicon atom is covalently bonded to four oxygen atoms (Fig. 21.1). As will be seen in Fig. 21.1(ii) each oxygen atom in this unit has an 'unused' valency bond so

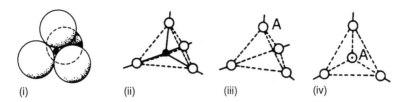

(i)   (ii)   (iii)   (iv)

Fig. 21.1   *The structure of the SiO$_4^{-4}$ unit in silicates. (i) The small silicon atom ● surrounded by and covalently bonded to four oxygen atoms ○. (ii) A 'spatial' representation of the unit. (iii) Here the silicon atom is omitted to simplify the diagram of the unit. (iv) A 'plan' view of the SiO$_4^{-4}$ unit, i.e. with the 'apex' oxygen atom, A, on 'top', again with the silicon atom omitted.*

that these units link up rather like the 'mers' in a polymer structure (19.22) to form a continuous structure.[1] A number of different basic structures can be formed in this way but two of the more important are described here.

21.21   Fig. 21.2 represents a 'chain' type arrangement in which silicon-oxygen units are found combined in a large number of naturally occurring minerals. However, although the basic tetragonal form of the 'SiO$_4$' unit is retained we are no longer dealing with a series of 'SiO$_4$ ions' linked together since oxygen atoms along the chain are shared between successive units so that in a single-chain structure (Fig. 21.2(i)) we are dealing with 'backbone chains' of a formula $n[(SiO_3)^{-2}]$, whilst in 'double-chain structures', since the sharing of oxygen valencies goes a stage further, the backbone chains will have a general formula $n[(Si_4O_{11})^{-6}]$ (Fig. 21.2(ii)). Those 'unused' oxygen valencies along the 'edges' and at the 'apices' of the units are in fact bonded *ionically* to either metallic or −OH ions. When the metallic ions concerned are of metals such as calcium or magnesium (both with valencies of two) the spare valency links the chain to a neighbouring chain.

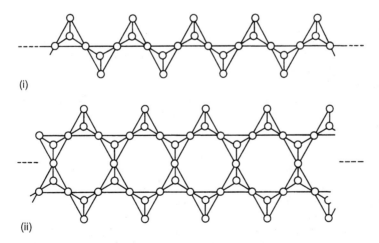

(i)

(ii)

Fig. 21.2   *This shows how the structure of chain-type 'giant ions' can be formed from 'SiO$_4$ groups'. In fact because oxygen atoms are* shared *between two units in a chain the basic formula of the chain is* $n[(SiO_3)^{-2}]$. *'Unused' oxygen valencies along the chain attach themselves to metallic ions.*

[1]   The 'family connection' between the elements carbon and silicon shows here.

21.22   However the point of all this is that the greatest strength of the structure is *along the covalently bonded backbone* of tetragonal silicon-oxygen units, long strands of which are held together by weaker ionic bonds. This shows some resemblance to the structure of thermoplastic polymers except that the van der Waals forces operating in polymers are replaced by ionic bonds here. This type of structure is present in the mineral *asbestos* and is convincingly reflected in its mechanical properties. Asbestos is a naturally occurring mineral consisting of giant chains based on Si_4O_{11} units ionically bonded to Ca^{+2}, Mg^{+2} and $-OH$ ions. It is mined in a rock-like form but a series of crushing and milling operations can separate the mineral in the form of fibres because of the basic structure described above. Fracture occurs of the weaker ionic bonds which hold the strong fibres together.

21.23   In another very large group of naturally occurring silicates the tetragonal silicon-oxygen units are covalently linked in a two-dimensional *sheet* form (Fig. 21.3) and the basic – or empirical – formula becomes $n[(Si_2O_5)^{-2}]$. This type of structure is found in naturally occurring minerals such as *talc* and some clays.

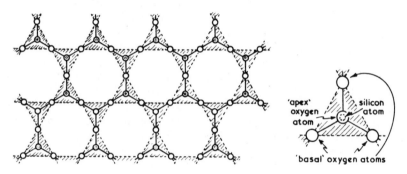

'apex'
oxygen
atom

silicon
atom

'basal' oxygen atoms

Fig. 21.3   *The 'sheet' structure of silicate tetrahedra.*

21.24   Other minerals of a similar type contain aluminium as a consistuent of the giant sheet-like units in which a silicon atom in the basic unit is replaced by an aluminium atom. These are classed as alumino-silicates. *Mica* is such a mineral and is characterized by the extremely thin sheets (less than 0.025 mm thick) into which the mineral can be cleft because the bonding forces between the sheets are very small compared with the strong covalent bonding within the sheets.

21.25   Clays are alumino-silicates in which the units can best be regarded as *three*-dimensional blocks rather than sheets and built up from silicon, oxygen and aluminium atoms. However, their outer surfaces contain $-OH$ groups which will attract water molecules by van der Waals forces operating between the two (1.31). These layers of water molecules (Fig. 21.4) separating the alumino-silicate blocks give to the raw clay its plasticity since one block can slide over another with the water molecules acting rather like a film of lubricating oil. When the clay is dried the water molecules evaporate as they receive enough thermal energy to break the weak van der Waals bonds and so escape. The clay then ceases to be plastic. On heating to a higher temperature further chemical changes occur in the clay and strong cross-links are formed between alumino-silicate blocks producing a

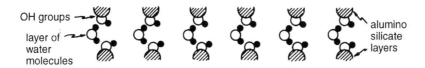

OH groups →

layer of →
water
molecules

alumino
silicate
layers

Fig. 21.4 *Water molecules acting as a weak link between alumina-silicate blocks in clay due to van der Waals forces operating between them.*

continuous hard, brittle structure. As with the reaction which occurs in thermosetting plastics materials, this chemical change is *non-reversible*. Clay, once fired, cannot be softened again with water.

Some silicate-based ceramics of commercial importance

21.30   Asbestos   A number of different minerals containing varying amounts of silicon, oxygen, calcium and magnesium qualify as forms of 'asbestos' but all have one feature in common, a structure based on long-chain molecules containing the repeating unit: $(Si_4O_{11})^{-6}$. This is responsible for the fibrous structure (21.22) common to all compositions of asbestos, which is a flexible, silky material from which fibres as fine as 1.8×10^{-6} mm diameter can be separated. The fibres are strong and with suitable compositions tensile strengths in the region of 1500 N mm^{-2} can be obtained. Strength begins to fall at temperatures above 200 °C though the mineral does not melt until a temperature of about 1500 °C is reached.

Asbestos is chemically inert and is not attacked by chemical reagents. It is little wonder therefore that *once its dust has reached the lungs it remains there to do its deadly work during the lifetime of the sufferer*.

21.31   Asbestos as a health hazard   Evidence began to accumulate at the beginning of the twentieth century pointing to a connection between exposure to asbestos dust and the incidence of some diseases of the human body. It became firmly established that the inhalation of the dust can lead to lung cancer and other serious respiratory diseases such as asbestosis – a chronic inflammation of the lungs. Likewise the ingestion (in food or water) of asbestos particles may lead to serious deterioration of other organs.

Naturally the use of asbestos in the raw fibrous form has been forbidden for many years but such large amounts of it were used in the past that it may be prudent to mention situations where it is likely to have been used.

21.32   When confronted by asbestos products the following precautions should be taken:

- If dealing with the raw fibre and woven forms of it, protective clothing should be worn including a respirator, rubber gloves and gum boots.
- Drilling and cutting of asbestos-cement and other composites should only be undertaken if the surface can be kept wet and continual damping down is possible to prevent the formation of air-borne dust.
- Asbestos composite sheets should be kept wet whilst handling to suppress dust formation.
- Machining processes should only be permitted if adequate exhaust systems are available to take off the dust-laden air.

An appreciation of the dangerous nature of asbestos dust must not of course blind us to the existence of other equally lethal substances. One remembers a very acrimonious meeting held in a school hall to protest about the alleged use of asbestos fibre as an insulator in its construction – a public meeting at which the air was literally blue with tobacco smoke. But enough of this cynicism!

21.33 Obsolete products which contained asbestos These are products used mainly where flame-resistance, thermal and electrical insulation, or resistance to chemical attack are involved. They consist principally of asbestos to which small quantities of a suitable binder were added to effect cohesion.

The object of listing these obsolete materials here is to alert the young engineer – particularly if he is engaged on maintenance work – to the possibility that *there may still be some of these materials around* and to deal with them carefully as outlined above.

Raw asbestos fibres were used as insulation, in particular as a packing for walls and floors and in underground conduits. Since it resists corrosion it was also used for the insulation of batteries.

Asbestos textiles were produced by spinning and weaving processes basically similar to those used for other fibres. They were used in belting for conveying hot materials; friction materials; and as electrical and thermal insulation. Other products included rope for caulking; seals; gaskets; theatre safety curtains; fire-fighting suits and blankets.

Asbestos paper was used as electrical and thermal insulation and for fire protection in military helmets; armoured car roofs; automobile silencers and linings for filing cabinets. Asbestos millboard was virtually a thick form of asbestos paper and could be drilled, nailed and screwed. It was used in the linings of stoves and ovens; fireproof linings of switch boxes, safes and doors; as thermal insulation in incubators; and as fire-proof wall boards.

Asbestos-cement products contained up to 75 per cent Portland cement. Here the strong asbestos fibre was used to strengthen the product as well as to improve thermal insulation properties. Asbestos-cement sheet was used for roofs and walls of small buildings, fire-protection booths, linings of gas and electrical cookers and switch boards. Asbestos-cement pipes were used for carrying water, sewage and gas, whilst large quantities are still in use as domestic roof guttering. Most of the pipe material has been replaced by plastics.

Asbestos reinforced and filled plastics materials included both thermoplastic and thermosetting types to which asbestos was added as loose fibre, woven textile or paper. Typical applications included circuit breakers; pulleys; castors; bearings and washing machine agitators.

21.34 Mica is an alumino-silicate which forms giant sheet-type molecules (21.24). Since the sheets are held together by relatively weak ionic bonds it is possible to separate thin films of mica along the basal planes, strength *along* the planes being relatively high because of the covalent bonds operating within them. The covalent bonding also explains why the electrical resistance of mica is very high. This fact, along with the ability to separate mica into sheets as thin as 0.025 mm, led to its use in small electrical capacitors in the early days of the domestic radio receiver. Mica is used in a large number of electrical industries as an insulating material.

Clay products

21.40 'Delhi Belly', 'the Turkey Trot' and 'Montezuma's Revenge' are all epithets used to describe the transient but uncomfortable (and often inconvenient) malady from which all of us who have travelled in hot climes have suffered at one time or another. The simple remedy was to take what the local pharmacist – if there was one – wrote up as 'kaolin et morph' (a suspension of kaolin in a dilute solution of morphine hydrochloride). However it may surprise the reader to learn that the same 'kaolin' which helps to 'soothe a troubled gut' is a purified form of the alumino-silicate which forms the basis of both pottery and other industrial clays.

21.41 The mineral *kaolinite* from which high-quality porcelain is manufactured was originally imported from a region in the Kaoling Mountains of China after which it was named. Such products became known as 'china' and kaolinite as 'china clay'. Fortunately supplies of good-quality china clay were later mined in Cornwall. Pure kaolinite is an alumino-silicate of the formula $Al_2O_3.2SiO_2.2H_2O$ but natural clays vary widely in composition. China clay as mined in Cornwall contains up to 95 per cent kaolinite whilst those clays used in the manufacture of ordinary bricks may contain only 30 per cent.

21.42 Fireclay used for furnace linings, firebricks and some crucibles is high in kaolinite. Such bricks can be used at temperatures up to 1500 °C but increased amounts of other constituents like silica, lime and iron oxides all reduce the melting point by chemical reaction. Since raw fireclay is relatively expensive, crushed used firebrick (known as 'grog') is added to new clay partly for reasons of economy but also to reduce shrinkage of the product during firing. Once fired, clay will not shrink a second time.

21.43 Shaping clay products Whilst modified forms of the potter's wheel – known as 'jolleys' and 'jiggers' – are still used to make a variety of domestic and ornamental products, articles such as dinner plates and the like are pressed in shaped moulds.

21.44 Lavatory basins, laboratory porcelain, pipe channels, as well as vases, porcelain figures and similar objects of re-entrant or complex shape, can be made by *slip-casting* (Fig. 21.5). Here a split mould made from some water-absorbent material such as Plaster of Paris is filled with liquid *slip* – a

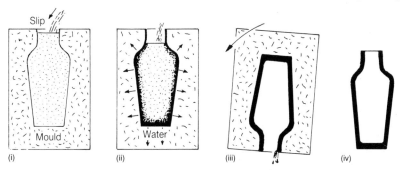

Fig. 21.5 *Slip casting of pottery. (i) The plaster mould is filled with slip; (ii) a layer of clay particles collects on the mould surface as water is absorbed by the mould; (iii) when the shell is thick enough excess slip is poured away; (iv) the resultant 'casting' is removed from the dry mould before being further dried and 'fired'.*

suspension of clay in water. Water is absorbed by the porous plaster mould, the previously suspended clay particles collecting as a uniform layer on the mould surface. When this layer has attained the required thickness the excess slip is poured away. The mould is allowed to dry out and the halves of the mould separated when the clay form is strong enough to be handled.

21.45 Hydroplastic forming of wet clay is employed in the manufacture of simple shapes such as bricks. The clay, softened with water, is extruded using a 'screw pump' of similar design to that shown in Fig. 20.10 for the extrusion of plastics materials. The extruded stock is 'parted off' as individual bricks.

21.46 The heat-treatment of clay products is, of necessity, a slow process since it must avoid cracking and distortion of the clay which would occur with rapid or uneven loss of water vapour. Two stages are involved:

1. *Drying* in which the work is loaded on to racks in large ovens and the temperature then raised to about 120 °C in an atmosphere *saturated with water vapour*. The humidity is then very gradually reduced so that water evaporates slowly and evenly from the work.
2. *Firing* is usually carried out in long gas- or electrically-heated tunnel kilns through which the work moves very slowly towards the hottest zone (at about 1500 °C), then into the cooling zone through which the work again moves slowly before being discharged. The slow, uniform temperature changes avoid cracking and distortion of the product.

During the initial drying operation *absorbed* water is removed but during the firing stage definite chemical changes occur. First, $-OH$ ions are lost so that they are no longer there to attract (by van der Waals forces) water molecules which cause raw clay to be plastic and 'mouldable'. Secondly, some vitrification occurs between the metallic oxides and silica, present in the original clay, forming a glassy phase which cements together the alumino-silicate particles forming a strong though rather brittle structure. Though brittle under conditions of shock, porcelain has a high crushing strength as was illustrated by Messrs Josiah Wedgwood and Sons Ltd., in a photograph showing an 11-tonne double-decker bus supported by six of their fine bone-china coffee cups.

Engineering ceramics

21.50 A few ceramics materials, for example carbon and glass fibres, are important for their high tensile strengths but most ceramics used in engineering are chosen for their great hardness. Only one of these, diamond, the hardest of them all, is of purely natural origin though small synthetic diamonds are now important as an abrasive grit.

Many readers will be familiar with 'cemented carbide' cutting tools in which particles of hard metallic carbides are held in a tough metallic matrix. Some modern cutting materials however are composed entirely of crystalline oxides, nitrides, carbides or borides. Thus sintered crystalline aluminium oxide (alumina) has been used as a cutting tool material for some years. Such tool tips are bonded to an ordinary metal shank, being discarded when they become worn. Not only do alumina and other similar materials retain hardness and compressive strength at high temperatures but they generally have low friction properties coupled with a high resistance to abrasion and chemical attack. Consequently they are often used for cutting at higher speeds than is possible with 'hard metal' tools. They are useful for cutting tough materials including plastics, rubber, aluminium and ceramics.

Yet other such materials, because of their great hardness coupled with a relative density which is low compared with that of metals, are finding increasing use in military armour plating.

21.51 Magnesium oxide (magnesia), MgO, is obtained by heating the mineral *magnesite* ($MgCO_3$) at about 1000 °C:

$$MgCO_3 \xrightarrow{\text{heat}} MgO + CO_2$$

The cubic structure of magnesium oxide (Fig. 21.6) is basically similar to that of lithium chloride (1.21) in that the positively charged metallic ions (Mg^{+2}) and negatively charged non-metallic ions (O^{-2}) are ionically bonded to each other. Each Mg^{+2} ion is surrounded by six O^{-2} ions. Similarly each O^{-2} ion is surrounded by six Mg^{+2} ions and the formula of magnesium oxide is MgO. As with most ionically bonded crystalline materials magnesium oxide is relatively brittle.

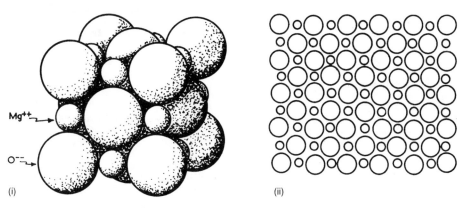

Mg^{++}

O^{--}

(i)   (ii)

Fig. 21.6   *The crystal structure of magnesium oxide ($Mg^{+2} O^{-2}$): (i) shows a single unit of the structure; (ii) shows a 'plan' view of the three-dimensional structure in which each oxygen ion is shared by neighbouring units.*

Magnesium oxide is used mainly as a refractory material, particularly in linings of furnaces used for steel-making by the now more common *basic* processes. Here slags are chemically basic since they contain lime so that a basic furnace lining must be used to avoid any chemical reaction between slag and lining. Magnesium oxide can withstand working temperatures up to 2000 °C.

21.52 Aluminium oxide (alumina), Al_2O_3, is produced from bauxite which is an impure form of alumina and the main ore from which metallic aluminium is manufactured.

In the crystalline structure of alumina (Fig. 21.7) a mixture of ionic bonds and covalent bonds operate between the resultant ions and atoms of aluminium and oxygen. Although the structure is basically close-packed hexagonal one third of the aluminium sites remain vacant in accordance with the valencies of the two elements, which result in a chemical formula of Al_2O_3.

Alumina is used as a cutting tool material but its poor thermal

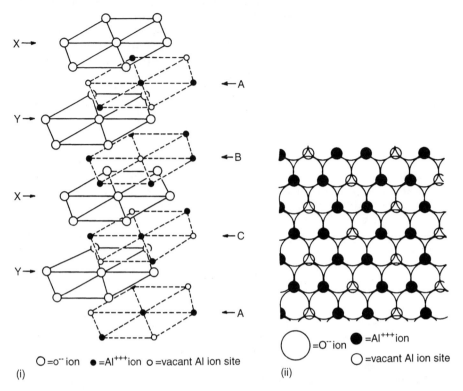

Fig. 21.7 *The crystal structure of α alumina. (i) represents a spatial arrangement of the positions of aluminium and oxygen atom/ion sites; (ii) represents a 'plan' view of the structure along plane X showing that only two-thirds of the aluminium atom/ion sites are occupied.*

conductivity makes it susceptible to overheating and hence rapid wear when used at high machining speeds. An alumina-titanium nitride mixture has a greater resistance to thermal shock and is also harder, making it more suitable for machining harder steels. Both materials are shaped by the usual pressing-sintering methods (7.40).

Much of the alumina now produced is used in military armour plating. Although, at 3.8, its relative density is fairly high it is comparatively inexpensive. Most of this armour is in the form of moulded breast plates used by anti-terrorist units who need protection against missiles fired from high-powered rifles. In the manufacture of such armour, alumina is compacted as a dry or slightly damp powder at high pressure. In this condition it is strong enough to be handled. The compact is then sintered for a long period at a temperature below the melting point which densifies and consolidates the crystal structure.

The hardness, wear resistance and *chemical inertness* of alumina have led to an increase of its use in hip-joint replacements, whilst a high electrical resistivity has favoured its use for the ceramic bodies of sparking plugs.

21.53 Silicon nitride, Si_3N_4, is a stable covalent compound. It cannot be sintered successfully since a very porous compact is the result so the required shapes are made in silicon. This produces porous compacts which are then heated at a temperature just below the melting point of silicon (1412 °C) in an atmosphere of nitrogen to produce the nitride.

As Si_3N_4 cannot be sintered small amounts of basic metallic oxides such as Al_2O_3 or MgO are added. On heating, a layer of SiO_2 forms on the surface of the Si_3N_4 grains due to some oxidation of the latter. This SiO_2 (being acidic) reacts with the metallic oxides (basic) to form an intergranular film of liquid which, on cooling, forms a glassy network cementing the Si_3N_4 grains into a solid mass. This glass can be devitrified to give a mainly crystalline structure either by slow cooling or by further heat-treatment.

Silicon nitride tool properties depend to a large extent on the composition and amount of the intergranular phase, particularly at high temperatures.

21.54 Sialons Pure Si_3N_4 can be modified in composition and properties by replacing some of the nitrogen in the structure by oxygen and at the same time substituting some aluminium for silicon. In each case the proportions added are adjusted to keep a balance of valencies. The resultant compounds are known as 'sialons' (an acronym on the chemical symbols of the main ingredients: Si–Al–O–N).

The main property of sialons is that they retain their hardness at higher temperatures than does alumina. They are tough but rather less so than a cemented carbide of equal hardness.

Sialons are used for cutting tool materials; dies for drawing wire and tubes; rock-cutting and coal-cutting equipment; nozzles and welding shields. Because they have a good combination of high-temperature and thermal-shock resistance sialons are used for the manufacture of thermocouple sheaths; radiant heater tubes; impellors; small crucibles and other purposes involving temeperatures up to 1250 °C.

21.55 Zirconia, ZrO_2, has been used for many years as a high-temperature crucible and furnace refractory where temperatures up to 2500 °C must be sustained. ZrO_2 is a polymorphic substance (1.40) which at 1170 °C undergoes a change in crystal structure similar to that in steel which leads to the formation of martensite. As with steel a sudden volume increase is involved on cooling and since ZrO_2, like martensite, is brittle severe distortion and cracking are liable to occur, leading to spalling and disintegration. Fortunately additions of yttria (yttrium oxide, Y_2O_3), suppress the martensite-type transformation to below ambient temperature – rather like the effect of nickel in 18/8 stainless steel (13.41). Lime and magnesia have a similar effect but the more expensive yttria is used because, unlike lime and magnesia, it does not lower the melting-point of zirconia.

A fully stabilized zirconia (FSZ) contains 18 per cent yttria, but partially stablized zirconia (PSZ) containing about 5 per cent yttria is generally used particularly as a temperature-resistant coating, since its *thermal conductivity is low*, for superalloy rotor blades in jet turbines. Here the *high* coefficient of *thermal expansion* of PSZ matches that of the metal it is coating so that cracking and exfoliation of the coating is eliminated. PSZ is manufactured by sintering *ultrafine* grade ZrO_2 powder with 5 per cent Y_2O_3 at 1450 °C. The resultant structure is fine grained because fine-grain powder was used.

Since zirconia is 'environmentally friendly' *inside* the human body it is finding use in implantology. As it is tougher and has a better 'bending strength' than alumina it is replacing the latter in North America and France for the manufacture of artificial hip joints.

21.56 Some other engineering ceramics As in the case of those engineering ceramics described above, these are materials which are used either because of their durability at high temperatures or because of their great hardness coupled with adequate strength and toughness. Of the former group beryllia (beryllium oxide, BeO), silicon carbide (SiC) and zirconium boride will withstand temperatures in excess of 2000 °C.

Among cutting tool materials the hardest and most effective of all is naturally occurring *diamond*. Its great hardness is due to the strong covalent bonds which bind all carbon atoms comprising the giant molecules (1.42) of this scarce and expensive precious stone. Diamonds are mined mainly in South Africa, Australia, Russia, Zaire and Botswana.

Synthetic diamonds were first produced in the 1950s by subjecting graphite to a pressure of some 7000 N mm^{-2} at a temperature of 3000 °C in the presence of a catalyst. Though not the material from which diamond tiaras could be fashioned these small black particles are very useful as a 'grit' for many cutting, grinding and polishing purposes.

Boron nitride, BN, is second only to diamond in respect of hardness. Manufactured in a similar way to synthetic diamond it is consequently as expensive. Its main fault is that above 1000 °C it will transform to a much softer crystalline form. It is therefore used to machine chill-cast irons and hardened steels at moderate speeds rather than for high-speed processes where the tool may overheat and consequently soften.

Other very hard ceramics are used for military armour plating rather than as cutting materials. They include *boron carbide*, B$_4$C, the low relative density of which makes it useful as armour plating for battlefield helicopters – and their crew; and *titanium diboride*, TiB$_2$, which, with a higher relative density is confined to the protection of ground warfare vehicles against anti-tank projectiles.

Properties of ceramics

21.60 Prior to the manufacture of modern engineering ceramics most materials of ceramic origin such as pottery, furnace refractories, bricks, tiles and electrical insulators consisted of materials containing particles of alumino-silicate origin cemented together by a brittle 'glassy' network formed by chemical reactions during the high-temperature firing process. Modern processes based on the technology of 'powder metallurgy' (7.40) where high-pressure compacting is followed by sintering, produce structures which are completely crystalline. Many of these materials therefore maintain high strength at high temperatures since the low melting-point glass network is absent. Toughness at ambient temperatures is also generally improved because of the absence of the brittle glass.

21.61 **Strength** The high *compressive* strength of many ceramics is typified by Josiah Wedgwood's bone china cups (21.46) and the retention of these properties at high temperatures is a useful feature of some ceramics. For example, titanium diboride will maintain a compressive strength of 250 N mm^{-2} at 2000 °C so that it is one of the strongest materials at such a temperature.

In terms of *tensile* strength however ceramics are generally less effective. Since many ceramics suffer from the presence of microcracks these act as stress raisers and the lack of ductility means that stresses within the material cannot be relieved by the kind of plastic flow which occurs in metals where the positively charged ions are held in a 'flexible' electron cloud, which accounts for both the *elasticity* and *plasticity* of metals, properties generally absent in ceramics. Consequently as stress increases catastrophic failure ultimately occurs, precipitated by the microcracks.

In most ceramics ductility is, for all practical purposes, zero. This is largely due to the presence of small voids in the structure as well as the lack of elasticity and plasticity mentioned above. These voids result from the methods of manufacture – *ceramics do not receive the amount of mechanical working from which wrought metals benefit*. In general most ceramics can only be classed as 'brittle'.

21.62   Creep   only takes place in ceramics at relatively high temperatures, and those which are totally crystalline in structure suffer the least creep. In those ceramics which contain glassy phases, particularly as intergranular networks, creep will take place much more easily and at relatively lower temperatures.

21.63   Hardness   Most ceramics are relatively hard and it is this property which makes them useful as abrasive grits and cutting-tool tips. This hardness is largely due to the operation of strong covalent bonds between atoms in their crystal structures. Materials like silicon carbide, boron nitride and of course diamond are examples.

Table 21.1   *Hardness of some ceramic materials.*

Material	Hardness index (Knoop)
Diamond	7000
Boron nitride	6900
Titanium diboride	3300
Boron carbide	2900
Silicon carbide	2600
Silicon nitride	2600
Aluminium oxide	2000
Beryllium oxide	1220
(Hardened steel)	700

21.64   Refractoriness   is the ability of a material to withstand the effects of high temperatures without serious deterioration in its mechanical properties.

Fireclays which are based on alumino-silicates soften gradually over a range of temperature and may collapse at a temperature below that at which fusion is expected to begin. Fig. 21.8 shows only the liquidus-solidus range of the Al_2O_3–SiO_2 equilibrium diagram. The line *ABCD* represents temperatures at which 'mixtures' of *pure* Al_2O_3 and SiO_2 would begin to melt on being heated. In practical terms this means that any fireclay containing in the region of only 5.5 per cent Al_2O_3 is best avoided since it would collapse completely and suddenly to form a liquid at 1595 °C. As the Al_2O_3 content increases the amount of liquid formed at 1595 °C decreases so that the brick would be less likely to collapse. However for bricks to be usable at temperatures well *above* 1595 °C a high-alumina mixture (above 71.8 per cent Al_2O_3) should be used so that the brick should be stable up to 1840 °C. It must be pointed out nevertheless that this is a purely theoretical assessment – using the Al_2O_3-SiO_2 equilibrium diagram – of the expected melting range of a *pure* clay mixture of known composition. In practice clays are rarely mixtures containing only kaolinite and either excess silica or

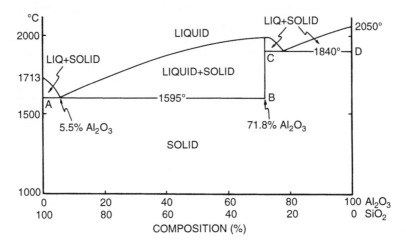

Fig. 21.8   *Part of the Al$_2$O$_3$-SiO$_2$ equilibrium diagram showing the liquidus-solidus range.*

excess alumina. Natural clays will contain other metallic oxides all of which will *lower* its melting range to a greater or lesser degree depending upon the type and quantity of metallic oxide present. Consequently some fireclays are unsuitable for use above 1200 °C whilst the more expensive clays containing above 70 per cent Al$_2$O$_3$ can be used in the range 1700–1800 °C.

Special ceramics for use at very high temperatures are far more expensive than fireclays. A few such materials are listed in Table 21.2.

Table 21.2   *Melting-points of some high-temperature ceramics.*

Material	Melting point (°C)
Hafnium carbide	3900
Tantalum carbide	3890
Thorium oxide	3315
Magnesium oxide (magnesia)	2800
Zirconium oxide (zirconia)	2600
Beryllium oxide (beryllia)	2550
Silicon carbide	2300 (decomposes)
Aluminium oxide (alumina)	2050
Silicon nitride	1900 (sublimes)
Fused silica glass	1680 (approx.)

Cement

21.70 It was the Romans, those pioneers of engineering, who manufactured the first effective cement from small lumps of limestone found in clay beds. These they heated to produce lime, which, when mixed with water, will absorb carbon dioxide from the atmosphere and slowly harden as crystalline calcium carbonate is produced. Later the Romans found that by mixing volcanic ash with lime they obtained a quick-hardening strong hydraulic cement, that is, one which will harden in water. Lime-based cements of the type developed by the Romans were in use into the twentieth century, being slowly replaced by modern Portland cement which was introduced during the nineteenth century.

A number of different types of cement exist but possibly the best known and certainly the most widely used is Portland cement so named in 1824 by its developer Joseph Aspdin since it resembles Portland Stone (a white limestone of the Island of Portland). The principal ingredients are ordinary limestone – or even dredged sea shells – and clay-bearing materials comprising suitable clays or shales. The raw materials are pulverized separately and then mixed in the correct chemical proportions before being fed as a paste into a long rotary kiln where the mixture is calcined at 1500 °C. The resultant 'clinker' is then ground along with a small amount of gypsum (calcium sulphate) to produce the fine greenish-grey powder – the well-known Portland cement – a typical composition of which is given in Table 21.3.

Table 21.3   *A typical Portland cement composition.*

Constituent	% by weight
Calcium silicates	73
Calcium aluminate	10
Calcium alumino-ferrite	8
Gypsum (calcium sulphate)	3
Magnesium oxide (magnesia)	3
Other metallic oxides	balance

When mixed with water Portland cement 'sets'. The chemistry of the process is quite complex but is one in which a reaction – usually termed 'hydration' – between water and the silicates, aluminates and gypsum produces a hard rigid crystalline mass. Portland cement is rarely used as a ceramic material by itself but generally mixed with sand (mortar) or with stone 'aggregate' (concrete).

21.71 Cement as an engineering material In building, cement and its products are widely used because they are relatively cheap to produce, being made from the most abundant constituents of the Earth's crust. Cement can be readily moulded into effective shapes without heating by simply mixing with water and pouring into a mould. Solidified cement has very useful properties; it does not burn, dissolve or rot and possesses a moderate compressive strength.

Nevertheless in other respects its mechanical properties are rather poor unless it is reinforced with steel (24.100). Ordinary hydraulic cement is very weak in tension or bending, particularly under impact loading when cracks propagate readily through the material. These cracks are formed from minute cavities which are present in ordinary commercially cast cement. Such cavities can be seen when a micro-section of such a cement is examined under a microscope and their action as 'stress raisers' is comparable to that of the graphite flakes in ordinary grey cast iron (15.60), another relatively brittle material.

Recent research into the production of Macro-Defect-Free (MDF) cement, in which the number and size of the cavities are greatly reduced, indicates that much improved mechanical properties are available. For example, a 3 mm thick sheet of MDF cement could not be broken by hand; and for cladding purposes 0.7 mm thick steel sheet could be replaced by 2 mm thick MDF cement *of the same weight* but *of vastly improved mechanical properties*. A coiled spring – similar in form to that used in automobile suspension – moulded from MDF cement was found to withstand a strain of 0.5 per cent before failure. The replacement of plastics materials for such applications as bottle tops is also suggested. MDF cement would neither melt nor burn and is twice as stiff as reinforced plastics materials.

MDF cement is prepared by mixing large and fine particles of Portland cement with a gel of polyacrylamide in water. Thorough mixing produces a stiff dough which is then moulded to shape under pressure.

CHAPTER 22

Glasses

22.10 Probably the first glass to be *seen* by Man was in the form of 'stones' produced by lightning strikes but little is known of its early manufacture, though bright blue glass ornaments dating from 7000 BC have been discovered. The Roman writer Pliny the Elder in the first century AD, tells us that glass was made accidentally by the Phoenicians from the fusion of sea-sand and soda. By Pliny's time glass making was a profitable industry and Phoenician glass blowers were producing beautiful vases, jugs and phials in decorated glass. By AD 300 the first window-panes were being made by pouring molten glass on to a flat stone but the glazing of windows was not widespread in England until the sixteenth century though window glass had reached us in Norman times. Progress in this field, as in so many others involving early science and engineering, had been halted by the Fall of Rome and the Dark Ages which followed, culminating in the Black Death. In those days the home was either very draughty – or very dark.

22.11 Most of us still think of 'glass' as the stuff from which window-panes and beer bottles are manufactured. We are all aware of its brittleness under mechanical shock and its less predictable behaviour in conditions of thermal shock – for some of us the first time we heard Mother use 'unacceptable' language was when she poured hot marmalade from her 'jam pan' into an inadequately heated glass jar only to see it shatter.

Now that the fundamental nature of glass is clearly understood materials scientists and engineers use the term 'glass' to describe the structure of a whole range of substances from boiled sweets and toffee to some plastics materials, other sophisticated engineering materials and even '*metallic* glasses', as well as the material of the well-known beer container we automatically describe as a 'glass'.

The composition and structure of glass

22.20 It was mentioned above that Pliny describes how, centuries before his time, the Phoenicians had developed glass manufacture by melting together a mixture of sea-sand and soda. Common glass is still manufactured from roughly the same ingredients today. The chemical reaction upon which this process relies is of a simple 'acid-basic' type. Non-metallic oxides (many of which combine with water to form acids) are 'acidic' in nature whilst metallic oxides are generally strongly 'basic'. Acidic oxides usually react with basic oxides to form neutral 'salts'. Thus silicon dioxide (silica), SiO_2, generally in the form of sand will combine with a number of metallic oxides when the two are heated strongly together. In addition to the sodium oxide, Na_2O, derived from 'soda ash' (commercial sodium carbonate) when the latter is heated, lime, CaO, is also used in the manufacture of ordinary 'lime-soda' glass. These substances react together to give a mixture of sodium and calcium silicates – neutral salts – which will melt at temperatures much lower than those of silica or any of the metallic oxides used.

22.21   When the liquid mixture of silicates is allowed to cool the viscous 'melt' becomes increasingly viscous and finally very hard and brittle – but, *crystallization does NOT take place* as it does with most materials such as metals. To explain this situation we will deal with a very simple 'glass' made from pure silica, SiO_2, only. This would need to be heated to a temperature above 1720 °C in order to produce a viscous liquid which would flow. If we then wished to produce *crystalline* silica we would need to cool it very slowly indeed to a temperature just below its freezing-point (1713 °C) in order to allow sufficient time for the SiO_2 molecules to link up forming the three-dimensional giant molecules $(SiO_2)n$, which are present in the crystalline structure of quartz and other rocks containing silica.

22.22   'Geological rates' of cooling when Mother Earth was born were of course very slow indeed and so vast amounts of crystalline quartz (Fig. 22.1(i)) were formed in the Earth's crust. The action of waves and weather has ground much of this to tiny particles – grains of sand.

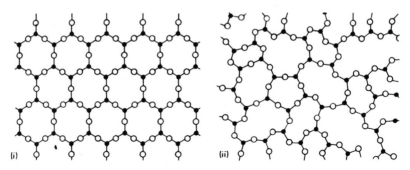

Fig. 22.1   *This is a two-dimensional representation of the three-dimensional cage-like structure of silica in (i)* crystalline *silica (quartz) and (ii) silica glass. In each case only three of the oxygen atoms (○) surrounding each silicon atom (●) are shown. The fourth oxygen atom is 'vertically above' the silicon atom (Fig. 21.1) and will be covalently bonded in turn to another silicon atom – and so on. In (i) there is a repetitive pattern throughout the orderly crystalline structure but in (ii) there is no repetitive pattern except inside the $(SiO_4)^{-4}$ units themselves – the structure is amorphous.*

If we allow molten silica to cool more quickly however there is insufficient opportunity for the complex SiO_2 network to 'arrange itself' completely. As the temperature continues to fall SiO_2 units lose more thermal energy and their movement becomes so sluggish that at ambient temperature it has ceased altogether. Thus we have failed to produce a crystalline silica structure – the material is still in its *amorphous* (or liquid) state (Fig. 22.1(ii)) and will remain so.

22.23   Although still retaining its amorphous or liquid state the silica has, at ambient temperature, become so viscous that it is extremely hard and brittle. Many polymer materials also retain amorphous – or 'glassy' – structures. Here the long chain-like molecules are so cumbersome that what thermal energy they possess is insufficient to overcome the van der Waals forces acting between them in order to permit them to 'wriggle' quickly enough to form a completely crystalline pattern. Hence most polymers contain some completely amorphous regions (Fig. 20.1). If we compare this state of affairs in glasses and polymers with that in metals we see that the single ions in a molten metal, accompanied by loosely attached *fast-moving* tiny electrons, can move very quickly by comparison with the massive cumbersome molecules in glasses and polymers. Hence metallic ions, along with electrons, slip quickly and easily into their appropriate positions in the crystal lattice.

22.24 This high mobility of metallic ions in a liquid metal makes it difficult to produce amorphous glass-like structures in metals. However it *can* be done with some metals and alloys – if we try hard enough! Generally it is necessary to cool the molten metal at a rate of at least one million degrees Celsius per second past its freezing point. In recent years this has been accomplished and 'metallic glasses' are now being manufactured as very valuable engineering materials (22.70).

Glass-transition temperature

22.30 Although silica does not crystallize at its freezing-point, T_m (1713 °C), under normal cooling conditions it does undergo a change in properties at some point below T_m. This is shown by a reduction in the rate at which it contracts in volume on cooling (Fig. 22.2), that is there is a reduction in the coefficient of expansion/contraction and this coincides with a change in properties from a viscous or plastic state to one which is hard and rigid. The point at which this change occurs is termed the *glass-transition temperature*, T_g.

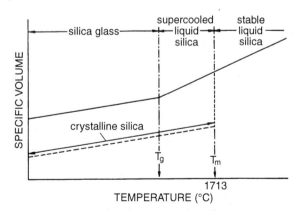

Fig. 22.2 *The glass transition temperature*, T_g, *of silica. (The broken line indicates the crystallization curve for SiO_2 which is cooled* extremely slowly *from above* T_m.)

22.31 There is no apparent change in the structure of the glass between T_m and T_g but clearly the glass has become a super-cooled liquid below T_m. On reaching T_g movement of the molecules relative to one another ceases and the glass as a result becomes hard and rigid – the molecules have become 'frozen' in their amorphous positions and a 'glassy' state has been reached. As the temperature falls below T_g the rate of contraction becomes smaller and is due to a reduction in thermal vibrations of the molecules.

22.32   It is not possible to give a precise value for T_g as it depends upon the rate at which the molten material is cooling when it passes through T_m. The more rapid the cooling rate the higher is the value of T_g because the molecules have less time to 'reorganize' themselves.

We have been discussing here the relationship between T_m and T_g for a simple silica glass because the latter has a single melting-point T_m (1713 °C). Commercial glasses, which are mixtures of silicates of more than one metal, do not melt at a single definite temperature but – like metallic alloys – over a range of temperatures. Nevertheless the relationships between T_m and T_g are basically similar.

22.33   Because glass has been cooled too rapidly to allow it to reach equilibrium by crystallizing at T_m it is an unstable amorphous material at ambient temperatures. It does not usually crystallize because it lacks the thermal energy to allow this to happen. However some glasses, presumably of unsuitable composition, will crystallize partially at ambient temperatures if given sufficient time – generally a few hundred years. This phenomenon is called *devitrification* and leads to a loss of strength of the glass and a tendency to become extremely brittle and opaque. Plans to remove some ancient stained-glass windows in order to protect them from bombing during the Second World War had to be abandoned because of the extreme brittleness of the glass.

The manufacture of glass

22.40   Ordinary 'soda' glass is made from a mixture of silica-sand, 'soda ash' (crude sodium carbonate) and lime (from limestone). Since glass is an easy product to recycle large amounts of scrap glass, known as 'cullet', are used in glass manufacture. Large 'tank' furnaces, usually gas-fired and operating at 1590 °C, hold up to 250 tonnes of molten glass produced from the mixture of raw materials and up to 90 per cent cullet. The cullet comes mainly from 'bottle banks' operated by municipal authorities, though sadly only one bottle in every six finds its way to these 'banks'. As a small boy I earned extra pocket money by collecting empty jam jars. For these the local grocer paid me 1/2d, i.e. 0.21p per jar. There was a similar market for empty bottles. Apparently it is now unprofitable, in these allegedly 'green' days, to collect, clean and re-use 'empties'. Worse still, few of these 'empties' end up even as cullet.

22.41   Sheet glass such as is used for windows is made by the Float Process in which a continuous ribbon of molten glass up to 4 m wide issues from the tank furnace and floats on the surface of a bath of molten tin (Fig. 22.3). The forward speed of the ribbon is adjusted so that the molten glass remains in the high-temperature zone of the float tank long enough for a smooth flat surface to be obtained. The surface of the glass will be parallel since both glass and tin retain horizontal surfaces. By the time that the glass leaves the float tank it will have cooled sufficiently so that it is stiff enough to be drawn into the annealing *lehr* without the use of rollers which might disfigure the surface of the glass. Slow, uniform cooling in the lehr limits internal stresses

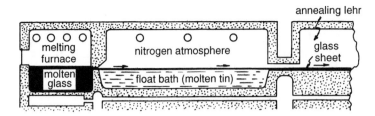

Fig. 22.3 *An outline of the float process for the manufacture of sheet glass.*

which might arise from rapid cooling. The glass is cut into sheets of the required length as it leaves the annealing lehr. Window glass (3 mm thick) and plate glass (6 mm thick) are made by this process, though flat glass between 2.5 mm and 25 mm in thickness can be produced.

22.42 Glass blowing Although glass blowing was probably introduced by the Phoenicians, modern glass-blowing processes were developed in Venice in the twelfth century from whence it moved to the nearby island of Murano in 1292 to avoid 'industrial espionage'. By the middle of the nineteenth century Flemish glass blowers had settled in the West Bromwich-Smethwick area to ply their craft at the Spon Lane works of Messrs Chance Brothers – sadly now long gone. In those days the requirements of a successful glass blower were very powerful lungs and the acquired skill, but rewards were high. As a glass blower at 'Chances' my maternal great-grandfather was able to retire at forty and buy his own pub!

Mouth-blowing is now restricted to the making of high-quality tableware and some artistic glass but most glass hollow-ware is machine blown into metal moulds using a press-and-blow sequence of operations. A 'gob' of molten glass is first pressed in a simple mould to form a cup-shaped blank. This blank is then machine-blown to shape in a mould using a set-up very similar to that shown in Fig. 20.14 which illustrates the shaping of a plastics container.

The properties of glass

22.50 In an explanation of the differences between crystalline and amorphous structures as they apply here (22.21), pure silica was used as an example. Common glass however contains sodium (added as soda ash) and calcium (as lime) in addition. These form Na^{+1} and Ca^{+2} ions which become bonded to oxygen atoms in the silicate structures. The 'silica network' then loses its continuity to an even greater degree and in general this makes the hot glass even more malleable and easily formed than is the case with pure silica glass.

22.51 Most metals are ductile and malleable because dislocations are able to move through the *crystals* comprising the metallic structure. Glasses have no such crystal structure and any deformation can therefore only occur as a result of viscous flow of the molecules. However a 'glass' is, by definition, a substance existing below the temperature T_g so that very little movement of molecules is possible. This means that a glass at ambient temperature is extremely brittle.

Nevertheless some movement of molecules in a glass can occur under stress at ambient temperatures but such movement is extremely slow. If a glass rod is supported at one end, cantilever fashion, and a weight placed at the free end deflection of the rod of course occurs immediately. This is elastic deformation which would disappear immediately if the load were removed. If however the load were allowed to remain acting over a very long period and then removed the elastic deformation would again disappear but the rod would probably be found to have acquired a 'permanent set' or deflection due to some movement of molecules into new positions having taken place during the long period of time in which the rod was in stress.

22.52   When glass is drawn to a fine fibre and cooled quickly in the process a high tensile strength is produced. Special glasses used in fibre-reinforced composites (24.31) can, under ideal conditions, reach strengths up to 15 kN mm^{-2} but in practice a lower strength of about 3.5 kN mm^{-2} would be obtained since surface damage of the fibre is caused by contact with other materials. These microscopic scratches on the surface of the glass act as stress-raisers.

Some special glasses

22.60   The coefficient of thermal expansion of ordinary soda glass is relatively high, whilst its thermal conductivity is low and this combination of properties makes it unsuitable for use in the majority of domestic situations involving sudden contact with boiling water. If boiling water is poured into a cold soda-glass container the latter usually shatters as the surface of contact expands quickly whilst the colder glass beneath does not. This sets up higher stresses within the structure than soda-glass can cope with.

22.61   **'Pyrex'**   is a well-known heat-resistant glass. It is of boro-silicate composition (Table 22.1) containing boron oxide, B_2O_3, which produces a material of very *low* coefficient of thermal expansion. This makes 'Pyrex' suitable for kitchen-ware, laboratory and industrial uses, particularly since its resistance to chemical attack is also good.

22.62   Some modern alumino-silicate glasses are heat-treated after fabrication to give them a partly crystalline structure. Such glasses have physical properties intermediate between those of true glasses and a ceramic like alumina. They are used in cookery-ware, heat exchangers and telescope mirror blanks. These and some other special glasses are listed in Table 22.1.

Metallic glasses

22.70   Although it is relatively easy to retain a glassy amorphous state in many materials which exist as large ungainly molecules, it is very difficult to preserve a glassy structure in metals for reasons stated earlier (22.24). Nevertheless a number of crafty methods have been developed to achieve the very high cooling rates (in the region of 10^6 °C per second) necessary to retain metals in an amorphous state. The most successful is based on some form of 'metal spinner' (Fig. 22.4). This consists of a copper wheel rotating at a peripheral speed of up to 50 m s^{-1} and a jet of molten metal under pressure which impinges against the rim of the wheel. The 'pool' of metal which forms

Table 22.1 *Compositions and uses of some glasses.*

Type of glass	Typical composition (%)	Properties and uses
Soda glass	SiO_2 – 72 Na_2O – 15 CaO – 9 MgO – 4	Window panes, plate glass, bottles, jars, etc.
Lead glass	PbO – 47.5 SiO_2 – 40 Na_2O K_2O } – 7.5 Al_2O_3 – 5	High refractive index and dispersive power. Lenses, prisms and other optics. 'Crystal' glass tableware.
Boro-silicate glass	SiO_2 – 70 B_2O_3 – 20 Na_2O K_2O } – 7 Al_2O_3 – 3	Low coefficient of expansion and good resistance to chemicals. Used for heat-resistant kitchen-ware and laboratory apparatus ('Pyrex').
Alumino-silicate glass	SiO_2 – 35 CaO – 30 Al_2O_3 – 25 B_2O_3 – 5 MgO – 5	High softening temperature (T_g up to 800 °C). A glass/ceramic ('Pyrosil' and 'Pyroceram') – cooking ware, heat exchangers, etc.
High silicon glass	SiO_2 – 96 B_2O_3 – 3 Na_2O – 1	'Vycor' – low coefficient of expansion. Missile nose cones, windows for space vehicles.
Silicon-free glass	B_2O_3 – 36 Al_2O_3 – 27 BaO – 27 MgO – 10	Sodium-vapour discharge lamps.

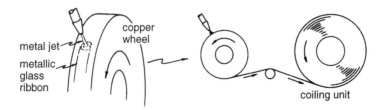

Fig. 22.4 *The principles of 'metal spinning' for the manufacture of 80Fe-20B metallic glass.*

on the rim cools extremely rapidly since copper is a good conductor of heat, forming a continuous ribbon which passes to a coiling unit.

22.71 In a metallic glass produced in this way the structure is amorphous and there is no regular arrangement of the atoms as exists in a normal crystalline metal below its freezing point. Here the atoms are still in the irregular array of the amorphous liquid, except that a little less space will be occupied because thermal movement has ceased. Imagine a plastics bag filled randomly with ball bearings – that is the type of 'structure' in which atoms are present in a metallic glass.

22.72   Some metals and alloys form glasses more readily than do others. Generally a mixture containing 80 per cent metal and 20 per cent 'metalloid' (for example boron, silicon, etc.) seems to be the most effective. Like other glasses they are very strong but in tension elongation is usually below 1 per cent. The most important of these glasses to date is 80Fe-20B which is mechanically stronger than the best carbon fibre. However by far the more important property is its extreme magnetic 'softness'. That is, it has a very low remanence and a high magnetic permeability (13.62).

This has led to its wide use in the USA as a core material for electric power transmission transformers, resulting in a very significant energy saving whilst at the same time being cost effective in the medium term. With traditional transformer cores there is a 3 per cent loss in the electricity generated. This is used up in magnetizing the transformer cores and can be reduced to about 0.75 per cent when 80Fe-20B cores are used. In the UK this translates to a saving of some 1.3 Mtonnes of oil (or its equivalent in gas or coal) per annum, which would lead to a reduction of 4.5 Mtonnes of CO_2 and 120 ktonnes of SO_2 being released as 'greenhouse gases'. Surely the universal adoption of such glass transformer cores should be an option to be taken seriously. Or is the threatened 'Carbon Tax' now overdue?

CHAPTER 23

Composite materials

23.10 Those engineering materials loosely referred to as 'composites' include a wide range of products varying from those used in high-strength aircraft components to road-building tarmacadam and concrete. Generally composites are manufactured by mixing together two separate components one of which forms a continuous matrix whilst the other, present either as particles or fibres, provides the strength or hardness required in the composite material. Of these materials *fibre-reinforced composites* are the most significant in the modern engineering world and the whole of the next chapter will be devoted to them.

23.11 The remainder, 'particle composites', can be divided into three groups:

- composites containing particles of a very hard constituent embedded in a tough, shock-resistant matrix, e.g. hard metallic carbide particles in a tough metallic matrix, used for tool and die materials;
- composites containing finely dispersed hard but *strong* particles which will raise the strength of the parent material, e.g. Al_2O_3 particles in specially prepared metallic aluminium;
- composites containing particulate material of very low cost which has been added as a 'filler' to 'bulk-up' the matrix material. Bakelite mouldings have long been 'filled' with sawdust, wood flour or finely ground minerals such as sand or limestone.

23.12 Some cohesion between the particles and the matrix is necessary and this may be achieved by either:

- *mechanical bonding* which will operate when the surface of the particle material is rough or irregular in texture and the matrix is added as a liquid, e.g. particles of aggregate in concrete;
- *physical bonding* which depends upon the operation of van der Waals forces acting between surface molecules in both materials;
- *chemical bonding* at the interface between particle and matrix; sometimes this can have a deleterious affect if the reaction product is in the form of a brittle film;
- *solid-solution bonding*. The particle material may dissolve in the matrix to a limited degree forming a solid solution. Such a situation generally produces a strong positive bond.

23.20 Particle-hardened composites These are generally the products of powder metallurgy (7.40) in which extremely hard particles of a ceramic material are held in a tough ductile matrix of some metal. Such materials are usually known as *cermets* and have been popular for many years as cutting tool and die materials.

23.21   The most widely used cermets consist of particles of hard tungsten carbide held in a tough matrix of cobalt (Fig. 23.1). The two components, in the form of fine powders, are thoroughly mixed and the mixture then compacted at high pressure in a die of the required shape. The application of high pressure causes the cobalt particles to slide over each other so that a degree of cold-welding occurs between the particles and the resultant compact is strong enough to permit handling. This stage of the process is followed by 'sintering' – that is, heating the composite at some temperature high enough above the recrystallization temperature of the cobalt so that a continuous, tough matrix of cobalt is formed. The heating process takes place in an atmosphere of hydrogen to protect the compact from oxidation.

Fig. 23.1   *The type of structure in a tungsten carbide/cobalt cermet. Particles of tungsten carbide (white) in a cobalt matrix, (black) (×1000).*

The proportions of tungsten carbide and cobalt in such a cermet are varied in accordance with the properties required in the tool. For machining hard materials cutting tools containing 95 per cent tungsten carbide and 5 per cent cobalt are used, whilst for cold-drawing dies, where greater toughness is required, a cermet containing 75 per cent tungsten carbide and 25 per cent cobalt would be employed.

23.22   Some cermets are manufactured by allowing a *liquid* matrix metal to infiltrate around particles of a solid ceramic. A strong positive bond is more likely to be produced when the ceramic forms a solid solution with the matrix metal. If the result is an intermetallic compound then the bond is likely to be weak and brittle.

Typical cermets are based on hard oxides, borides and nitrides as well as upon carbides. The more important groups are shown in Table 23.1.

Dispersion-hardened materials

23.30   A material can be strengthened by the presence of small roughly spherical particles of another material which is *stronger* than the matrix material itself. The function of these small strong particles is to impede the

Table 23.1 *Some cermet materials.*

Cermet group	Ceramic	Bonding matrix	Uses
Carbides	Tungsten carbide	Cobalt	Cutting tools.
	Titanium carbide	Cobalt, molybdenum or tungsten	
	Molybdenum carbide	Cobalt (or nickel)	
	Chromium carbide	Nickel	Slip gauges, wire-drawing dies.
Oxides	Aluminium oxide	Cobalt or chromium	Rocket motor and jet-engine parts – other uses where high temperatures are encountered.
	Magnesium oxide	Magnesium, aluminium or cobalt	
	Chromium oxide	Chromium	'Throw-away' tool bits.
Borides	Titanium boride	Cobalt or nickel	Cutting tool tips.
	Chromium boride	Nickel	
	Molybdenum boride	Nickel	

movement of 'dislocation fronts' which are passing through the structure of the material. (Fig. 6.4). Since the dislocation fronts are unable to move through the particles – because the latter are stronger – 'loops' are formed. These loops will provide an even more effective barrier to any further dislocation fronts following along the same plane. Thus the strength of the material increases as the 'dislocation jam' builds up.

23.31 An example of such a dispersion-hardened material is 'sintered aluminium powder' (SAP) in which the dispersed strong particles – or 'dispersoid' – is aluminium oxide. SAP is manufactured by grinding fine aluminium powder in the presence of oxygen under pressure. Aluminium oxide forms on the surface of the aluminium particles and much of it disintegrates during grinding to form fine powder intermingled with the aluminium particles. The mixture is then compacted and sintered by powder metallurgy techniques to produce a hommogeneous aluminium matrix containing about 6 per cent aluminium oxide particles. As Fig. 23.2 indicates SAP retains a much higher strength then does pure aluminium *at all temperatures*. More important still it is stronger than some of the precipitation-hardened aluminium alloys at temperatures above 200 °C –

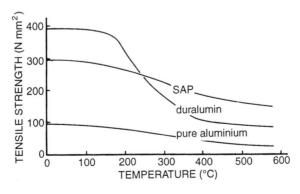

Fig. 23.2 *The relationship between tensile strength and temperature for duralumin, SAP and pure aluminium.*

temperatures at which the precipitation-treated alloys lose strength rapidly as reversion (17.74) takes place.

23.32 Orthodox melting/casting processes are unsuitable for the manufacture of dispersion-hardened products since the dispersoid is usually of lower relative density and would float to the surface of the melt. Instead both dispersoid and matrix metal in powder form are treated in a high-speed ball mill where the high impact energy of the hard steel balls causes some 'mechanical alloying' between particles of both dispersoid and matrix metal. The final consolidation and shaping is by normal powder-metallurgy processes or by extrusion.

23.33 Some modern superalloys used in aerospace projects are dispersion-hardened by yttrium oxide particles. Thus *Incoloy MA956* contains 4.5 per cent Al, 2.0 per cent Cr, 0.5 per cent Ti, 0.5 per cent Y_2O_3. bal.-Fe, and is used in combustion chambers and turbine casing sections where resistance to creep is essential. Both aluminium oxide and aluminium carbide are used as dispersoids in some aluminium alloys. *Inco MAP Al9052* is an aluminium base alloy containing 4 per cent Mg, 1.1 per cent C and 0.6 per cent O and has a tensile strength of $450\,N\,mm^{-2}$. It is used in military aerospace where a combination of low relative density, high strength and resistance to corrosion is required.

Mortar and concrete

23.40 These are products in which relatively expensive 'cement' is used to stick together particles of low-value sand, gravel, stones and other 'aggregates' to form composites of fairly high compressive strengths.

23.41 Mortar is the adhesive material used between bricks in the building industry. Before the advent of Portland cement (21.70) mortar consisted of a mixture of slaked lime, sand and water. This is still used occasionally when a *slow* hardening process can be tolerated. The slaked lime contained in the mortar reacts slowly with carbon dioxide present in the atmosphere:

$$Ca(OH)_2 + CO_2 \longrightarrow CaCO_3 + H_2O$$
$$\text{slaked}$$
$$\text{lime}$$

The calcium carbonate ($CaCO_3$) forms as a hard interlocking network so that the slaked lime/sand 'paste' is replaced by a hard rigid solid.

Modern mortar usually consists of a mixture of Portland cement and clean sand with sufficient water to make the mixture workable. Ideally the proportion of sand : cement in the dry mixture is 3 : 1 but higher proportions of sand are generally used.

23.42 Concrete is produced from a mouldable mixture of Portland cement, stone aggregate, sand and water. When hard it has many of the characteristics of natural stone, and is extensively used in building and civil engineering. In the 'wet' state, it can be moulded easily, and is one of the cheapest constructional materials in Britain, because of the availability of suitable raw material, and also the low maintenance costs of the finished product.

The aggregate may be selected from a variety of materials. Stone and gravel are most widely used, but in some cases other substances available cheaply, such as broken brick and furnace slag, can be employed. Sand is also included in the aggregate, and, in order to obtain a dense product, a correct stone/sand ratio is essential. Aggregate materials should be clean, and free from clay.

The proportion of cement to aggregate used depends upon the strength required in the product, and varies from 1 : 3 in a 'rich' mixture to 1 : 10 in a 'lean' one. The cement/aggregate ratio commonly specified is 1 : 6, and this produces excellent concrete, provided that the materials are sound, and properly mixed and consolidated. Other things being equal, the greater the size of the large stones, the less cement is required to produce concrete of a given strength.

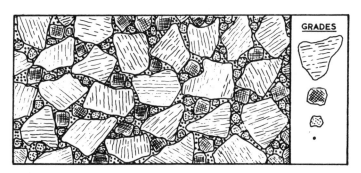

Fig. 23.3 *A satisfactory concrete structure. The aggregate should consist of different grades (or sizes) in the correct proportions. In this way, the smaller particles fill in the spaces between the larger ones, whilst particles of sand occupy the remaining gaps. The whole is held together by a film of cement.*

The cement and its aggregates are mixed in the dry state, water is then added, and the mixing process is continued until each particle of the aggregate is coated with a film of cement paste. The duration of the hardening period is influenced by the type of cement used, as well as by other factors such as the temperature and humidity of the surroundings. Concrete made with Portland cement will generally harden in about a week, but rapid-curing methods will reduce this time. Concrete structures can be 'cast *in situ*' (as in the laying of foundations for buildings), or 'pre-cast' (sections cast in moulds, and allowed to harden before being raised into their final positions).

23.43 Plain concrete   This is suitable for use in retaining walls, dams, and other structures which rely for their stability on great mass. Other uses include foundations where large excavations have to be filled; whilst vast quantities of plain concrete are employed in modern motorway construction. At the other end of the scale, many small pre-cast parts are manufactured; for example, 'reconstituted stone' for ornamental walls, as well as sundry other articles of garden 'furniture'.

Tarmacadam

23.50   Modern methods of road making were originated early in the nineteenth century by the Scots engineer, John McAdam. The method he used – coating suitable hard aggregate material with tar – is roughly similar to the process used today except that the tar (obtained from the gasworks where coal was destructively distilled) has been replaced largely by bitumen (residues from the refining of crude petroleum). Some asphalts also occur naturally, e.g. 'Trinidad Lake'.

The bituminous material is mixed with a suitable aggregate such as crushed blast-furnace slag for the coarse foundation work, or fine gravel for the finishing layers. The resultant mixture is tough and crack-resistant because of the bituminous matrix, whilst it is hard-wearing because of the exposed surface of hard aggregate material. Its structure and properties resemble very closely those of a bearing metal in which hard, low-friction particles standing 'proud' of the surface are held in a tough ductile shock-resistant matrix. However whilst slip at a very low coefficient of friction is the objective in a bearing the reverse is true in a road surface, so that the properties of the hard rubber tyre must be designed to provide maximum adhesion between tyre and road surface.

CHAPTER 24

Fibre-reinforced composite materials

24.10 Wood is our oldest structural material and is still employed in almost every industry. Much timber, generally of inferior quality, is also used in the production of paper pulp. This may be regarded as being acceptable procedure when the end-product is a textbook but one may feel less enthusiastic about the slaughter of trees to make possible the printing of the average daily 'tabloid', which, in these days of increasing awareness of the importance of hygiene, is deemed unfit even for the wrapping of fish and chips!

Trees are the largest and most ancient living things on Earth – some of them are 5000 years old and have lived through the rise and fall of empires since the days of the Pharaohs. When tramping through the forests of towering Redwoods along the Canadian Pacific coast one cannot but be impressed by the awesome silence of the place. It is as though the rest of Nature Herself is in reverence of these majestic immemorial giants.

A tree is also an example of very good engineering design. The cynic may suggest that it took a few millions of years of trial-and-error by blundering Evolution to achieve this end. Nevertheless though Man had been using wood since the time when he began making spears and building shelters it was not until the 1940s that he sought to make materials whose structures were based on that of wood but with enhanced mechanical properties.

24.11 The structure of wood In metallic structures the building unit is the crystal whilst a superpolymer is an agglomeration of large numbers of long thread-like molecules. A glass consists of a mass of fairly large silicate units which are much too sluggish in their movements to be able to crystallize. In living matter, both plant and animal, the simplest unit is the cell. As a tree grows the wood tissue forms as long tube-like cells of varying shapes and sizes. These are known as *tracheids* but we may regard them as 'fibres' which are arranged in roughly parallel directions along the length of the trunk (Fig. 24.1). They vary in size from 0.025 to 0.5 mm in diameter and from 0.5 to 5.0 mm in length and are composed mainly of cellulose (19.33). The fibre-like cells are cemented together by the natural resin *lignin*; so wood can be regarded as a naturally occurring composite material in which the matrix of lignin resin is reinforced and strengthened by the relatively strong tracheids or fibres of cellulose.

24.12 Most amateur carpenters and DIY experts soon come to realize that the strength of a piece of timber is 'along the grain', i.e. in the same

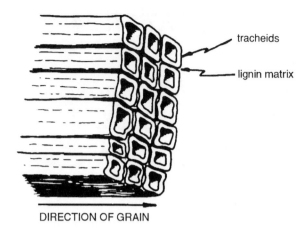

Fig. 24.1   *The structure of wood – the cellulose fibres (tracheids) are cemented in a matrix of lignin resin.*

direction as the cellulose fibres, whilst 'across the grain' the timber is relatively weak and brittle. Thus wood is a very anisotropic[1] material; for example, European redwood may have a tensile strength of 9 N mm^{-2} 'along the grain', whilst 'across the grain' (i.e. perpendicular to it) the strength may be no more than 1.5 N mm^{-2}.

Reference has already been made to 'fibre' in rolled metal sections (6.50), and it was stated that the maximum strength of such a material was in the direction of the fibres. The same principle applies in the case of wood, but to a much greater extent. Timber has an excellent specific strength (strength/weight ratio) and when considered in this context, some types of wood can compete with steel. Unfortunately, a bulky cross-section is necessary if wood is to sustain a tensile load; consequently, steel, with its correspondingly small cross-section, will generally be used instead. In compression and bending, provided there is no bulk disadvantage, wood may be preferable, as indeed it is for domestic building. Wood has a high modulus of elasticity, and, when vibrated, it produces strong resonance; hence it is used for the sounding-boards of many stringed musical instruments, such as violins and guitars – the traditional or 'acoustic' types of course, not the electrically amplified gadgets 'manipulated' by the perpetrators of ear-splitting cacophony of 'rock music'. Sorry chaps! but my youth was spent in the era of real jazz and the days of Django, Stephan and their timeless 'Quintette'.

Man-made fibre-reinforced composites

24.20   The concept of fibre-reinforced materials had its origin in nature in the structure of wood. Although Man had been using wood for a long time it was not until the 1940s that strong fibre materials and suitable bonding resins became available so that he could hope to surpass the tensile and stiffness properties of natural wood. The development of polyester resin/glass fibre composites then began.

---

[1]   A material in which the properties vary according to the *direction* in which they are measured in the material.

24.21   In general fibre-reinforced composites include:

- Matrix materials such as thermosetting or thermoplastics polymers and some low-melting point metals; reinforced with fibres of carbon, glass or organic polymer.
- Polymers (usually thermosetting) reinforced with fibres or laminates of woven textile materials.
- Vehicle tyres in which vulcanized rubber is reinforced with woven textiles or steel wire.
- Materials such as concrete reinforced with steel rods.

Obviously when the reinforcing fibres are uni-directional, as are the tracheids in a tree trunk, then maximum strength is also uni-directional. In many fibre-reinforced composites where woven textiles are used instead of uni-directional fibres then maximum strength is available in more than one direction. A case in point is the woven textile or steel mat used to reinforce a vehicle tyre. However we shall be dealing here mainly with composites containing uni-directional fibres so that we can more easily assess the advantages of such reinforcement.

24.22   In successful composites there must be adequate bonding between fibre and matrix and this bonding may be either physical or chemical. The main function of the matrix material is to hold the fibres in the correct position so that they carry the stress applied to the composite as well as to provide adequate rigidity. At the same time the matrix protects the fibres from surface damage and from the action of the environment. The fibres should be long enough so that the bonding force between the surface of the fibre and the surrounding matrix is greater than the force necessary to break the fibre in tension – short fibres may slip inside the matrix because the interface surface is too small so that bonding fails before the fibre breaks in tension.

The overall relative density of such a composite is important since it affects both *specific* strength and *specific* modulus of elasticity – both particularly relevant when materials for use in air or land transport vehicles are involved.

24.23   We will now consider a composite rod consisting of long parallel fibres of some strong material held in a matrix of a rigid substance (Fig. 24.2), e.g. glass or carbon fibres in an epoxy resin matrix.

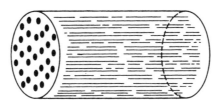

Fig. 24.2

The relative density, ρ_c, of such a composite can be calculated using the simple 'rule of mixtures':

$$\rho_c = \rho_f V_f + \rho_m (1 - V_f)$$

where ρ_c, ρ_f and ρ_m are the relative densities of composite, fibre and matrix respectively, and V_f is the volume fraction of the fibre (clearly the volume fraction of the matrix will then be $(1 - V_f)$).

Q A glass-reinforced polyester resin contains 60 per cent by volume of glass fibre. If the relative densities of glass and polyester are 2.1 and 1.3, respectively, calculate the relative density of the composite.

A Using

$$\rho_c = \rho_f V_f + \rho_m (1 - V_f),$$
$$\rho_c = (2.1 \times 0.6) + (1.3 [1 - 0.6])$$
$$= \underline{1.78}.$$

24.24   If the fibres and matrix are firmly bonded to each other then:

$$\text{force acting on the composite} = \text{force acting on the fibres} + \text{force acting on the matrix.}$$

Since stress, σ = force/cross-section, in each case,

then:

$$\text{total force} = \sigma_f A_f + \sigma_m A_m$$

where A_f and A_m are the cross-sectional areas of fibre and matrix, respectively. If we now divide throughout by the total cross-sectional area of the composite, A_c, we get:

$$\frac{\text{total force}}{A_c} = \frac{\sigma_f A_f}{A_c} + \frac{\sigma_m A_m}{A_c}$$

but total force/A_c is equal to the stress, σ_c, on the composite whilst if we consider a unit length of the composite, V_f is the volume fraction of fibres and V_m the volume fraction of the matrix. Therefore

$$\sigma_c = \sigma_f V_f + \sigma_m V_m;$$

but $V_f + V_m = 1$ so

$$\sigma_c = \sigma_f V_f + \sigma_m (1 - V_f).$$

Q A composite material in rod form consists of a polyester reinforced with 40 per cent by volume of long parallel glass fibres. If the tensile strengths of the glass and polyester are 3.4 kN mm^{-2} and 60 N mm^{-2}, respectively, calculate the tensile strength of the composite.

A Using

$$\sigma_c = \sigma_f V_f + \sigma_m (1 - V_f)$$

then

$$\sigma_c = (3400 \times 0.4) + (60 \times 0.6)$$
$$= \underline{1.396 \text{ kN mm}^{-2}}.$$

24.25   When the composite is under stress the resultant strain, ϵ, will be the same in both fibres and matrix (since we have assumed that they are firmly bonded). Hence:

$$\frac{\sigma_c}{\epsilon} = \frac{\sigma_f V_f}{\epsilon} + \frac{\sigma_m (1 - V_f)}{\epsilon}.$$

But stress/strain = Young's modulus of elasticity (E) in each case, therefore:

$$E_c = E_f V_f + E_m (1 - V_f).$$

Q A composite material is made by adding 15 per cent by volume of glass fibre and 5 per cent by volume carbon fibre to an epoxy resin matrix, the fibres being long and parallel in direction and firmly bonded to the matrix. If the values of Young's modulus, E, for glass, carbon and epoxy resin are 75 kN mm^{-2}, 320 kN mm^{-2} and 4 kN mm^{-2}, respectively, calculate E for the composite.

A Using

then

$$E_c = E_f V_f + E_m (1 - V_f)$$

$$E_c = (75 \times 0.15) + (320 \times 0.05) + (4 \times 0.8)$$

$$= \underline{30.45 \text{ kN mm}^{-2}}.$$

Fibres used in composites

24.30 Cold-setting resins strengthened by woven textile fibres, e.g. 'Tufnol', were developed many years ago for use in the electrical trades as tough panels with high electrical insulation properties, but glass fibre was the first man-made material to be used for reinforcement purposes to produce a strong light-weight material. Now a number of other man-made fibres, in particular carbon, boron, some ceramics and polymers, all possessing the necessary high strength, stiffness and low relative density, are in use. Research continues with the object of producing materials of increasingly high *specific* strength for the aerospace industries and since relative density is important many fibres are made from compounds based on the elements boron, carbon, nitrogen, oxygen, aluminium and silicon – all elements with low atomic mass which often, though *not* inevitably, means that relative density is also low. Moreover compounds based on these elements are likely to be covalently bonded – the strongest of chemical bonds which is generally reflected in the mechanical strength of the compound.

24.31 **Glass fibre** was the first of modern fibres to be used in composites. A number of different compositions are now in production. The best known of these are S-glass (which is a high strength magnesia-aluminosilicate composition) and E-glass (a non-alkali metal boro-silicate glass developed originally for its electrical insulation properties). Whilst glass fibre is strong and relatively cheap it is less stiff than carbon fibre, the relative density of which is lower than that of glass.

High-quality glass fibre is spun at a high temperature and then cooled very rapidly before being given a coating of 'size'. This is a mixture, the main object of which is to protect the glass surface from contamination which would lead to a loss in strength. It also contains a derivative of silane, SiH_4, which acts as a bonding agent between glass and matrix.

24.32 Carbon fibre is made up of layers of carbon atoms arranged in a graphite-type structure (Fig. 1.11) but whilst graphite consists of small 'plates' of atoms which slide over each other (hence graphite is a lubricant), in carbon fibre the graphite-type structure persists to form long fibres. Whilst strong covalent bonds join the carbon atoms along a layer, only weak van der Waals forces operate *between* layers. Thus the carbon fibre structure is anisotropic in many of its properties, for example, in a direction parallel to the fibres Young's modulus reaches 1000 kN mm^{-2}, whilst perpendicular to the fibres it is only 35 kN mm^{-2}.

The bulk of carbon fibre is manufactured by the heat-treatment of poly(acrylonitrile) filament. The process takes place in three stages in an inert atmosphere:

1. A low-temperature treatment at 220 °C. This promotes cross-linking between adjacent molecules so that filaments do not melt during subsequent high-temperature treatments.
2. The temperature is raised to 900 °C to 'carbonize' the filaments. Decomposition takes place as all single atoms and 'side groups' are 'stripped' from the molecules, leaving a 'skeleton' of carbon atoms in a graphite-type pattern.
3. The heat-treatment temperature is then raised to produce the desired combination of properties. Lower temperatures (1300–1500 °C) produce fibres of high tensile strength and low modulus, whilst higher temperatures (2000–3000 °C) provide fibres of low strength but high modulus.

The fibres are kept in tension during the heat-treatment processes to prevent curling and to favour good alignment. Carbon fibres are between 5 and 30 μm in diameter and are marketed in 'tows' containing between one and twenty thousand filaments. Carbon fibre is the most promising of composite reinforcements particularly in terms of specific strength. Although initially carbon fibre was expensive, production costs have fallen as a result of improved manufacturing techniques and seem likely to fall further.

24.33 Boron fibre has a very high Young's modulus but only a moderate tensile strength. It is used in the aerospace industries because of its low relative density but mainly because of its high transverse strength, which, along with a large fibre diameter, enables it to withstand high compressive forces. However since it is difficult to fabricate it is very expensive.

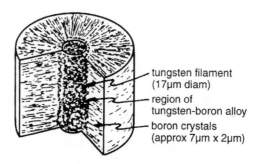

tungsten filament
(17μm diam)
region of
tungsten-boron alloy
boron crystals
(approx 7μm x 2μm)

Fig. 24.3   *Section through a boron fibre, grown on to a tungsten filament.*

Boron fibre is grown direct from the vapour phase on to a tungsten filament only 15 μm in diameter, the resultant boron fibre being between 100 and 200 μm in diameter. Since boron reacts chemically with tungsten there is a continuous bond between boron and the thin tungsten core.

24.34 'Aramid' fibre Most of the plastics materials consisting of long-chain molecules have limited strength and stiffness because the molecules are *not straight* but generally coiled or folded. Hence, under stress, extension takes place as the molecules straighten and also slide past each other. Nevertheless if the molecules are stretched and straightened during a fibre manufacturing process much higher strength and stiffness can be developed. Aromatic[2] polyamide – hence 'Aramid' – fibres are the most successful of these and are manufactured under the trade name of *'Kevlar'*.

After being spun the fibres are heated *whilst in tension* in a nitrogen atmosphere at temperatures up to 550 °C. The 'stretched' properties are then retained. The resultant material is strong in a direction parallel to the straightened molecular chains but since the chains are held together only by weak van der Waals forces the transverse and compressive strengths are relatively low. However this does mean that Kevlar has a good capacity for absorbing energy so Kevlar-reinforced composites are useful for resisting impact damage, e.g. bullet-proof vests and the like where low relative density is important. For the latter reason aramids are also used in aerospace industries. Their main disadvantages are a sensitivity to ultra-violet light from which they must be shielded and the fact that they absorb water.

Table 24.1 *Mechanical properties of some fibre materials.*

Fibre type	Tensile strength (kN mm^{-2})	Young's modulus (kN mm^{-2})	Relative density	Specific strength (kN mm^{-2})	Specific modulus (kN mm^{-2})
Aramid (Kevlar 49)	2.75	125	1.44	1.9	87
Silica glass	5.9	74	2.2	2.7	34
E-glass	1.7	70	2.5	0.7	28
S-glass	4.6	84	2.6	1.8	32
Carbon (high strength)	3.5	200	1.7	2.0	118
Carbon (high modulus)	2.5	400	2.0	1.3	200
Boron	3.9	400	2.6	1.5	154
Silicon carbide	7.0	400	3.2	2.2	125
Alumina	1.7	380	4.0	0.43	96
Bulk materials for comparison					
Steel	1.3	260	7.8	0.2	26
Wood (spruce)	0.1	10	0.46	0.22	22

---

[2] The term 'organic' as applied to the chemistry of some carbon compounds was adopted early in the nineteenth century when it was thought that compounds such as alcohol, urea and glucose could be produced only by living organisms (plant or animal). These so-called 'organic' chemicals were then divided into two groups – *aliphatic*, those compounds having a mainly chain-like structure; and *aromatic*, those compounds based on the 'benzene ring'. However do not be misled by the use of 'aromatic' in this context – some of the compounds have a completely disgusting smell whilst some 'aliphatics' are quite fragrant!

24.35   Other fibres   Some ceramics, most importantly silicon carbide and alumina, are being developed as reinforcement fibres in metal-matrix composites (MMCs). These are important for future developments in the aerospace industries where both cast and extruded aluminium alloys can be further strengthened by fibre reinforcement.

Matrix materials

24.40   Polymers are the most widely used of matrix materials in fibre-reinforced composites. The main function of the matrix is to provide a rigid base which holds the strong fibres in position. To provide a successful composite the matrix material must bond adequately with the fibre surface either frictionally or, as is more usual, chemically. When a chemical bond is formed any reaction between the matrix material and the fibre must not be so vigorous as to damage the surface of the latter. The matrix material must flow easily during the investing process, so that the spaces between the fibres are filled, and then 'set' quickly.

24.41   The main *service* requirements of a matrix material are:

* it should be stable to a temperature at which the properties of the fibre begin to deteriorate;
* it must be capable of resisting any chemical attack by its environment;
* it should not be affected by moisture.

24.42   Thermosetting resins   are polymers which cross-link during 'curing' (19.11) to form a hard glassy solid. The most popular are polyesters and epoxy resins. Polyesters were the first thermosetting resins to be used, with glass-fibre reinforcement, during the Second World War in the manufacture of housings for aircraft radar antennae, the composite being 'transparent' to radio transmissions. These resins are still widely used since they are relatively cheap and are easy to work with and will 'cure' at relatively low temperatures. Unfortunately polyesters do not bond well with fibres so that the transverse strength of the composite depends mainly on weak van der Waals forces. Moreover since polyesters shrink considerably on curing this further reduces adhesion to the fibres. For both reasons polyesters are not generally used in high-strength composites.

For high-performance composites based on either continuous carbon or aramid fibres, epoxide resins are extensively used. The principles of the manipulation of these resins will be familiar to readers who have used DIY epoxide glues (26.21). The resin is mixed with a 'hardener' and heated for up to eight hours at about 170 °C to form a hard, insoluble, cross-linked structure with a tensile strength of 80–100 N mm^{-2} and a Young's modulus of 2–4 kN mm^{-2}. Such a material can be used at temperatures up to about 160 °C.

Other thermosetting matrices include the well-known and inexpensive phenolic resins (19.41) which are fire-resistant but unfortunately rather brittle. Some high-performance polyimides can withstand temperatures to 315 °C continuously but they are very expensive and require long curing times following the fibre-impregnation stage.

24.43   Thermoplastic polymers   used as matrices are usually long-chain polymers of high molecular mass which develop strength at ambient temperatures, even though they remain amorphous, because the bulky units in the chains become entangled with each other, e.g. polycarbonates. Others develop strength by crystallizing partially, e.g. nylon. Many thermoplastics have the advantage over thermosets of greater toughness and also shorter fabrication times (since 'curing' is not involved). Their main disadvantage is that high viscosity at the moulding temperature makes impregnation of the fibres more difficult.

Many thermoplastics can be used as matrices but the most popular are nylon 66, polyethylene terephthalate (PET), polyamideimide (PAI), poly-sulphone (PS) (150–170 °C) and polyether ether ketone (PEEK) (250 °C). (Figures are maximum working temperatures.) PEEK-based composites have good strength and excellent resistance to failure under impact.

Metal matrices

24.50 Since the main object in making a composite is to produce a material of high *specific* strength the most suitable metals are those of low relative density, i.e. aluminium (2.7), magnesium (1.7), titanium (4.5) and their respective alloys. Metal matrix composites (MMCs) are of interest mainly because of the higher temperatures at which some of them can operate (around 500 °C). Higher impact, transverse and shear properties are also important whilst they can be joined by most metallurgical processes. Their main disadvantages are high relative density and difficulty of fabrication.

Mechanical properties of fibre composites

24.60 A composite containing fibres in a single direction will be extremely anisotropic. In fact in a direction transverse to that of the fibres strength may be only 5 per cent or less than that measured in the direction of the fibres. Often the transverse strength is less than that of the matrix itself because of the presence of discontinuities and incomplete bonding between fibres and the matrix material. Nevertheless when measured in the fibre direction many fibre-reinforced composites have greater *specific* strength and *specific* modulus than can be found in other materials.

24.61 Properties of some *uni*-directional composites are shown in Table 24.2 which indicates that, although strengths and stiffnesses of composites are not much different from those of metals, these composites have much higher *specific* properties because their *relative densities* are *lower*. Savings in weight of up to 30 per cent are usually achieved over aluminium alloys.

Table 24.2 *Mechanical properties of some uni-directional fibre composites.*

Material (60% by volume of fibre in epoxy resin matrix)	Relative density	Tensile strength (N mm⁻²) Longitudinal	Transverse	Young's modulus (kN mm⁻²) (longitudinal)	Specific strength (N mm⁻²) (longitudinal)	Specific modulus (kN mm⁻²) (longitudinal)
Aramid	1.35	1600	30	80	1185	59
S-glass	2.0	1800	40	55	900	28
High-strength carbon	1.55	1770	50	174	1140	112
High-modulus carbon	1.63	1100	21	272	675	167
Bulk materials for comparison						
Duralumin	2.7	460	–	69	170	26
High-tensile (Ni-Cr-Mo) steel	7.83	2000	–	210	255	27

Fibre-composite manufacture

24.70   Fibre materials are supplied in the following commercial forms:

- *Rovings*. A 'roving' of glass fibre, which may be several kilometres in length, consists of 'strands', or bundles of filaments wound on to a 'creel'. A 'strand' contains some 200 filaments each about 10 μm in diameter. Bundles of continuous carbon fibres are known as 'tows'.
- *Woven fabrics* in various weave styles.
- *Chopped fibres* usually between 1 mm and 50 mm long.

24.71   Composites are manufactured either as:

1. Continuously produced sections (rod, tube or channel), or sheet, from which required lengths can be cut. Such a process can only produce composites which are anisotropic in their properties, strength being in a direction parallel to the fibre direction.
2. Composites manufactured as individual components. Here the fibre may be woven into a 'preform' which roughly follows the mould or die contours. In this case the mechanical properties will tend to be multidirectional.

The following are typical processes:

24.72   Poltrusion   This process is used in the manufacture of continuous lengths of composites of fixed cross-section – rods, tubes, channel profiles and sheet – but window-framing, railings, building panels and the like are also produced. Most poltrusions are of glass-fibre/polyester composites. Such material provides good corrosion resistance at low cost as well as satisfactory mechanical properties.

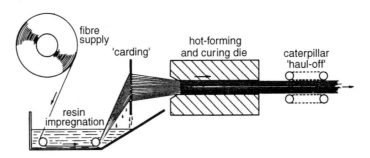

Fig. 24.4   *The principles of the poltrusion process. The reinforcement fibres virtually pull the composite through the system.*

In this process (Fig. 24.4) the fibre material is drawn through the liquid resin, the surplus of which is removed during the 'carding' process. It then passes through the hot-forming die which compacts the composite. Curing then follows this stage. The finished poltrusion is cut into the required lengths as it leaves the 'haul off' unit.

24.73   'Hand-and-spray' placement   probably accounts for the largest proportion of glass-reinforced plastics (GRP) made in Britain at present. It is a labour-intensive process in which, in its simplest form, woven mats of glass-fibre rovings are laid over a mould which has previously been coated with a non-stick agent. The liquid resin is then worked into the fibre material by hand using either a brush or roller. Polyester resins are most commonly used. This type of process is widely used in the manufacture of boats and other pleasure craft but larger vessels such as mine sweepers up to 60 m long – where a non-magnetic hull is required – have been 'hand layered' by this process.

24.74 Press moulding is the most popular process for mass producing fibre-reinforced plastics components. It is similar in principle to that shown in Fig. 20.11, the charge to the mould cavity consisting of fibres and resin.

24.75 Resin-transfer moulding follows the same principle as the process shown in Fig. 20.13. The fibre, usually as a woven pre-form, is placed in the die cavity and the resin then injected.

24.76 Metal matrix composites (MMCs) are relatively difficult to manufacture because high temperatures and/or pressures need to be used. Both may cause damage to the fibre reinforcement. The bulk of MMCs are based on aluminium alloy matrices which are reinforced with fibres of alumina, silicon carbide, carbon or boron. A few low melting-point metals can be gravity die cast to form the matrix but generally it is necessary to use pressure in order to get adequate infiltration of the fibres.

In *powder metallurgy* processes a metal powder is mixed with chopped fibres, pressed to shape and then sintered using basically the method described in 7.40.

Uses of fibre-reinforced composites

24.80 The most important of these materials commercially are polymer matrix composites reinforced with either glass, carbon or aramid fibres. Of these the greatest volume produced consists of GRP, mainly because of its low cost relative to the others.

24.81 Such composites constitute a large proportion of the interior fittings and trim of most airliners along with some parts of their primary structure. Sporting and leisure goods – gliders, boats, sailboards, skis and rackets – are now made largely of fibre-reinforced composites. The plastics bodies of many domestic appliances are reinforced with short chopped fibres incorporated during the moulding process.

24.82 In addition to the aerospace industries large quantities of GRP are used in land transport vehicles in the interests of fuel economy. In motor car manufacture the use of fibre-reinforced plastics continues to grow in the form of panels, bumpers and many interior fittings. Leaf springs made in GRP have good fatigue properties and lead to a saving in weight.

E-glass was developed originally as an electrical insulation material and is now used as a fibre reinforcement for much equipment in the electrical industries – switch casings, cable and distribution cabinets, junction boxes, etc. are manufactured in GRP. Since these materials are also very corrosion-resistant they are used in the chemical industries for storing aggressive chemicals and also in silos, wine vats and pipelines for water and sewage.

Where more sophisticated applications are involved requiring maximum specific strength and specific modulus, either carbon or aramid fibres are used for reinforcement. Aerospace and military aircraft design, helicopter blades and the like employ CFRP.

24.83   The following characteristics of fibre composites commend their use:

- low relative density and hence high specific strength and modulus of elasticity;
- good resistance to corrosion;
- good fatigue resistance, particularly parallel to the fibre direction;
- generally low coefficient of thermal expansion, especially with carbon or aramid fibres.

Reinforced wood products

24.90   The development of strong synthetic resin adhesives some years ago resulted in much progress in the use of timber as a constructional material.

24.91   Laminated wood   Glued lamination of timber ('gluelam') has many advantages over conventional methods of mechanical jointing, such as nailing, screwing, and bolting. Thin boards or planks can be bonded firmly together in parallel fashion, using these strong adhesives, and members be thus manufactured in sizes and shapes which it would be impossible to cut directly from trees. Timber members can be made indefinitely long by gluing boards end to end on a long sloping 'scarf' joint, and laminae can be 'stacked' to produce any necessary thickness and width. Moreover, thin laminae can be bent to provide curved structural members which are extremely useful as long-span arches, to cover churches, concert-halls, gymnasia, and similar large buildings. Arches a hundred metres or more long, with members up to two metres deep, are not uncommon. The main problem is in transporting them, since assembly 'on site' is unsatisfactory.

Adhesives used in glued laminations include casein and urea formaldehyde for interior work, and the more durable resins phenol formaldehyde and resorcinol formaldehyde for timbers exposed to weather, or other damp conditions. All synthetic-resin adhesives, many of which are available from 'do-it-yourself' stores, consist of two parts – the resin and an acid hardener, supplied separately or premixed in powder form. Setting time is often several hours, though special fast-setting hardeners are available.

24.92   Plywood, blockboard and particleboard   An inherent weakness of natural timber is its directionality of properties, as shown by a lack of strength perpendicular to the grain. This fault is largely overcome in laminated board – or plywood, as it is commonly called – by gluing together thin 'veneers' of wood, so that the grain direction in each successive layer is perpendicular to that in the preceding layer. These veneers are produced by turning a steam-softened log in a lathe, so that the log rotates against a peeling knife. Most softwoods and many hardwoods can be peeled successfully, provided that the log is straight and cylindrical.

Plywood varies in thickness from 3 mm to 25 mm. The thinnest material is three-ply, whilst thicker boards are of five-ply or multi-ply construction. An odd number of plies is necessary, so that the grain on the face and the back will run in the same direction, otherwise the product will be unbalanced, and any change in the moisture-content during service may lead to warping.

Blockboard resembles plywood, in that it has veneers on both face and back. The core, however, consists of solid wood strips up to 25 mm wide,

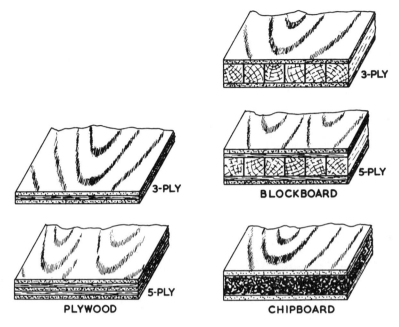

Fig. 24.5 *Plywood, blockboard and chipboard.*

assembled and glued side by side (Fig. 24.5). Adhesives used in the production of these materials are chosen to suit service conditions. Plywood and blockboard for interior use are glued with casein, which is resistant to attack by pests, but which will not withstand moisture for long periods. Consequently, plywood and blockboard destined for exterior use are bonded with urea-formaldehyde or melamine-formaldehyde resins. Better still, a mixture of phenol formaldehyde and resorcinol-formaldehyde is used to produce material which is 'weather- and boil-proof'.

Another important modern introduction in timber sheet material is *particleboard* or *chipboard*. This consists of a mass of wood chips, ranging from coarse sawdust to flat shavings, bonded with a resin, and sandwiched between veneers. Here the strength is dependent mainly on the resin bond, but there is no directionality of properties, since the particles constituting the wood filling are orientated at random.

24.93 Corrugated cardboard is derived from wood pulp and is a laminated product. It consists of corrugated paper sandwiched between outer layers of paper (Fig. 24.6). This introduces increased stiffness along

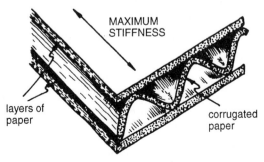

Fig. 24.6 *The structure of corrugated cardboard.*

the direction parallel to the corrugation ridges. In fact corrugated cardboard is stiff for the same reason as is a RSJ which consists of two load-bearing *flanges* separated by the connecting *web*. The latter ensures that the flanges are at appropriate distances from the neutral axis of the joist. Corrugated cardboard is still widely used as a packaging material for a large amount of domestic and other equipment, often in conjunction with foamed polystyrene (20.45). The latter is much lighter and is stiff enough to offer good protection against mechanical shock. Foamed PS is, in a sense, a composite consisting of a mass of air bubbles cemented by thin films of PS – which brings to mind a schoolboy 'howler' of my childhood describing a fishing net as 'a lot of holes tied together with string'.

Reinforced concrete

24.100 In common with many ceramic materials, concrete is stronger in compression than in tension. For structural members such as beams, where stresses are both compressive and tensile, the use of plain concrete would be most uneconomical, as indicated in Fig. 24.7 (i) and (ii). In reinforced concrete, the tensile forces are carried by fine steel rods. Full advantage can then be taken of the high compressive strength of the concrete, and the cross-section can be reduced accordingly. The rods are so shaped (Fig. 24.7 (iii)) that they are firmly gripped by the rigid concrete. Thus, reinforcement allows concrete members to be used in situations where plain concrete of adequate strength would be unsuitable because of the bulk of material necessary.

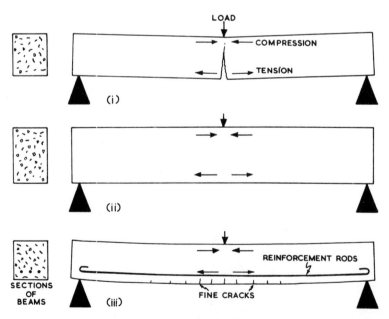

Fig. 24.7 *The advantage of reinforcement in concrete. In (i), a plain concrete beam fails at the edge which is in tension. A beam strong enough in this respect (ii) will be uneconomically bulky. In (iii), steel reinforcement rods support the portion of the beam which is in tension.*

I began this chapter by indicating how the principles of fibre-reinforced composite structures are based on those which have existed in trees for millions of years. Perhaps it is appropriate to finish with one of the early fibre-reinforced composites to be developed by Man – reinforced concrete. However I am reminded of Pharoah's directive to the taskmasters of the Israelite slaves some 3400 years earlier: '*Ye shall no more give the people straw to make brick, as heretofore; let them go and gather straw for themselves*' (Exodus V, 7). In those days bricks made from sun-dried mud were strengthened by the incorporation of straws. One may still see such bricks in use in some of the more remote mountain regions of the Balkans.

CHAPTER 25

The environmental deterioration of materials

25.10 Sooner or later all engineering materials – whether as components of the Pyramids or the family 'tin lizzie' – decay in their surrounding environments. Much of this degradation is due to chemical reactions typified by the 'rusting' of steel and the corrosion of metals in general. Many plastics materials also degenerate when in contact with the environment. Such 'weathering' is due in some degree to simple oxidation but, more seriously, to the action of ultra-violet radiation which of course can have disastrous effects on our own *epidermis* if we are foolish enough to court a suntan in the belief that it makes us appear more glamorous!

Generally, attack by living organisms, whether from the animal or botanical kingdoms, is confined to those materials used by Man which are part of the biological world. Thus timber is attacked by both insects and fungi. Rarely rodents will chew their way through man-made plastics materials, whilst it was reported recently that a pair of woodpeckers had set up home in one of America's latest space projects by drilling a neat hole in it.

The corrosion of metals

25.20 Probably the reader's first serious confrontation with the problem of corrosion was when he became the proud possessor of his first second-hand motor car. What left the dealer's show-room as a 'mint-condition' vehicle began to show those ubiquitous rust spots within a few weeks. It is a melancholy reflection that, whilst the engine and any other worn mechanical parts of a motor car can be replaced almost indefinitely, once the body – which is also the 'chassis' – of a modern car begins to deteriorate badly, little can be done save consign the car to the scrap yard.[1]

25.21 However, the corrosion of steel is a problem which in some degree must be faced by all of us, and it is estimated that the annual cost of the fight against corrosion on a world-wide basis runs into some thousands of millions of pounds. Other engineering metals corrode when exposed to the atmosphere, though generally to a lesser extent than do iron and steel. Because of their high resistance to all forms of corrosion, plastics are replacing metals for many applications where corrosion-resistance is more important than mechanical strength.

[1] Since this was written some quarter of a century ago there has been a considerable improvement in the durability of the bodywork of the average motor car, but one wonders how much this has been due to the very serious competition from manufacturers in the Far East.

Chemical corrosion

25.30 Metals may corrode by a process which we can describe as simple chemical attack. Thus oxygen, ever present in the atmosphere, can combine with some metals to form a film of oxide on the surface. If this film is porous, or if it rubs off easily, the process of oxidation can continue, and the metal will gradually corrode away. Aluminium oxidizes very easily, but fortunately the thin oxide film so formed is very dense, and sticks tightly to the surface, thus effectively protecting the metal beneath. Other metals, notably titanium, tantalum and chromium, 'seal' themselves from further attack in a similar way.

At higher temperatures the process of oxidation takes place more rapidly and as the temperature increases so does the rate of oxidation. Thus iron oxidizes – or scales – readily at temperatures above 650 °C. This process of oxidation involves a transfer of electrons from the metallic atom to the oxygen atom (1.21) so that positively charged metallic ions and negatively charged oxygen ions are formed. Since these oppositely charged ions attract each other a crystalline oxide layer is formed on the metal surface. Often this scale is not coherent with the metal surface so that it flakes away exposing the metal to further oxidation.

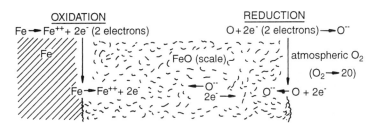

Fig. 25.1   *Reactions at and near the surface of iron heated in contact with the atmosphere.*

Both metal ions and oxygen ions are able to diffuse through the scale and so thicken the oxide layer. When the crystal lattice of this layer is not coherent – that is, it does not 'match up' – with the crystal lattice of the metal beneath, it flakes away so that the metal surface is exposed to more rapid oxidation. If the oxide layer is coherent – as with aluminium, titanium, etc. – the metal is protected by this closely adherent oxide layer.

25.31 Some metals, notably nickel, are readily attacked at high temperatures by gases containing sulphur. Therefore heat-resisting steels used in the presence of sulphurous vapours should not contain large amounts of nickel. Although in this case oxygen itself is not involved in the chemical attack, nickel is said to have been 'oxidized'. In fact the term 'oxidation' is applied in chemical jargon to any reaction in which metallic atoms lose electrons and so become ions.

25.32 The rusting of iron and steel is not a case of simple oxidation, and is associated with the presence of both air *and moisture*. Iron will not rust in dry air, nor will it rust in pure water; but, when air and moisture are present together, iron, and particularly steel, will begin to rust very quickly. Rusting continues unabated, because the layer of corrosion product formed is loose and porous, so that a fresh film of rust will form beneath and lift upwards the upper layer.

Electrolytic corrosion

25.40   Essentially, this is also a form of chemical corrosion, though a little more complex in its origin than the simple chemical corrosion described above.

25.41   Most readers will be familiar with the principle of a simple electric cell. It consists of a plate of copper and a plate of zinc, both of which are immersed in dilute sulphuric acid. If the plates are not touching each other in the solution, and are not connected to each other outside the solution, then no action takes place; but, as soon as they are connected (Fig. 25.2), a current of electricity, sufficient to light a small bulb, flows through the completed circuit. At the same time, bubbles of hydrogen form at the copper plate, whilst the zinc plate begins to dissolve in the acid, to form a salt, zinc sulphate. In this way, the 'chemical potential energy' of the zinc is being converted to electrical energy.

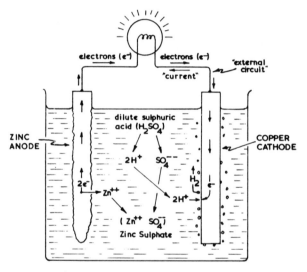

Fig. 25.2   *The 'corrosion' of zinc in a simple voltaic cell. In this and the following diagrams e*⁻ *denotes electrons.*

As the circut is closed atoms on the surface of the zinc plate begin to form ions (Zn^{++}), each zinc atom losing two electrons. These electrons pass round the external circuit to the copper plate which as a result becomes negatively charged. Since hydrogen ions (H^+) are present as a result of the ionization[2] of dilute sulphuric acid these will be attracted towards the negatively charged copper plate along with the positively charged zinc ions (Zn^{++}). However the hydrogen ions have a greater attraction for electrons than do the zinc ions so that hydrogen ions combine with these electrons to form hydrogen atoms which immediately combine forming hydrogen molecules (H_2):

$$2H^+ + 2e^- \longrightarrow 2H \longrightarrow H_2$$

Hence small bubbles of hydrogen gas form on the copper plate. In the meantime the Zn^{++} ions 'pair up' with the SO_4^{--} ions (from the sulphuric

---

[2]   All mineral acids and salts *ionize* to some degree on contact with water. That is, they go into solution as ions. Thus sulphuric acid (H_2SO_4) forms hydrogen ions (H^+) and 'sulphate' ions (SO_4^{--}).

acid) to form zinc sulphate. The 'current of electricity' which causes the bulb to light up is the stream of liberated electrons from the dissolving zinc. These electrons lose energy – as heat – as they force their way through the bulb filament.

The dilute sulphuric acid is termed an *electrolyte*. An electrolyte is any substance which in the liquid state – or in this case as a solution in water – contains separate positive and negative ions. The crystals of lithium fluoride mentioned in 1.21 will, as they dissolve in water, separate into Li^+ and F^- ions and so constitute an electrolyte. Similarly if lithium fluoride crystals were heated to 848 °C they would melt and Li^+ and F^- ions would separate forming an electrolyte. Any electrolyte will 'conduct electricity' in the manner described above.

25.42   If the reader examines Fig. 25.2, he will notice that the terminology used is opposite to that which may have been used in his school science lessons. The usual electrical terminology shows the zinc plate as being the negative pole (or cathode), whilst the copper plate is shown as being the positive pole (or anode). Similarly, the current is usually shown as flowing from copper to zinc in the external circuit, and not from zinc to copper as indicated in Fig. 25.2. This apparent confusion arises from the fact that, in the early days of electricity, the true nature of an electric current was not really understood. It was thought that 'positive' electricity flowed from copper to zinc in the external circuit; whereas we now know that an electric current consists of a stream of *negatively charged* particles (electrons) flowing from zinc to copper.

However, the important point to remember is that zinc is anodic towards copper; so that, when these metals are connected and immersed in any electrolyte, the zinc will dissolve – or corrode – far more quickly than if immersed in the electrolyte by itself. This phenomenon of electrolytic action is not confined to copper and zinc but will apply to any pair of metals, one of which will always be anodic to the other. Electrolytic action occurs as a result of a difference in *electrode potential* between two metals. This electrode potential is related to the amount of energy required to remove the outer-shell electrons from the atom (1.20). Thus zinc atoms lose their outer-shell electrons more easily than do copper atoms and so zinc is said to be *anodic* to copper. The electrode which supplies electrons to the external circuit is called the *anode* whilst the electrode which receives electrons from the external circuit is called the *cathode*.

In the Electrochemical (or Galvanic) Series (Table 25.1) any metal in the table will replace from solution any metal above it. This explains why in the simple cell described above, hydrogen ions are discharged from the electrolyte instead of zinc ions. Similarly, if a steel (iron) pen-knife blade is immersed in a copper sulphate solution it becomes coated with metallic copper because iron atoms have formed ions to replace the copper ions in solution. As a result electrons pass from the iron atoms to the copper ions which are then deposited as copper atoms. This happens because iron is anodic to copper as indicated by their relative positions in Table 25.1.

It follows that any pair of metals in the table when immersed in an electrolyte and also in electrical contact with each other will form a voltaic cell, and the metal which is lower in the table will be anodic to the other and will go into solution as its atoms form ions and release electrons into the external circuit. The further apart these metals are in the table the greater the electrode potential between them and the greater the tendency of the anode to dissolve or corrode.

Table 25.1   *The Electrochemical (Galvanic) Series.*

Metal (ion)		Electrode potential (volts)	
(Noble metals)	Gold (Au^{+++})	+1.5	Cathodic
	Silver (Ag^{+})	+0.8	↑
	Copper (Cu^{+})	+0.52	
	(Cu^{++})	+0.34	
	Hydrogen (H^{+})	0.00	(Reference)
	Iron (Fe^{+++})	−0.05	
	Lead (Pb^{++})	−0.13	
	Tin (Sn^{++})	−0.14	
	Nickel (Ni^{++})	−0.25	
	Iron (Fe^{++})	−0.44	
	Chromium (Cr^{+++})	−0.74	
	Zinc (Zn^{++})	−0.76	
	Aluminium (Al^{+++})	−1.66	
	Magnesium (Mg^{++})	−2.37	↓
(Base metals)	Lithium (Li^{+})	−3.04	Anodic

(Some metals e.g. copper and iron, form more than one type of ion, depending upon the number of electrons lost by the atom under the electrochemical conditions prevailing.)

Often the rate of electrolytic action is extremely slow and consequently the flow of current (i.e. electrons) between the two metals so small as to be undetected. Nevertheless electrolytic action will be taking place and will lead to the accelerated corrosion of that member of the pair which is anodic to the other. *Pure* water is only very slightly ionized but in industrial regions rain water dissolves sulphur dioxide from the atmosphere to form sulphurous acid giving rise to the 'acid rain' we hear so much about these days. This dilute acid is very highly ionized and so forms an electrolyte which allows corrosion to proceed more rapidly.

25.43   If the reader has been long engaged in practical engineering, he may have learned that it is considered bad practice to use two dissimilar metals in close proximity to each other and in the presence of even such a weak electrolyte as rain water. In spite of this, many examples of such bad practice are encountered, especially in domestic plumbing. Figure 25.3 (i) illustrates such a case. Here a section of mild steel pipe has been attached to one of copper, the system being used to carry water. Since the mild steel will be anodic to copper, it will rust far *more quickly than if the pipe were of mild steel throughout*. Rusting will be the more severe where it joins the copper pipe as Fe^{+++} ions go into solution more quickly there. These Fe^{+++} ions in solution will combine with hydroxyl ions (OH^{-}), formed by the ionization of water, and produce ferric hydroxide which is the basis of the reddish brown deposit we call *rust*.

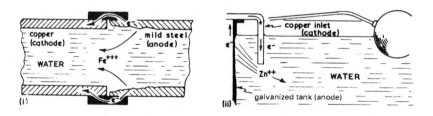

Fig. 25.3   *Examples of bad plumbing practice.*

25.44 A further example of bad practice was all too frequently encountered in domestic installations (Fig. 25.3 (ii)). Here the inlet pipe was invariably of copper whilst the storage tank was of galvanized (zinc coated) mild steel. Since most mains water contains sufficient dissolved salts to make it an electrolyte the inevitable corrosion took place. As a result the zinc coating in an area adjacent to the copper inlet pipe went into solution as Zn^{++} ions and the exposed mild steel, also anodic to copper, then dissolved rapidly as a result of the close proximity of the cathodic copper pipe. The disastrous leak which followed generally occurred during the annual holiday or on Christmas Day, and always just after the ground-floor rooms had been decorated. Enlightened plumbers used inlet pipes of inert plastics materials. Fortunately modern domestic installations use plastics tanks which are free from such problems, but if your house was built before 1970 take a quick look at the 'cold-water' tank. If it's of galvanized iron – start worrying!

25.45 Many pure metals have a good resistance to atmospheric corrosion. Unfortunately, these metals are usually expensive, and many of them are mechanically weak. However, a thin coating of one of these metals can often be used to protect mild steel. Pure tin has an excellent resistance to corrosion, not only by the atmosphere and by water, but by very many other liquids and solutions. Hence tinplate – that is, mild-steel sheet with a thin coating of tin – is widely used in the canning industry. Figure 25.4 illustrates what happens if a tin coating on mild steel becomes scratched. The mild steel

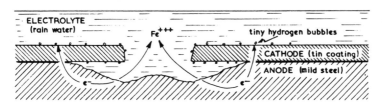

Fig. 25.4 *The corrosion of scratched tinplate.*

is anodic to tin and so it will go into solution as Fe^{+++} ions more quickly in the area exposed by the scratch than if the *tin coating were not there at all*. The electrons released by the ionization of iron atoms travel to the cathode (the tin coating) where they attract hydrogen ions (H^+) from the ionized water forming hydrogen molecules which may appear as minute bubbles but are more likely to remain dissolved in the water. In the meantime Fe^{+++} ions combine with hydroxyl ions (OH^-) also present in the ionized water, leading to the formation of rust. Consequently to be of any use in protecting mild steel a tin coating must be absolutely sound and unbroken.

25.46 Fig. 25.5 illustrates the effect of a zinc coating in similar circumstances. Here the zinc is anodic to mild steel so in this case it is the zinc which goes into solution (as Zn^{++} ions) in preference to the iron (mild

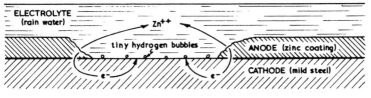

Fig. 25.5 *The 'sacrificial protection' offered by zinc to mild steel.*

to accelerated electrolytic corrosion of the more heavily cold-worked area. Similarly the presence of a *non-uniform electrolyte* can also cause electrolytic action in a material of uniform composition and treatment.

Consider the case of flaking paintwork on a motor car (Fig. 25.9 (i)). Rain water can seep beneath the loose paint film and since it becomes thus isolated from the atmosphere it will inevitably contain less dissolved atmospheric oxygen, as time goes on, than that water which remains exposed to the atmosphere. The composition of the electrolyte is therefore non-uniform and that metal covered by the oxygen-depleted water becomes anodic to the metal covered by oxygen-rich water. Thus Fe^{+++} ions go into solution more quickly *beneath* the paintwork leading to the build up of corrosion products there. This is why blisters form and ultimately push off the paint film.

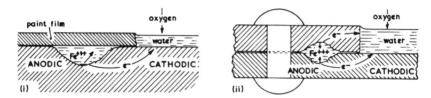

Fig. 25.9   *Corrosion due to a non-uniform electrolyte.*

This type of corrosion can also be caused by faults in design. Consider two steel plates of similar composition rivetted as shown (Fig. 25.9 (ii)) with rivets also of composition similar to that of the plates. Water seeping into the fissure between the plates will become depleted in oxygen as compared with water which is in continuous contact with the atmosphere. Thus non-uniformity in the composition of the electrolyte will cause metal within the fissure to become anodic to the remainder and corrosion will take place there. This is generally termed *crevice corrosion*. Piles of leaves, sand or other forms of inert sediment collecting at the bottom of a metal water storage tank will cause corrosion of the metal beneath in a similar manner.

The protection of metal surfaces

25.50   The corrosion of carbon steel is due partly at least to electrolytic action between different phases in the structure of the steel. Pearlite consists of microscopically thin layers of ferrite and cementite, arranged alternately in the structure (11.43). Ferrite is anodic to cementite, and so it is this ferrite which corrodes away, leaving the cementite layers standing proud. Being very brittle, these cementite layers soon break away.

In order to protect the surface of steel, therefore, it must be coated with some impervious layer which will form a mechanical barrier against any electrolyte which is likely to come into contact with its surface. In all cases, the surface to be coated must be absolutely clean and rust free.

25.51   Painting   This is used to coat vast amounts of mild steel, not only to protect it against corrosion by the atmosphere, but to provide an attractive finish. Optimum results are obtained by first 'phosphating' the surface of the steel. This involves treating it with a phosphoric acid preparation, which not only dissolves rust, but also coats the surface of the steel with a dense and slightly rough surface of iron phosphate. This affords some protection against corrosion, but also acts as an excellent 'key' for the priming paint and the undercoat of subsequent paint.

In the automobile industry undercoat paint is applied using a process called *electrophoresis* or 'electro-painting'. The articles to be painted are made the anode in a bath which contains the paint 'resin' suspended as particles in a soap solution in water. The negatively charged ions from the dissolved soap become attached to the 'resin' particles so that the negatively charged combination is then attracted to the anode (the article being painted). A very uniform paint film is produced because the charged particles 'seek out' any areas not already insulated by paint.

25.52 Stove-enamelling This finish is used to provide a hard-wearing corrosion-proof coating for many domestic appliances, such as washing-machines, refrigerators, and cooking-stoves.

25.53 Coating the surface with another metal A thin coating of a corrosion-resistant metal can be applied to one which is less corrosion-resistant, in order to protect it. The aim is *always* to provide a *mechanical* barrier against possible electrolytes or corrosive atmospheres, but it must be remembered that, whilst zinc and aluminium will offer sacrificial protection should coatings of these metals become damaged, the presence of damaged coatings of most other metals will *accelerate* corrosion.

The metallic coating can be applied in a number of different ways.

25.53.1 *Hot-dipping* can be used to coat the surface of iron and steel components with both tin and zinc. Tin plate is still manufactured in South Wales, where the industry was established some three hundred years ago. Clean mild-steel sheets are passed through a bath of molten tin, and then through squeeze rolls, which remove the surplus tin.

Galvanizing is a similar process, whereby articles are coated with zinc. Buckets, dustbins, wheelbarrows, cold-water tanks, and barbed-wire are all coated by immersion in molten zinc. Many of these galvanized items are now largely superseded by plastics materials products, though barbed-wire festoons the British countryside in increasing amounts as it replaces hedgerows either grubbed-out or systematically killed off by the mechanical flail.

25.53.2 *Spraying* can be employed to coat surfaces with a wide range of molten metals, though zinc is most often used. In the Schoop process, an arc is struck between two zinc wires within the spray gun, and the molten metal so produced is carried forward in an air blast. This type of process is useful for coating structures *in situ*, and was employed in the protection of the Forth road-bridge.

25.53.3 *Sherardizing* is a 'cementation' process, similar in principle to carburizing. Steel components are heated in a rotating drum containing some zinc powder at about 370 °C. A very thin but uniform layer of zinc is deposited on the surface of the components. It is an ideal method for treating nuts and bolts, the threads of which would become clogged during ordinary hot-dip galvanizing.

25.53.4 *Electroplating* is used to deposit a large number of metals on to both metallic and non-metallic surfaces. Gold, silver, nickel, chromium, copper, cadmium, tin, zinc, and some alloys can be deposited in this way. Electroplating is a relatively expensive process, but provides a very uniform surface layer of very high quality, since accurate control of the process is possible at all stages. Moreover, heating of the component being coated is not involved, so that there is no risk of destruction of mechanical properties which may have been developed by previous heat-treatment.

25.53.5 *Cladding* is applicable mainly to the manufacture of 'clad' sheet. The basis metal is sandwiched between sheets of the coating metal, and the sandwich is then rolled to the required thickness. During the process, the coating film welds on to the base metal. 'Alclad', which is duralumin coated with pure aluminium, is the best known of these products.

25.54   Protection by oxide coatings In some cases the film of oxide which forms on the surface of a metal is very dense and closely adherent. It then protects the metal beneath from further oxidation. The 'blueing' of ordinary carbon steel during tempering produces an oxide film which offers some protection against corrosion.

25.54.1 *Anodizing* is applied to suitable alloys of aluminium, in order to give them added protection against corrosion. The natural oxide film on the surface of these materials is an excellent barrier to further oxidation, and, in the anodizing process, this film is thickened by making the article the anode in an electrolytic bath. As current is passed through the bath, atoms of oxygen are liberated at the surface of the article, and these immediately combine with the aluminium, thus thickening the natural oxide film. Since aluminium oxide (alumina) is extremely hard, this film is also wear-resistant; it is also thick enough to enable it to be dyed an attractive colour.

Metals and alloys which are inherently corrosion-resistant

25.60   Stainless steels are resistant to corrosion partly because the tenacious chromium oxide film which coats the surface behaves in much the same way as does the oxide film on the surface of aluminium. They are also corrosion-resistant, because of the uniform structure (13.51) which is generally present in such steels. If a structure consists of crystals which are *all of the same composition*, then there can be no electrolytic action between them, as there is, for example, between the ferrite and cementite in the structure of an ordinary carbon steel.

Pure metals are generally corrosion-resistant for the same reason; though, of course, particles of impurity present at crystal boundaries can give rise to intercrystalline corrosion. For this reason, extremely pure metals have very high resistances to corrosion. This applies particularly to iron and aluminium, which can be obtained as much as 99.9999 per cent pure. Unfortunately, such metals are generally so expensive in this state of purity that their everyday use is not possible.

Galvanic protection

25.70   The sacrificial protection of a ship's steel hull from electrolytic corrosion, which would be accelerated by the presence of a 'manganese bronze' propellor near by, has been described. In a similar way underground steel pipelines can be protected from corrosion by burying slabs of zinc (or magnesium) near to the pipe at suitable intervals. Alternatively a small EMF can be used (Fig. 25.10) making the pipe *cathodic* to its surroundings. Either a battery or a low-voltage dc generator (operated from a small low-maintenance windmill) can be used to supply the EMF.

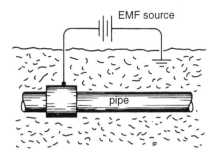

Fig. 25.10   *Using an impressed EMF to protect a steel pipeline from corrosion.*

The stability of plastics materials

25.80   Those metals which corrode in contact with the atmosphere do so as their atoms lose electrons and the ions so formed go into solution in the electrolyte (moisture on the surface of the metal). Many plastics materials are relatively inert and will resist chemical attack by those reagents which would lead to the severe corrosion of most engineering metals. Thus polythene is unaffected by prolonged contact with concentrated acids (including hydrofluoric acid which readily dissolves glass) yet it is not completely stable when exposed to outdoor atmospheres, tending to become opaque and brittle with the passage of time. Similarly rubber will 'perish', that is become brittle and useless after varying periods of time unless steps have been taken to stabilize it.

25.81   Weathering of plastics materials   Almost all polymers, but in particular thermoplastics materials, deteriorate in appearance at a significant rate when exposed to the atmosphere unless they have been 'stabilized'. This deterioration in appearance coupled with an increase in brittleness continues to the extent that some plastics eventually become useless. At ambient temperatures this weathering process appears to be due to the combined effects of oxygen and ultra-violet (UV) light, the effects of oxygen by itself being negligible. Most polymers absorb UV light and the energy held by this radiation excites the polymer molecules causing them to vibrate so that chemical bonds break and chains thus shorten. At the same time the presence of oxygen leads to the formation of other chemical groups and possibly cross-links. Both events will reduce flexibility and strength of the polymer whilst increasing its brittleness.

The simplest method by which plastics can be protected from weathering is to include some substance in the original moulding mixture which will *screen* the material from UV radiation by making it opaque. 'Carbon black' and other pigments will absorb UV radiation and so act as an effective screen. A large number of organic compounds are used as compatible absorbers of UV radiation when transparency to visible wavelengths of light is required in the polymer. Their function is to absorb UV radiation without decomposing. A further type of UV absorber converts UV radiation into light of longer wavelength which has no deleterious effect. Such a substance is hydroxyphenylbenzotriazole.

Other stabilizers and anti-oxidants may be added. Thus 0.1 to 0.2 per cent phenol or an amine will effectively absorb oxygen and so prevent it from reacting with the polymer chains.

Other forms of short-wavelength radiation such as γ-rays or X-rays will cause degradation – or *scission* – of polymer molecules. High-speed particles such as electrons and neutrons may have a similar effect. Scission refers to a breaking of the linear 'backbone' chains. As might be expected strength is reduced because the sum total of van der Waals forces holding molecules together will be less if these molecular chains are now shorter.

Weathering is of course more serious with polymer materials exposed to intense UV radiation outdoors, but often the cost of stabilizing the material must be balanced against the cost of frequent replacement. Weathering is inevitable with materials like the thin LDPE sheet used for covering horticultural greenhouses. Such sheet will not last for more than two seasons but the overall running cost of replacement is lower than it would be using a glass house which would, in addition, require a stronger frame and still involve some maintenance.

The phenomenon of weathering was previously known as 'ageing' but this term is now generally used to describe an 'annealing' process whereby the structure of a polymer attains equilibrium and, consequently, stability in mechanical properties. For example a PVC moulding which has been cooled rapidly may have a tensile strength of 40 N mm^{-2} but if immediately 'aged' (annealed) at 65 °C the tensile strength will be increased to 60 N mm^{-2}. 'Natural ageing' at ambient temperature (20 °C) would achieve a similar result but only after a number of years.

25.82 Perishing of rubbers Both natural rubber and some synthetic elastomers such as SBR are prone to attack by ozone in the atmosphere. Ozone, an allotrope of oxygen in which the molecule contains three atoms of oxygen, is much more reactive than 'ordinary' oxygen in which the molecule contains two atoms, and can provide oxygen atoms which form cross-links between rubber molecules at points not used up during the vulcanization process. At the same time some degradation (scission) of the linear chain results. The extra cross-linking reduces elasticity whilst scission reduces strength so that the rubber becomes both weak and brittle.

Fig. 25.11   *The perishing of rubber. (i) Original sulphur (S) cross-links only; (ii) scission of linear molecules and also extra oxygen (O) cross-links.*

In rubber for car-tyre production the moulding mixture contains about 3.5 per cent amines added as anti-oxidants and anti-ozonants. The effect of both oxygen and ozone in promoting perishing of rubber is accelerated by UV radiation but the presence of some 30 per cent carbon black in the mix acts as a screen to UV penetration.

25.83 Stress cracking and crazing of polymers Some active chemical reagents have an adverse effect on the surface finish of some plastics materials when the latter are in a state of mechanical stress. Thus water solutions of detergents can produce brittle cracking in polythene utensils which contain residual stresses from the moulding process. *Environmental stress cracking*, as it is termed, is therefore a result of the combined effects of stress and chemical attack. The surface cracking of stressed PVC gas pipes takes place in the presence of some hydrocarbon impurities, whilst some organic liquids and gases may also promote the formation of a network of minute cracks (crazes) in amorphous polymers such as clear polythene. Although such crazed material has a poor visual appearance it generally retains its strength quite well. Nevertheless there is the danger that these minute surface cracks may act as stress-raisers and so promote ultimate failure.

25.84   Stability to solvents   Some plastics materials dissolve in or absorb water. Nylon, for example, will absorb considerable amounts of water either when immersed or held in a humid atmosphere. Unexpectedly this 'wet nylon' has a much higher impact value than does 'dry nylon'. Most of the common plastics, however, are insoluble in water and the old practical rule that *like dissolves like* applies (lubricating oil – an organic compound – will not dissolve in water but it will dissolve in many organic solvents such as acetone, carbon tetrachloride and many others).

Solution occurs when the van der Waals forces operating between the solvent molecules and the polymer molecules are greater than the van der Waals forces operating between the adjacent polymer molecules within the solid polymer. Individual polymer molecules are thus pulled away from the solid and so go into solution. Thermosetting and other cross-linked polymers cannot be dissolved in this way because the van der Waals forces exerted by the solvent molecules are much too small to break the strong covalent cross-links holding the polymer chains together. For this reason a cross-linked polymer does not dissolve or lose its original overall shape. Instead it may absorb molecules of the solvent into itself because of the action of van der Waals forces. This often causes a polymer to swell considerably. Some rubbers (cross-linked elastomers) swell when in contact with certain organic solvents though they are impervious to water.

Many thermoplastics polymers are dissolved by various organic solvents. Conversely the polymer will absorb these solvents and small quantities may lead to a softening of the polymer. This can often be utilized in *plasticization*, as an important aid in polymer technology but is an unwelcome property in that it limits the range of organic liquids with which many polymers can be permitted to make contact.

The preservation of timber

25.90   In 1628, the Swedish warship *Wasa* capsized and sank at the start of her maiden voyage. Until 1961, she remained at the bottom of Stockholm harbour, when it was discovered that she was in remarkably good condition, and was subsequently raised. This great oak ship, with all its beautifully carved decorations and wooden sculptures, has since had a museum built around it. The excellent state of preservation was explained by the fact that the water in which the wreck lay was too saline to support freshwater organisms, but not sufficiently salt to allow 'sea-worms' to survive. Thus, due to the absence of any predatory organisms, the wood did not decay.

Metals are subject to corrosion by fairly simple chemical attack in the presence of moisture, but timber, which is obtained from a living organism, is attacked by other living organisms – insects and fungi. Normally, when a tree dies, it is, in the course of time, reduced to its original chemical ingredients by the scavengers of the forest – boring insects, fungi, and bacteria. The chemical ingredients return to the soil, and so the cycle of nature repeats itself. When timber is used for constructional purposes, these scavengers are regarded as enemies, and steps must be taken to repel them.

25.91 Insect pests In Britain, various types of beetle constitute the most important insects which attack wood. In each case, the female lays her eggs on or just beneath the surface of the wood, and the larvae (or grubs) which ultimately hatch eat their way through the timber until, considerably grown, they become dormant, and form pupae or chrysalids. Eventually, the adult beetles emerge, find their way to the surface via bore-holes, and the life-cycle repeats itself.

The *common furniture-beetle* attacks both softwoods and hardwoods, but seems to prefer old furniture. The *death-watch beetle*, on the other hand, generally attacks hardwoods, and is most frequently found in the roof timbers of old churches. It thrives in damp situations arising from a lack of ventilation, and derives its name from the tapping sound it makes during the mating season – a not unusual behaviour pattern among certain higher forms of life.

These two are the most common insect pests which attack wood, though there are several others encountered less frequently. They often reveal their presence by the little mounds of wood powder scattered around the scenes of their crimes, though the death-watch beetle differs from the others in that its larvae produce little bun-shaped pellets. If timber is badly infected, the only sure way to eradicate the pest is to cut out such timber and burn it. A number of proprietary preservatives are available to ward off such attack. To be effective, such preservatives should be permanent in their effect, poisonous to both insects and fungi, but non-toxic to human beings and animals.

25.92 Fungus attack A very large number of fungi and mould (or microfungi) attack timber in their search for food and lodgings. However, the best known are those which cause the effects generally known as *dry-rot* and *wet-rot*.

The fungus which causes dry-rot thrives in damp, poorly ventilated situations where a temperature between 16 °C and 20 °C prevails. It consists of a somewhat disgusting, dark, feathery mass, with branching tendrils, and is able to penetrate brickwork. An unpleasant mouldy smell generally prevails when dry-rot is present. The timber becomes discoloured, and develops a dry, shrunken appearance. This is the origin of the term 'dry-rot', but it must be emphasized that the fungus which causes it operates only in damp situations.

The fungi responsible for wet-rot cause internal rotting of the wood, and thrive only in very wet conditions. A pale green scum first appears, but this soon changes to dark brown, and ultimately to black.

Whilst these forms of fungal attack occur most frequently in the basements and ground floors of buildings, they may occur elsewhere, if the combination of dampness and poor ventilation produce conditions under which they can thrive. All infected timber must be cut out and burnt, and the surviving sound wood treated with a preservative.

25.93 Wood-preservatives fall into three main groups:

• tar-oil derivatives (e.g. creosote) – applied by brushing, dipping, or by pressure;

- water-soluble materials – applied by pressure;
- materials in organic solvents – applied by brushing, dipping, or spraying.

Creosote is probably the most widely used wood preservative. It is both economical and effective, and is suitable for the treatment of fencing, railway-sleepers, marine timbers, telegraph-poles, and wooden buildings. It does not attack metals, and is not washed out by rain. Its main disadvantage is that it is inflammable. Inhalation of the vapours as well as contact with the skin should be avoided with many of these organic preservatives as they are thought to be carcinogenic.

A very effective preservative was tri-butyl tin or 'TBT' as it was dubbed by BBC newsreaders (they being reluctant to risk the mispronunciation one has come to expect by these people of the names of even simple organic compounds). Tri-butyl tin was widely used for the preservation of the wooden hulls of yachts until it was found to be entering our food chain via oyster beds sited near to pleasure-craft marinas. It was promptly withdrawn and replaced by a solvent-based organic compound, acypetacs-zinc, considered to be less toxic.

CHAPTER 26

Methods of joining materials

26.10 Metal welding, as carried out by the blacksmith, is without doubt one of the most ancient of metal-working processes; yet the modern technology of welding has undergone revolutionary change during the last two or three decades. In many instances where expensive riveting processes were once employed, steel sheets are now successfully joined by welding. The fabrication of the American 'Liberty' ships, during the Second World War, helped to establish this technique.

26.11 Many methods are used for joining materials. Wood is screwed or nailed, as well as being joined by glue. For many purposes, metals are still riveted, or joined by bolts and nuts. However, in this chapter we shall consider only those methods of joining materials where a 'continuous' joint is employed; that is, adhesion is produced by forces of attraction acting between the fundamental particles of the materials involved.

Glues and adhesives

26.20 Adhesion signifies the joining together of two dissimilar materials, and is dependent upon forces of attraction which operate between molecules in the surface of the adhesive and molecules in the surfaces of the materials being joined. Generally speaking, the larger the molecules involved, the greater are these forces of attraction, and hence the greater the adhesion. Consequently, most adhesives are organic compounds, being composed of very large and complex molecules.

26.21 Another important feature is that a successful adhesive must mould itself perfectly to the surfaces being joined, in order that its molecules will remain in close contact with the molecules of the surfaces being joined. Adhesives bond two 'substrates'[1] together either by chemical attraction for the surfaces involved, or by a mechanical intertwining action, by which the adhesive is carried *into* the two substrates, and sets as a tough film. Adhesives used in industry generally work on a combination of both methods, and they normally consist of solutions or suspensions. In these forms, the adhesive material can conveniently be applied to a substrate. Such adhesives are 'cured by solvent loss', that is, they harden as the water (or other solvent) evaporates or is absorbed by the substrates. Some of the solvents employed

[1] This term signifies the surface and those regions near to the surface of the work-piece.

have become notorious by their abuse in 'glue sniffing'. Direct heat is sometimes used to accelerate solvent loss. Unfortunately, this means that, to heat the glue-line, the whole assembly must be heated, and then cooled before further handling; a decrease in production-rate would result.

Radio-frequency waves are a means of applying heat only to the point where it is required, namely the glue-line. By means of very high-frequency vibrations acting on the molecules of the adhesive, and trying to vibrate them, considerable heat is generated by the friction caused by molecules resisting the process. This energy is converted to heat, and dries off the solvent from the adhesive.

In many of the modern adhesives, e.g. epoxy resins such as Araldite, a two-part system is used. Two liquids, stored separately, are mixed and immediately begin to polymerize so that cross-linking (19.40) between adjaent molecules occurs, leading to solidification of the adhesive film and,

Table 26.1 *Glues and adhesives.*

Group	Adhesive substance (or raw material)	Materials joined
Animal glues	Animal hides or bones; fish; casein (from milk); blood albumen	Wood; paper; fabrics and leather
Vegetable glues	Starch; dextrine	Paper and fabrics
	Soya beans	Paper-sizing
Natural resins	Bitumens (inc. asphalt)	Laying floor blocks; felt
	Gum arabic	Paper and fabrics
Inorganic cements	Sodium silicate	Foundry moulds
	Portland cement; plaster of Paris	Building industries
Elastomer materials	Natural rubber (latex/solvent)	Rubber; sealing strips
	Synthetic rubbers (neoprene; nitrile)	Footwear industries; polythene; PVC
Synthetic polymer materials	Polyvinyl acetate and vinyl co-polymers	Wood; paper; fabrics; book binding
	Cellulose derivatives (solvent release)	Glass; paper; balsa wood
	Acrylics	Acrylics; polycarbonates
	Anaerobic acrylics	Metals
	Cyanoacrylates – cure in presence of moisture	Metals; rubbers; PVC; polycarbonates; polystyrene; polyimide
	Epoxy/amine	Metals; glass; ceramics; wood; reinforced plastics. Very wide range of uses
	Epoxy/polyamide	
	Phenol, urea, melamine and resorcinol formaldehydes	Weather-proof plywoods; fabrics and paper
	Polyurethane – hot- or cold-curing liquid	Polyurethane; PVC; polycarbonates; paper and fabrics
	Polyimide – hot-curing film	Metals; glass; ceramics; polyimide
	Silicones	Silicone rubbers; sealing seams and joints in other materials.

consequently, bonding of the substrates. This reaction sometimes takes place in the cold but with other adhesives the application of heat is necessary. With DIY materials of the 'Araldite' variety a stronger bond is usually achieved if the bonded assembly is heated. Placing it on or near a domestic radiator for an hour will lead to co-polymerization proceeding to a more advanced state. Other materials are blended to produce a better adhesive, particularly in the case of synthetic resins and rubbers. For example, well known adhesives like Bostik and Evo-stick are basically rubber-resin mixtures.

26.22   Service requirements must be considered in the choice of an adhesive. The most important requirements include strength, temperature-range, and resistance to water or moisture. The working properties of the adhesive are also important, and include the method of preparation and use, the storage-life, drying-time, odour, toxicity, and cost of bonding. Many adhesives require some type of solvent, in order to make them fluid: the solvent should evaporate reasonably quickly. Other adhesives, such as animal glues, need to be heated to make them fluid.

As with other joining processes, surface preparation is important. Surfaces must, above all, be thoroughly degreased, whilst the removal of dust and loose coatings is essential. Sand-blasting, wire-brushing, or grinding may have to be used.

Soldering and brazing

26.30   A solder must 'wet' – that is, alloy with – the metals to be joined, and, at the same time, have a freezing-range which is much lower, so that the work itself is in no danger of being melted. The solder must also provide a mechanically strong joint.

Table 26.2   *Soft solders.*

BS specification EN29453:	Composition (%)			Freezing range (°C)	Uses
	Sn	Pb	Others		
1 (S–Sn63Pb37)	63	37	–	183 (eutectic)	Tinman's solder – mass soldering of printed circuits.
3 (S–Pb50Sn50)	50	50	–	183–215	Coarse tinman's solder – general sheet-metal work.
14 (S–PB58Sn40Sb2)	40	58	Sb–2	185–213	Heat exchangers, automobile radiators, refrigerators.
–	15	85	–	227–288	Electric lamp bases.
28 (S–Sn96Ag4)	96	–	Ag–4	221 (eutectic)	For producing capiliary joints in all copper plumbing installations, particularly when a lead-free content is required in domestic and commercial situations.
29 (S–Sn97Ag3)	97	–	Ag–3	221–230	
23 (S–Sn99Cu1)	99	–	Cu–1	230–240	
24 (S–Sn97Cu3)	97	–	Cu–3	230–250	
27 (S–Sn50In50)	50	–	In–50	117–125	Soldering glazed surfaces.

26.31 Alloys based on tin and lead fulfil most of these requirements. Tin will alloy with iron and with copper and its alloys, as well as with lead; and the joints produced are mechanically tough. Suitable tin–lead alloys melt at temperatures between 183 °C and 250 °C, which is well below the temperatures at which there is likely to be any deterioration in the materials being joined. Best-quality tinman's solder contains 62 per cent tin and 38 per cent lead, and, being of eutectic composition (9.11), melts at the single temperature of 183 °C. For this reason, this alloy will melt and solidify quickly at the lowest possible temperature, passing directly from a completely liquid to a completely solid state, so that there is less opportunity for a joint to be broken by disturbance during soldering.

Because tin is a very expensive metal as compared with lead, the tin content is often reduced to 50 per cent, or even less. Then the solder will freeze over a range of temperature, between 183 °C and approximately 220 °C. Solders are sometimes strengthened by adding small amounts of antimony.

When the highly toxic nature of lead was realized lead piping used in plumbing systems was replaced by copper. These copper pipes should be joined using a *lead-free* solder, e.g. 98 per cent tin – 2 per cent silver, which is molten between 221.3 and 223 °C (the tin-silver eutectic contains 3.5 per cent silver and melts at 221.3 °C). However the operative who instals our copper pipework is still known as a 'plumber' (Latin: *'plumbum'* – lead (Pb)).

26.32 For a solder to 'wet' the surfaces being joined, the latter must be completely clean, and free of oxide films. Some type of flux is therefore used to dissolve such oxide films, and expose the metal beneath to the action of the solder. Possibly the best-known flux is hydrochloric acid ('spirits of salts'), or the acid zinc chloride solution which is obtained by dissolving metallic zinc in hydrochloric acid. Unfortunately, such mixtures are corrosive to many metals, and, if it is not feasible to wash off the flux residue after soldering, a resin-type flux should be used instead. Aluminium and its alloys are particularly difficult to solder, because of the very tenacious film of oxide which always coats the surface.

Table 26.3   *Brazing alloys and 'silver solders'.*

BS specification 1845:	Composition (%)			Freezing range (°C)	Uses
	Cu	Zn	Others		
CZ6	60	Bal.	Si–0.3	875–895	Copper, steels, malleable irons, nickel alloys.
Ag7	28	–	Ag–72	780 (eutectic)	'Silver solders – copper, copper alloys, carbon and alloy steels.
Ag1	15	Bal.	Ag–50 Cd–19	620–640	

26.33 Brazing is fundamentally similar to soldering, in that the jointing material melts, whereas the work-pieces remain in the solid state during the joining process. Brazing is used where a stronger, tougher joint is required, particularly in alloys of higher melting-point. Most ferrous materials can be brazed successfully. A borax-type flux is used, though, for the lower temperatures involved in silver soldering, a fluoride-type flux may be employed.

Arc-welding processes

26.40   Early arc-welding processes made use of the heat generated by an arc struck between carbon electrodes, or between a carbon electrode and the work. 'Filler' metal to form the actual joint was supplied from a separate rod. The carbon arc is now no longer used in ordinary welding processes, and has been replaced by one or other of the metallic arc processes.

26.41   Metallic-arc welding, using hand-operated equipment, is the most widely used fusion-welding process. In this process, a metal electrode serves both to carry the arc and to act as a filler-rod which deposits molten metal into the joint. In order to reduce oxidation of the metal in and around the weld, flux-coated electrodes are generally used. This flux coating consists of a mixture of cellulose materials, silica, lime, calcium fluoride, and de-oxidants such as ferro-silicon. The cellulose material burns, to give a protective shield of carbon dioxide around the weld, whilst the other solids combine to form a protective layer of fusible slag over the weld. Either a.c. or d.c. may be used for metallic-arc welding, the choice depending largely upon the metals being welded.

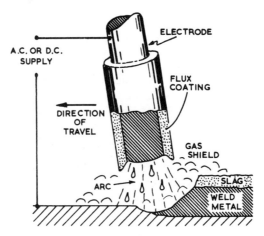

Fig. 26.1   *Metallic-arc welding.*

26.42   Submerged-arc welding is essentially an automatic form of metallic-arc welding which can be used in the straight-line joining of metals. A tube, which feeds powdered flux into the prepared joint, just in advance of the electrode, is built into the electrode holder. The flux covers the melting end of the electrode, and also the arc. Most of the flux melts, and forms a protective coating of slag on top of the weld metal. The slag is easily detached when the metal has cooled.

The process is used extensively for welding low-carbon, medium-carbon, and low-alloy steels, particularly in the fabrication of pressure-vessels, boilers, and pipes, as well as in shipbuilding and structural engineering.

26.43   Electro-slag welding   The main feature of the process, which was developed in the USSR, in 1953, is that heavy sections can be joined in a single run, by placing the plates to be welded in a *vertical* position, so that the molten metal is delivered progressively to the vertical gap, rather as in an ingot-casting operation. The plates themselves form two sides of the 'mould', whilst travelling water-cooled copper shoes dam the flow of weld

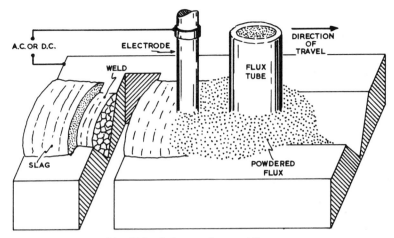

Fig. 26.2   *Submerged-arc welding.*

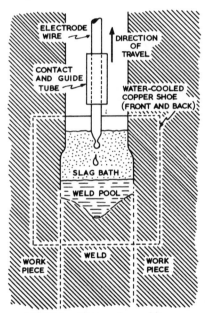

Fig. 26.3   *Electro-slag welding.*

metal from the edges of the weld, until solidification is complete (Fig. 26.3).

In this process, the arc merely initiates the melting process, and thereafter heat is generated by the electrical resistance of the slag, which is sufficiently conductive to permit the current to pass through it from the electrode to the metal pool beneath.

The process was originally developed for joining large castings and forgings, but its use has been extended to cover many branches of the heavy engineering industry. It can be used for welding a wide range of steels, and also titanium.

26.44 Gas-shielded arc-welding   Gases such as nitrogen and carbon dioxide, which are often referred to as being 'inert', will, in fact, react with some molten metals. Thus, nitrogen will combine with molten magnesium; whilst carbon dioxide will react with steel, oxidizing it under some circumstances. The only truly inert gases are those which are found in small quantities in the atmosphere, viz. argon, helium, neon, krypton, and xenon. Of these, argon is by far the most plentiful – comprising 0.9 per cent, by volume, of the atmosphere – and is therefore used for filling electric light bulbs, and also as a protective atmosphere in inert-gas welding. In the USA, substantial amounts of helium are derived from natural gas deposits, so it is used there as a gas-shield in welding. Since argon and helium are expensive to produce, carbon dioxide (CO_2) is now used in those cases where it does not react appreciably with the metals being welded.

In gas-shielded arc-welding, the arc can be struck between the work-piece and a tungsten electrode, in which case a separate filler-rod is needed, to supply the weld metal. This is referred to as the 'tungsten inert gas' (or TIG) process. Alternatively, the filler-rod itself can serve as the electrode, as it does in other metallic-arc processes. In this case, it is called the 'metallic inert gas' (or MIG) process.

26.44.1   *The TIG process*   was developed in the USA during the Second World War, for welding magnesium alloys, other processes being unsatisfactory because of the extreme reactivity of molten magnesium. It is one of the most versatile methods of welding, and uses currents from as little as 0.5 A, for welding thin foil, to 750 A, for welding thick copper, and other materials of high thermal conductivity.

Since its inception, it has been developed for welding aluminium and other materials, and, because of the high quality of welds produced, TIG welding has become popular for precision work in aircraft, atomic engineering, and instrument industries.

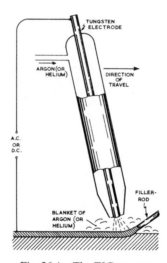

Fig. 26.4   *The TIG process.*

26.44.2   *The MIG process*   uses a consumable electrode which is generally in the form of a coiled, uncoated wire, fed to the argon-shielded arc by a motor drive. Most MIG welding equipment is for semi-automatic operation, in which the operator guides the torch or 'gun' (Fig. 26.5), but has little else to do once the initial control settings have been made.

In addition to the advantages arising from semi-automatic operation, the MIG process is generally a clean welding process, due to the absence of flux. Consequently, the MIG process is one of the most diversely used welding methods, in terms of the number of different jobs with which it can cope

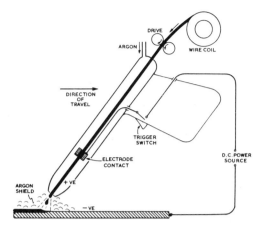

Fig. 26.5   *The MIG process.*

successfully. Some industries where it finds application include motor-car manufacture, shipbuilding, aircraft engineering, heavy electrical engineering, and the manufacture of tanks, pressure-vessels, and pipes.

26.44.3   *The CO_2 welding process*   is reasonably effective, provided the filler-rod used contains adequate deoxidizing agents to cope with any oxidation which may arise due to the weld metals reacting with carbon dioxide. The normal CO_2 process is a modified MIG process, in which argon has been replaced by carbon dioxide, and in which the electrode wire is rich in deoxidants.

26.45   Plasma-arc welding   In this relatively new process, a suitable gas, such as argon, is passed through a constricted electrical arc. Under these conditions, the gas ionizes, that is, its atoms split up into electrons and positively charged particles, the mixture being termed 'plasma'. As the ions recombine, heat is released, and an extremely hot 'electric flame' is produced. The sun's surface consists essentially of high-temperature plasma, though even higher temperatures of up to $15\,000\,°C$ can be produced artificially.

The use of plasma as a high-temperature heat source is finding application in cutting, drilling, and spraying of very refractory materials, like tungsten, molybdenum, and ceramics, as well as in welding.

Other fusion-welding processes

26.50   Some welding processes make use of heat obtained by a chemical reaction. Oxyacetylene welding is probably the best known of these, though the Thermit process also continues to have its uses. In addition, some sophisticated methods based on both electron and laser beams have been developed in recent years.

26.51   Oxyacetylene welding   Formerly, gas-welding ranked equally important with metallic-arc methods, but the introduction of argon-shielding to the latter process placed gas-welding at a disadvantage for welding metals where a flux coating is necessary. Consequently, the use of oxyacetylene welding has declined in recent years. However, it is still widely used where maintenance or general repair work is involved.

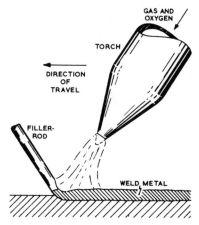

Fig. 26.6   *The principles of gas-welding.*

Both oxygen and acetylene can conveniently be stored in cylinders, and their flow to the torch can easily be controlled by using simple valves. Although acetylene normally burns with a smoky, luminous flame, when oxygen is fed into the flame in the correct proportions, an intensely hot flame is produced, allowing temperatures in the region of 3500 °C to be attained. Such a high temperature will quickly melt all ordinary metals, and is necessary in order to overcome the tendency of sheet metals to conduct heat away from the joint so quickly that fusion never occurs.

As with most other fusion-welding processes, a flux-coated filler-rod is used to supply the weld metal.

26.51.1 *Bronze-welding* is the term applied to the joining of metals of high melting-point, such as mild steel, by the use of copper-alloy filler-metals. An oxyacetylene flame is generally used to supply the heat. Bronze-welding differs from true welding in that little or no fusion of the work-pieces takes place.

26.52   Thermit welding is used chiefly in the repair of large iron and steel castings. A mould is constructed around the parts to be joined, and above this is a crucible containing a sufficient quantity of the Thermit powder. The parts to be welded are first preheated, and the powder then ignited.

Thermit powder consists of a mixture of powdered iron oxide and aluminium dust, in calculated proportions. The chemical affinity of aluminium for oxygen is greater than the affinity of iron for oxygen; hence the powders react so:

$$Fe_2O_3 + 2Al = Al_2O_3 + 2Fe + heat$$

The heat of reaction is so intense that molten iron is produced, and this is tapped from the crucible into the prepared joint.

26.53   Electron-beam welding When a stream of fast-moving electrons strikes a target, the kinetic energy of the electrons is converted into heat. Since the electron-beam can be focused sharply, to impinge at a point, intense heat is produced there. To be effective, the process must be carried out in a vacuum, otherwise electrons tend to collide with molecules of oxygen and nitrogen present in the air-space between the electron-gun and the target.

Consequently, electron-beam welding is at present used mainly for 'difficult' metals which melt at high temperatures, such as tungsten, molybdenum, and tantalum; and also for the chemically reactive metals beryllium, zirconium, and uranium, which benefit from being welded in a vacuum.

26.54 Laser welding The term 'laser' is an acronym; that is, a name coined from the initial letter of each important word of its descriptive title. In this case, 'laser' represents Light Amplification by Stimulated Emission of Radiation. The device consists of a suitable generator of light pulses – or 'optical pump' – and an active element, and it emits a beam of light of extremely high intensity.

At present, laser technology is probably still in its infancy, but already it has had great successes in the fields of medical research, and in space technology. In the field of metal manufacture, it will ultimately find application in cutting, drilling, shaping, and welding, particularly in *micro* spot-welding.

Pressure-welding processes

26.60 In order to weld together two pieces of iron, the blacksmith first heats them to a high temperature, and then brings them into contact under pressure. He does this by hammering them together on an anvil, thus perpetuating one of the most ancient of metallurgical processes used. Whilst it is highly probable that Tubal Cain used this process some 6000 years ago (Genesis, iv, 22), we know with certainty that smith-welding was carried out by the Ancient Greeks, at least 400 years BC.

Apart from smith-welding, which is still used widely in many different forms, modern pressure-welding processes are mostly of the electrical-resistance type, in which the passage of a heavy electric current generates sufficient heat to permit welding.

26.61 Spot-welding In this process, the parts to be joined are overlapped, and firmly gripped between heavy metal electrodes (Fig. 26.7). An electric current of sufficient magnitude is then passed, so that local heating of the work-pieces to a plastic state occurs. Since the metal at that spot is under pressure, a weld is produced. This method is used principally for joining plates and sheets, but in particular for providing a temporary joint.

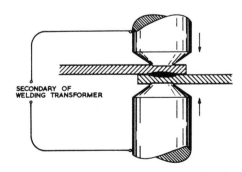

SECONDARY OF
WELDING TRANSFORMER

Fig. 26.7 *Spot-welding.*

26.62   Projection-welding   is a modified form of spot-welding, in which the current flow, and hence the resultant heating, are localized to a restricted area, by embossing one of the parts to be joined. When heavy sections have to be joined, projection-welding can be used where spot-welding would be unsuitable because of the heavy currents and pressures required. Moreover, with projection-welding, it is easier to localize the heating to that zone near the embossed projection. As Fig. 26.8 shows, the projection in the upper work-piece is held in contact with the lower work-piece (i). When the current flows, heating is localized around the projection (ii), which ultimately collapses under the pressure of the electrodes, to form a weld (iii).

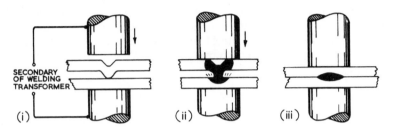

Fig. 26.8   *Projection-welding.*

26.63   Seam-welding   resembles spot-welding in principle, but produces a continuous weld by using wheels as electrodes (Fig. 26.9). The work-pieces are passed between the rotating electrodes, and are heated to a plastic state by the flow of current. The pressure applied by the wheels is sufficient to form a continuous weld.

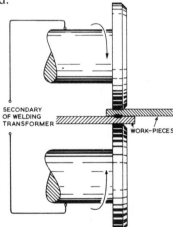

Fig. 26.9   *Seam-welding.*

26.64   Butt-welding   This process is used for welding together lengths of rod, tubes, or wire. The ends are pressed together (Fig. 26.10), and an electric current is passed through the work.

Since there will be a higher electrical resistance at the point of contact (it is most unlikely that the two ends will be perfectly square to each other), most heat will be generated there. As the metal reaches a plastic state, the pressure applied is sufficient to lead to welding.

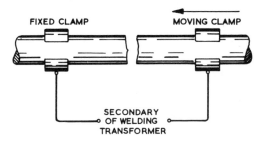

Fig. 26.10   *Butt-welding.*

26.65   Flash-welding   is somewhat similar to butt-welding, except that heat is generated by striking an arc between the two ends which are to be joined. Not only are the ends heated in this way, but any irregularities there are melted away. The ends are then brought together quickly, under pressure, so that a sound weld is produced.

26.66   Other pressure-welding processes   Several other welding processes which involve the use of pressure are worth a mention here.

26.66.1   *Induction-welding*   In this process, the parts to be joined are pressed together, and an induction coil is placed around the joint. The high-frequency current induced in the work heats it to welding temperature.

26.66.2   *Friction-welding*   can be used to join work-pieces which are in rod form. The ends of the rods are gripped in chucks, one rotating, and the other stationary (Fig. 26.11). As the ends are brought together, heat is generated, due to frition between the slipping surfaces. Sufficient heat is generated to cause the ends to weld. Rotation ceases as welding commences.

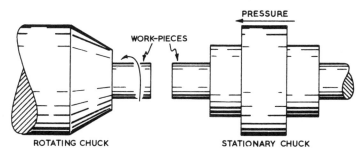

Fig. 26.11   *Friction-welding.*

26.66.3   *Cold-pressure welding,*   in the form of lap-, seam-, and butt-welds, generally involves making the two surfaces slip relative to each other under great pressure. Any oxide films are broken, and a cold-weld forms between the two surfaces. Spot-welds are produced in this way by punching the two work-pieces together.

Structures of welds

26.70   Since most welds are made at high temperatures, this inevitably leads to the formation of relatively coarse grain in the metal in and around the weld. A weld produced by a fusion process will have an 'as-cast' type of structure, and will be coarse-grained, as compared with the materials of the work-pieces,

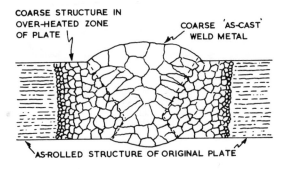

Fig. 26.12   *The crystal structure of a fusion-weld.*

which will generally be in a wrought condition. Not only will the crystals be large (Fig. 26.12), but other as-cast features, such as the segregation of impurities, may be present, giving rise to intercrystalline weakness.

When possible, it is of advantage to hammer a weld *whilst it is still hot*. Not only does this smooth up the surface, but, since recrystallization of the weld metal will follow this mechanical working, a tough fine-grained structure will be produced. Moreover, the effects of segregation will be significantly reduced.

Some alloys are difficult to weld successfully, because of structural changes which accompany either the welding process or the cooling which follows. Thus, air-hardening steels, particularly high-chromium steels, tend to become martensitic and brittle in a region near to the weld, whilst 18–8 austenitic stainless steels may ultimately exhibit the defect known as 'weld-decay' (13.42), unless they have been 'proofed' against it.

Welding of plastics

26.80   Thermoplastic materials can be welded together by methods which are fundamentally similar to those used for welding metals. All methods depend upon the application of heat, and sometimes pressure.

26.81   Hot-gas welding   In this process, a jet of hot air from a welding torch is used (Fig. 26.13). In other respects, this method resembles

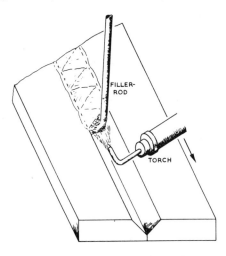

Fig. 26.13   *The hot-gas welding plastics.*

oxyacetylene welding of metals. A thin filler-rod is used, and is of the same material as that of the work-pieces. This process is used widely in the building of chemical plant, particularly in rigid PVC.

26.82 Seam- and spot-welding These methods are very similar to those used in metal-joining, except that, since plastics are non-conductors of electricity, the 'electrodes' must be heated by high-frequency coils.

26.83 Electrofusion-welding (Fig. 26.14) is a novel process used by British Gas for joining yellow polythene pipes up to 180 mm diameter. The abutting pipe ends are inserted into a connecting polythene collar which contains a moulded-in electric-heater element. Resistance heating causes a continuous joint to form in about half-an-hour.

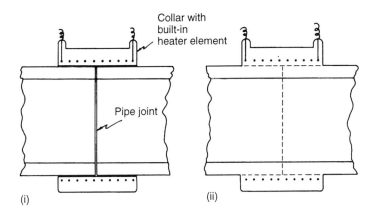

Fig. 26.14 *The principles of electrofusion-welding of polythene gas pipes.*

26.84 Stitch-welding is a method of welding thermoplastics, using a device similar to a sewing-machine, but fitted with two electrodes, which weld the material progressively. Again, the electrodes are heated by a high-frequency current supply.

26.85 Jig-welding In this method, the work-pieces are gripped in jigs. These jigs act as electrodes which are heated by high-frequency fields.

26.86 Friction-welding also known as spin-welding, is similar in principle to the process which has recently been developed for joining metals (26.66.2). One part is rotated at speed, whilst the other is pressed against it. This heats both of the work-pieces. Motion is then stopped, and the parts are held together until the joint hardens.

CHAPTER 27

The choice of materials and processes

27.10   At about the time I had acquired the status of a 'teenager' it was possible to buy a brand new, rather heavy 'sit-up-and-beg' pedal bicycle from the local branch of H*lf*r*s for the sum of £3. 19s. 11d (£3.99½p). The 'bike' is still with us and 'trail riders' are encountered in all sorts of unlikely places such as canal tow-paths, bridleways and steep mountain tracks. It seems that the fanatics – sorry, 'enthusiasts' – who ride these machines are willing to pay in excess of £1000 for the frame alone, provided that it offers an acceptable 'strength/weight ratio'.

27.11   Currently it seems that about half-a-dozen different materials are competing for this trade. The main contenders are:

1. A MMC in which aluminium is dispersion strengthened (23.31) by the presence of aluminium oxide particles (a development of SAP). Frames made with this MMC are TIG welded (26.44.1).
2. A MMC, again with an aluminium matrix, but this time stiffened with boron carbide in amounts up to 15 per cent. This composite is also TIG welded.
3. Another MMC in which aluminium is stiffened by the incorporation of 40 per cent beryllium, giving a material of low relative density about 2.3. The metal beryllium, though scarce and very expensive, has now come on the market for civil uses as a result of the end of the Cold War. (Recently a cycle with a frame constructed entirely from this 'modern miracle metal'[1] was marketed in the USA for around $25,000.) The use of beryllium under all but the most carefully controlled conditions is made more difficult because its dust (from machining operations) and vapour (from welding) are extremely toxic.
4. A type of maraging steel (13.27) based on iron, nickel, cobalt, chromium, molybdenum and carbon from which frame tubes with a wall thickness of only 0.5 mm can be drawn. This makes them extremely light and strong but also difficult to weld due to ensuing heat distortion. Silver soldering (Table 26.3) is used instead.
5. A novel carbon-fibre reinforced nylon matrix in which a woven mat of mixed carbon and nylon fibres is hot moulded to form the frame. During this process the nylon fibres fuse to form the *thermoplastic* matrix holding the carbon fibres.

---

[1]   In fact beryllium was discovered in 1797 and industrial production on a small scale began in 1916!

27.12   These then are processes and materials competing to manufacture very strong light-weight cycle *frames* at prices (1996) ranging from £900 to £1100. Here we are dealing with the 'pursuit of excellence' with few, if any, restraints on the cost of the product, but how often can this situation prevail in the real world of engineering? Everyday engineering production involves a highly competitive enterprise which, in the long run, nearly always leads to some sacrifice of quality, in favour of lower cost. The manufacturer is not necessarily to blame for this state of affairs, since he is usually satisfying an existing demand, even though that demand is often stimulated by a vigorous advertising-campaign. Often the cost of perfection is too high, and in most cases the effects of competition, coupled with a lack of power of discrimination by the consumer, lead to the manufacture of an article which is 'cheap and nasty' – or in many cases nasty without being cheap.

27.13   In an age when the development of new materials was extremely slow industrial philosophy favoured the manufacture of high-quality durable products. Thus a pocket watch in its silver case was handed down from father to son for several generations. Kettles, cooking pots and domestic equipment generally was made to last indefinitely. However, methods of mass production which followed in the wake of the Industrial Revolution meant that the petrol-driven 'tin lizzie' of the 1920s never quite reached the standard of craftsmanship of the horse-drawn carriage of earlier days. Of course mass production had to provide a commodity which was cheap enough to be purchased by the masses who were employed in making it. Invention, research and development now proceed so quickly that many products become obsolescent within a very short time of purchase (the market value of electronics gadgets like computers falls so rapidly that one purchased now becomes almost unsaleable in a couple of years).

Fortunately when technical development of a class of appliance stabilizes quality of production tends to improve as there is less incentive for the consumer to buy a new model because of some minor innovation in design and this in turn encourages him to hope that it will last longer without breaking down. Obviously there will always be the manufacturer who seeks to make something a little cheaper – and a little nastier. It is up to Joe Public to discriminate between products offered for sale in large choice.

Selection of materials

27.20   In choosing a material for a specific application the engineer must consider:

- the ability of the material to withstand service conditions;
- the method(s) by which it will be shaped; and
- the overall cost and in some cases the availability of the material and the cost of the shaping process(es).

In some cases, a number of materials may be satisfactory in respect of fulfilling the service requirements. Normally the engineer will choose the one which, when the cost of forming and shaping is taken into account, results in the lowest overall production cost. Thus, the cheapest material may not necessarily be the one which is used; a more expensive material may be capable of being formed very cheaply, and so give a lower overall production cost.

27.21   The service requirements of a material may involve properties which fall under one or all of the following three headings:

Mechanical properties	Physical properties	Chemical properties
(a) tensile strength	(a) relative density	(a) resistance to oxidation
(b) elasticity	(b) melting-point or	(b) resistance to electrolytic
(c) toughness (impact value)	softening temperature	corrosion
(d) stiffness (Young's	(c) thermal conductivity	(c) resistance to
modulus)	(d) coefficient of expansion	degradation by electro-
(e) hardness	(e) effect of temperature	magnetic radiation
(f) fatigue resistance	changes on properties	(d) resistance to biological
(g) resistance to creep	(f) electrical conductivity	attack by plants or
(h) frictional properties	(g) magnetic properties	animals

We will now deal with the influence of some of these properties on materials selection.

27.30   Tensile strength and 'specific strength'   It is an easy matter to make a list of materials in order of tensile strength. Such a list (Table 27.1) is headed by maraging steels and the well-known 'nickel-chrome-moly' constructional steels, followed by other steels, fibres, non-ferrous alloys, ceramics and plastics materials in appropriate order. The choice of a material however will not usually be made with reference to strength alone. Such criteria as the temperature and chemical nature of its working environment must be considered as well as the type of stress it must bear (i.e. 'static', 'live' or 'alternating').

Table 27.1   *Tensile strengths and specific strengths of some important engineering materials.*

Material	Tensile strength (approx.) ($N\,mm^{-2}$)	Specific strength (approx.) ($N\,mm^{-2}$)
maraging steel	2460	299
*S-glass	1800	900
*high-strength carbon fibre	1770	1140
Ni–Cr–Mo constructional steel	1700	215
beryllium bronze, heat treated	1300	146
titanium alloy T_A10	1120	280
'Nimonic 115'	1100	134
0.5% C steel, quenched and tempered	925	119
18/8 stainless steel, hard rolled	800	98
phosphor bronze, hard drawn	700	80
65/35 brass, hard rolled	695	84
0.2% C steel, quenched and tempered	620	79
duralumin, heat treated	420	155
'400' grade grey cast iron	400	52
zinc-base die-casting alloy	300	43
magnesium-base alloy, cast and heat-treated	200	118
aluminium alloy LM6, cast	190	73
thermoplastic polyester	170	120
PEEK	92	70
nylon 6:6	85	74
epoxy resin	80	69
bakelite	55	40
PVC (rigid form)	49	36
polythene (HD)	31	32

*60% by volume of fibre in epoxy resin matrix.

The choice of dimensions of the load-bearing member may also be relevant. Thus, if dimensions allow a material of *lower* tensile strength but *greater* cross-sectional area (CSA) to be used, then a cheaper material may be available but this may mean using a member of *greater* weight and so the factor £/kg is involved. For example although a medium-carbon steel is only about half as strong as a similar nickel-chromium constructional steel it is only about one-third of the cost of the latter in terms of £/kg. Consequently although a member in medium-carbon steel capable of bearing a similar load will be greater in CSA than one in the nickel-chrome steel, it will cost less.

The energy required to blast a space vehicle free of the gravitational 'pull' of Mother Earth is considerable so that it is important to ensure that its mass is kept to a minimum. Therefore constructional materials used in such a craft must have a high 'strength-to-weight ratio'. This is now more correctly known as *specific strength* and is the ratio (tensile strength)/(relative density). Of course it is not only when dealing with space vehicles that we must take into account energy used in working against gravity. Heavy road-haulage vehicles constructed from unnecessarily heavy component parts consume extra energy – and release unacceptable amounts of 'greenhouse gases' as a result. Undoubtedly fibre-reinforced composites will be used increasingly in such cases.

27.31 Young's modulus and 'specific modulus' Young's modulus of elasticity, E, (4.21) is derived from the ratio stress/strain for stresses applied to a material below its yield point. Thus E is equivalent to the slope of that part of the force/extension diagram below the yield point and is a measure of the 'stiffness' of the material. For example the slope of that part of the diagram for a steel is much 'steeper' than that for aluminium – the steel is much the stiffer material under the action of either tensile or compressive loads. Young's modulus can also be related to the relative density to give the value 'specific modulus', i.e.

$$\text{specific modulus} = \frac{\text{Young's modulus of elasticity}}{\text{relative density}} \quad \text{(Table 27.2)}$$

When considering the *specific* moduli of materials note that there is little to choose between metallic alloys and it is here where composites reinforced with carbon, boron or aramid fibres score heavily.

Table 27.2 *Young's modulus of elasticity and specific modulus for some engineering materials.*

Material	Young's modulus ($kN\,mm^{-2}$) *measured longitudinally*	Specific modulus ($kN\,mm^{-2}$)
high-tensile Ni–Cr–Mo steel	210	27
18/8 stainless steel	205	26
high-carbon steel	203	26
low-carbon steel	203	26
titanium alloy (6Al–4V)	115	26
duralumin	69	26
*S-glass	55	28
*aramid	80	59
*boron	210	95
*high-modulus carbon	272	167

*60% by volume fibre in epoxy resin matrix.

27.32   Toughness and impact value   The easy way to define 'toughness' is to say that it is the opposite of 'brittleness', but to the engineer it implies the ability of a material to absorb energy under conditions of mechanical shock without fracture. The Izod impact test (4.40) itself is a demonstration of this value measuring as it does the mechanical energy absorbed from a swinging hammer as it fractures a standard test-piece.

Strong materials are not necessarily tough ones. Thus, soft, annealed copper with a tensile strength of no more than $200 \, N \, mm^{-2}$ may well almost 'stop' the swinging hammer in the Izod test. Any treatment which increases the yield stress or the hardness of a material generally reduces its toughness. Thus a quenched and tempered low-alloy steel may be stronger and harder than the same steel in the normalized condition but the Izod value will usually be lower.

Whilst most ceramics are brittle when compared with metals there are exceptions. Thus crystalline alumina, though used in armour plating because of its great hardness, also possesses adequate toughness. Thermoplastic polymers are usually weak but tough whilst thermosetting polymers are generally stronger and harder but more brittle.

27.33   Fatigue resistance   In 27.30 we were comparing the tensile strengths of some materials under the action of *static* loads. In practice however we must often consider the behaviour of a material under the action of 'live' (fluctuating) or alternating stresses and we assess the *fatigue limit* as the maximum stress a material can sustain under the action of an 'infinite number' of reversals of stress. Since we can only work in 'finite' large numbers of stress reversals we base our fatigue data on these values (4.62).

For most metallic alloys the fatigue limit is usually between one-third and one-half of the static tensile strength. With many plastics materials it is difficult to assess fatigue properties because the heat generated by the energy used in overcoming van der Waals forces acting between the large molecules is not quickly dispersed as it is in metals which have a much higher thermal conductivity. The plastics material therefore tends to overheat on test and fail at a low stress. The higher the frequency of stress alternations the quicker the build up of heat and the more likely the polymer to fail at low stress.

27.34   Creep resistance   Creep (4.50) is the gradual extension which can take place in materials at a stress *below* the yield stress. Although measurable creep may occur in some materials such as lead and a number of polymers at ambient temperatures it is generally associated with engineering materials at high temperatures. Metallic alloys used under these conditions are usually dispersion hardened (23.30) by including in the microstructure tiny, strong particles which impede the progress of dislocations along the crystallographic planes. It is also necessary for high-temperature alloys to have a good resistance to oxidation.

The need, engendered in the early 1940s, for a material suitable for the manufacture of parts for jet engines and gas turbines led to one of the epic metallurgical research projects of the twentieth century and resulted in the production of the 'Nimonic' series of alloys (18.24) by Messrs Henry Wiggin. The basis of a Nimonic is an alloy containing 75 per cent nickel and 20 per cent chromium giving a material of high melting-point which will remain tough at high temperatures because of the grain-growth restricting influence of nickel and at the same time be oxidation-resistant because it becomes coated with a tenacious film of protective chromium oxide. Small amounts of carbon, along with titanium, aluminium and molybdenum, form the particles of 'dispersoid' which strengthen the alloy at high temperatures. Other alloys based on nickel and chromium along with dispersion hardening additions are now available. Some are wrought alloys whilst others are cast to shape.

27.35 Refractoriness Whilst most materials working at high temperatures require high creep resistance, others which are exposed to aggressive environments combining high temperatures and corrosive atmospheres need carry only light static loads. Such materials are used in furnace linings and other static parts, crucibles and the like. Materials which are used principally because of their ability to withstand high temperatures are known as *refractories*. Some of the 'heavy metals' like tungsten and molybdenum fall into this category. Table 27.3 lists some refractories along with other engineering materials of which maximum working temperatures are indicated.

Table 27.3 *Maximum working temperatures for some engineering materials.*

Material	Maximum working temperature (°C)
Refractories	
zirconia	2500
magnesia	2000
boron nitride	1800
silica	1700
fireclay	1400
'sialons' (silicon nitride based)	1250
molybdenum	1000
Other metals	
nimonic alloys	900
high nickel-chromium steels	800
12% chromium stainless steel	600
low Ni–Cr–Mo–V steel	550
titanium alloys	500
carbon steels (normalized)	400
cast iron	300
copper alloys	190
aluminium alloys	180
Polymers	
themosetting polymers	150
thermoplastic polymers	100

27.36 Friction and wear resistance When two surfaces rub together energy is consumed in proportion to the frictional forces which operate between them. At the same time there is a tendency for particles of the softer material to become detached – its surface wears away. In machines these problems are reduced to some extent by lubrication, the function of a lubricant being to *separate* the running surfaces and so reduce friction. Lubricants include both liquids (from water to heavy mineral oils) and solids (molybdenum disulphide, MoS_2, and graphite).

As far as load-bearing surfaces are concerned materials with high hardness values usually have a low coefficient of friction (μ) and a good resistance to wear. Unfortunately such materials are often intermetallic compounds (8.40) which are both weak and brittle. Nevertheless the need for a material which is hard, has a very low value of μ but is also reasonably tough and ductile is met by using an alloy with a complex microstructure as is found in bearing metals (18.60). Thus the 'white' bearing metals consist of hard particles of an antimony-tin intermetallic compound (SbSn) embedded in a tough matrix of tin-base or lead-base alloy; whilst bronze bearings consist of particles of the hard intermetallic compound δ ($Cu_{31}Sn_8$) embedded in a tough copper-tin solid solution (18.62). In both cases the softer matrix tends to wear away leaving the hard compound standing proud. This in itself reduces friction between the running surfaces but also provides channels which assist the flow of lubricants.

Similar properties are required in the journal which is running on the bearing surface provided, i.e. a strong, tough 'body' with a hard, low-μ surface. This is provided by using one of two classes of steel:

1. a low-carbon or low-alloy, low-carbon steel the surface of which is then either carburized and case-hardened (14.20) or nitrided (14.50);
2. a medium-carbon or low-alloy steel the surface of which receives separate heat-treatment by either flame- (14.70) or induction-hardening (14.80).

In each case a tough, strong core is provided with a very hard wear-resistant skin.

Cast irons tend to have a good wear-resistane – 'grey' irons because the presence of graphite flakes (15.21) has a lubricating effect, and 'white' irons because of the presence of very hard cementite. Among plastics materials which have a low μ value are nylon 6:6 and PTFE (19.32) both of which are used in low-duty bearings, gears and the like.

27.37   Stability in the environment   Iron is one of the most plentiful and inexpensive of engineering metals. Furthermore its alloy, steel, can either by cold-work (mild steel) or by heat-treatment (medium- and high-carbon steels) be made to provide us with a wider range of mechanical properties than any other material. It has one major fault – its corrosion-resistance is poor and, except when alloyed with expensive chromium, it will rust if exposed to damp atmospheres.

The cost of protecting steel structures – bridges, pylons, radio masts and the like – from corrosion in the UK alone runs into more than £10 billion per year, whilst of the 'unprotected' steel some 1000 tonnes 'escapes back into Nature' as rust on every single day.

Steel will rust more quickly in the annealed or normalized conditions since in the presence of an electrolyte (moisture containing dissolved atmospheric pollutants such as SO_2 from burning coal) the layers of ferrite and cementite present in pearlite constitute an electrolytic cell (25.41) leading to the 'rusting' of the ferrite component which is anodic to cementite. A quenched (hardened) carbon steel is slightly more corrosion-resistant because most of the carbon is in solution and electrolytic action is reduced.

Thus for carbon- and low-alloy steels which are to be used in an aggressive environment some form of coating whether by paint, paint assisted by electrophoresis (25.51), enamelling, or some form of metal coating must be used. Where mild-steel presswork is exposed to corrosive conditions it may be worthwhile considering alternative materials to painted mild steel. Some of the better-quality 'family cars' now have bodywork where the 'skirtings' – notorious corrosion traps – are largely of plastics materials.

Most of the engineering non-ferrous metals have much higher corrosion resistance than does carbon steel but, weight-for-weight, they are more expensive. Titanium and its alloys are exceptionally resistant to attack by air, water and sea water but are very expensive. Nevertheless when it can be considered in terms of £/m^3 instead of £/kg the low relative density of titanium may make it an attractive proposition (27.39.1). Nickel alloys too are very resistant but expensive whilst the amounts of nickel and chromium necessary to make steel 'stainless' also render it up to eight times more expensive, weight-for-weight, than mild steel. Copper and its alloys are moderately resistant to atmospheric corrosion but some brasses suffer dezincification in contact with some natural waters – small amounts of arsenic inhibit this fault if added to brasses used for condenser tubes. Aluminium, which in terms of cost per unit *volume* is less expensive than other non-ferrous alloys, has an excellent resistance to corrosion (17.30) but some of the high-strength alloys, particularly those containing zinc (2L88) require protection. Zinc-base die-casting alloys have a good corrosion resistance provided that the metal is of high purity – 'four-nines zinc', i.e. 99.99 per cent pure is required for this purpose.

Whilst plastics materials have a high resistance to chemical attack and can be used underground for conveying water and gas, some are prone to damage by ultra-violet (UV) light and must be made opaque by suitable additives (25.81). PTFE resists attack by the most aggressive reagents but is very expensive. Natural rubber may 'perish' due to exposure to intense UV radiation and the presence of atmospheric ozone.

Glass has an excellent resistance to all chemical reagents except hydrofluoric acid (used for etching stained glass) as does vitreous enamel. 'Glass' coated steel tanks are widely used in the chemical industries for containing corrosive liquids. If a heavy spanner is accidentally dropped on to such a vitreous-coated surface the chipped area can be 'drilled through' and the surface sealed by carefully hammering in a tantalum plug. Tantalum is a very soft, ductile heavy metal with an extremely high resistance to corrosion for, like aluminium and chromium, its surface seals itself with an adherent, impervious oxide film.

27.38   Electrical conductivity   A material has a high electrical conductivity if electrons are able to pass through it easily, that is with little loss of energy (dissipated as heat). Electrical conductors, e.g. power lines, must lose as little energy as possible (the 'i^2R loss') in this way. With electrical heater elements the opposite is required and these alloys should have a low conductivity (or high electrical *resistivity*) so that heat is generated without the element melting. Hence for this purpose high-temperature nickel-chromium alloys (18.22) are used.

Copper is the most widely used metal industrially where high electrical conductivity is required and the bulk of copper produced is for the electrical trades. In respect of electrical conductivity it is second only to silver which is slightly more conductive but when relative costs of the two metals are considered copper is of course the metal used. The International Annealed Copper Standard (IACS) relates to copper of which the resistance of a wire 1 m long and weighing 1 g is 0.15328 ohm at 20 °C. Such copper is said to have a conductivity of 100 per cent. Since this standard was adopted in 1913 the chemical purity of commercial copper has improved and this explains why seemingly impossible conductivities of 101 per cent or more are quoted.

The electrical conductivity of copper is adversely affected by the presence of impurities (Fig. 27.1). Some impurities like zinc, cadmium and silver have

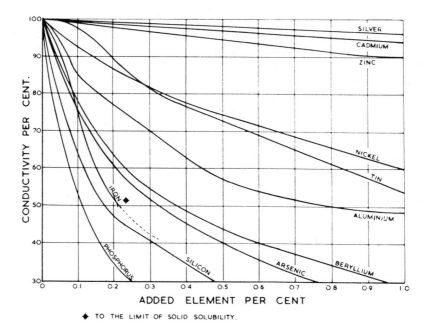

TO THE LIMIT OF SOLID SOLUBILITY.

Fig. 27.1   *The effect of impurities on the electrical conductivity of copper.*

little effect if present in small amounts. Thus about 1 per cent Cd is added to copper destined for use in overhead telephone lines in order to strengthen them sufficiently to support their own weight. The conductivity is still of the order of 95 per cent. The presence of only 0.1 per cent P however reduces electrical conductivity by 50 per cent so that phosphorus-deoxidized copper should not be used where high conductivity is required.

The electrical conductivity of aluminium is only 60 per cent that of copper but in terms of *specific* conductivity on a mass/unit length basis (Fig. 17.1) aluminium is a better conductor, for which reason it is used in the power grid network (17.32). The electrical conductivities of various grades of copper and other alloys likely to be used in electrical trades are shown in Table 27.4.

Table 27.4   *Electrical conductivities (IACS) of grades of copper and its alloys.*

Metal or alloy	Conductivity (%)
silver	105
IACS copper standard	100
oxygen-free high-conductivity (OFHC) copper	101.4
hard-drawn copper (1% Cd)	95
phosphorus-deoxidized copper (0.08% P)	54
phosphorus-deoxidized arsenical copper	45
70–30 brass	27
tin-bronzes	10–27

27.39   Relative costs of important engineering materials   Table 27.5 gives a very approximate idea of the relative costs of engineering materials. In terms of relating the costs of polymer materials to those of metals, 'weight-

Table 27.5   *Approximate relative costs of some engineering materials.*

Material	Cost per unit mass	Relative density	Cost per unit volume
Bar form			
mild steel	1.0	7.87	1.0
medium carbon steel	1.5	7.83	1.5
low-alloy steel	4.5	7.85	4.5
brass	6.5	8.45	7.0
aluminium	8.8	2.70	3.0
stainless steel	9.7	7.92	9.8
copper	12.0	8.94	13.6
phosphor bronze	14.6	8.93	16.6
titanium	24.0	4.51	13.7
Sheet form			
mild steel	1.0	7.87	1.0
lead	2.9	11.30	4.0
brass	4.0	8.45	4.3
stainless steel	6.3	7.92	6.3
aluminium	6.7	2.70	2.3
copper	7.0	8.94	8.0
titanium	20.0	4.51	11.5
Castings			
grey cast iron	1.0	7.4	1.0
brass	3.5	8.38	4.0
zinc-base die-casting alloy	4.0	6.8	3.7
aluminium alloy LM6	4.8	2.65	1.7
phosphor bronze	6.8	8.93	8.2

Polymer materials	Cost/unit mass or volume
polythene	1.0
PVC	1.4
natural rubber	1.7
phenolics	3.8
ABS	4.5
polycarbonate	11.0
nylons	12.0
PTFE	25.0

for-weight' polythene is very approximately three times as expensive as mild steel but when considered 'volume-for-volume' mild steel is roughly three times more expensive than polythene.

In the real world both the *quality* and the relative *costs* of materials change and the design engineer must keep under review such factors as:

- variations in the market value of a material relative to those of competitive materials;
- improvements in relevant properties of materials resulting from developments in production methods.

27.39.1   For example until recently there was no serious competitor for copper as a prestigious roofing material but in Japan, where corrosion from marine atmospheres is prevalent, it was decided to roof the Municipal Aqualife Museum in Kobe with titanium sheet. On the face of it this would seem to be an expensive project since, despite a gradual fall in the price of titanium as production methods improve, it is still several times more expensive than is copper when considered on a weight-for-weight basis. However the relative density of titanium is only 50.7 per cent that of copper, so, assuming equal thicknesses of cladding sheet are involved the price difference is already halved; but the comparison does not end there. Since the use of titanium cladding reduces by half the load to be carried by roof members, modification in the design of the latter on the side of economy is also possible. Since the price of titanium is likely to fall further relative to other metals, and particularly copper, we can expect to see increased use of titanium as a roofing material, particularly in buildings of high prestige.

27.39.2   Improvements in quality of some of the old well-established materials continues to take place. We may not regard prosaic mild steel as a technically 'glamorous' material but in the years since the Second World War considerable improvements have been achieved in its chemical 'cleanliness' and a reduction in the quantity of dissolved atmospheric nitrogen it contained hitherto. These achievements have led to a big improvement in the overall ductility of those high-quality mild steels produced in this way. So much so that they have largely replaced far more expensive materials like 70–30 brass, cupro-nickel and aluminium for the manufacture of *deep-drawn* (7.33) components. This 'revolution' has of course been assisted by improvements in tool design and methods of lubrication. Much of the practice ammunition used by the military under conditions where corrosion is no problem uses mild steel cases to replace those formerly deep drawn from 'cartridge' (70–30) brass.

The choice of a shaping process

27.40   The properties of a metallic material which affect its suitability for forming and shaping operations include:

- malleability
- ductility
- strength
- the effects of temperature on the above properties
- castability
- machinability
- capacity, if any, for heat treatment
- methods by which the material can be joined.

27.41   Forging processes can be carried out on metals and alloys which are malleable in either the hot or cold state but whilst a metal may be malleable at all temperatures those which are ductile lose their ductility at high temperatures where strength falls so that the metal tears apart in tension. Thus ductile metals can be drawn or deep drawn, spun or stretch formed (7.32–7.35). Some materials are neither malleable nor ductile and can only be cast to shape or shaped by powder metallurgy techniques. If great accuracy of dimensions is required in such castings then an expensive investment casting process (3.100) may be involved.

27.42   Thermoplastic polymers and glasses can be shaped by hot pressing and injection moulding but, since both of these materials also retain some ductility at high temperatures, they can also be blown to shape. Very often the choice of shaping process will depend upon the number of components required. Thus it may be desirable to use a die-casting process to produce a component, and thus eliminate machining operations which would otherwise be necessary if sand-casting were employed instead. Unfortunately, the use of a die-casting process will only be economical if a

Table 27.6 *The ductility of some metals and alloys (as measured by percentage elongation in the tensile test) for use in cold-pressing or deep-drawing processes.*

Metal or alloy	Elongation (%) in tensile test
70–30 brass	70
5% aluminium bronze	70
3% tin bronze	65
65–35 brass	65
aluminium (99.9%)	65
stainless steel (18% Cr–10% Ni)	65
copper (OFHC)	60
cupro-nickel (5% Ni)	50
mild steel	40
stainless iron (13% Cr)	40
aluminium alloy (1.2% Mn)	40

large number of castings, say 5000 or more, is required, thus covering the high initial cost of the steel dies (3.102). Similarly, drop-forging is only economical if very large numbers of components are required, since expensive die-blocks must be sunk. Sometimes a compromise may be struck, at the expense of some reduction in mechanical properties, by using a malleabilized-iron sand-casting in cases where only a few components are required.

Frequently, the shaping process is also used to develop strength in a component. Thus, an aluminium cooking-pot or the bodywork of an automobile become sufficiently rigid as a result of work-hardening which accompanies the shaping process.

27.43   Some metals and alloys have a very limited scope for shaping. These alloys such as 'Nimonics' (18.24; 27.34) and the Incoloy series, destined for service at high temperatures in gas turbines need of course to be both strong and stiff at those temperatures if they are not to suffer distortion. This hardness and stiffness in turn makes such alloys very difficult to forge even at high temperatures where they are intended to be stiff. Thus a slow, expensive processing schedule is involved with frequent reheating stages. Sometimes investment casting is the alternative in such cases but whichever process is chosen shaping is an expensive matter, the cost of which must be added to the initial high cost of such materials.

27.44   Conditions are continually changing, as new materials become available. During the last few decades, plastics have replaced metals and glass for innumerable applications. As an example, most curtain-rail material was extruded from a 60–40 type brass. Now this material is produced in a thermoplastic polymer which is extruded on to a steel core which provides the necessary strength. Many components which were once made in the form of metal die-castings are now manufactured as plastics mouldings, die-casting being retained as the shaping process where higher strength, rigidity and temperature resistance are required. In disposable tubes for toiletries, such as toothpaste, aluminium has been replaced by plastics materials but unfortunately it is now less easy to squeeze out the last 'blob' – though perhaps it is my inborn suspicion of manufacturers' motives that makes me ponder this fact!

The main barrier to the greater use of polymers is their low range of

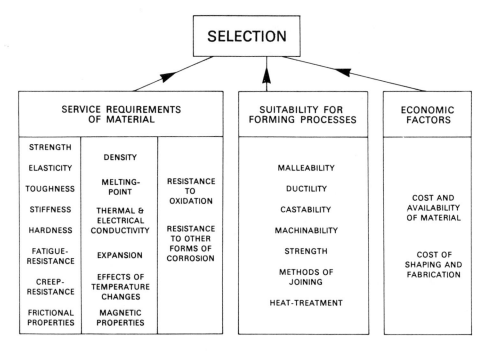

Fig. 27.2   *The selection of materials.*

softening temperatures. Nevertheless the corrosion-resistance of a plastics moulding is often very much higher than was that of the former die-casting, particularly if the latter was made from an alloy in a poor state of purity.

27.45 Without doubt the field in which most engineering materials development will take place in the near future is that of fibre-reinforced composites. Having spent a long life working mainly as a metallurgist I am of the opinion that the properties of metals and alloys have been exhaustively investigated and that there is now limited scope for outstanding developments in metallic alloys to take place – but I wonder if perhaps I made a similar remark, mistakenly, as a young whippersnapper in the 1930s?

Appendix

Properties of the important engineering metals.

Metal	Chemical symbol	Relative density	Melting-point (°C)	Tensile strength ($N mm^{-2}$)(*soft*)	Elongation (%)
Aluminium	Al	2.7	660	59	60
Antimony	Sb	6.6	630	10	0
Beryllium	Be	1.8	1285	310	2.3
Cadmium	Cd	8.6	321	80	50
Chromium	Cr	7.1	1890	220	0
Cobalt	Co	8.9	1495	250	6
Copper	Cu	8.9	1083	220	60
Gold	Au	19.3	1063	120	30
Iron	Fe	7.9	1535	500	10
Lead	Pb	11.3	327	18	64
Magnesium	Mg	1.7	651	180	5
Manganese	Mn	7.2	1260	500	20
Mercury	Hg	13.6	−39	molten at ordinary temperatures	molten at ordinary temperatures
Molybdenum	Mo	10.2	2620	420	50
Nickel	Ni	8.9	1458	310	28
Niobium	Nb	8.6	1950	270	49
Platinum	Pt	21.4	1773	130	35
Silver	Ag	10.5	960	140	50
Tin	Sn	7.3	232	11	60
Titanium	Ti	4.5	1667	230	55
Tungsten	W	19.3	3410	420	16
Uranium	U	18.7	1150	390	4
Vanadium	V	5.7	1710	200	38
Zinc	Zn	7.1	420	110	25
Zirconium	Zr	6.4	1800	220	25

The most widely used of the 'light metals'. Common in the Earth's crust.

A brittle, crystalline metal, used in limited amounts in bearing and type-metals.

A light metal, the use of which is limited by its scarcity. Used in beryllium bronze and in nuclear-power industries.

Used for plating steel, and to strengthen copper for telephone-wires.

A metal which resists corrosion, and is therefore used for plating, and in stainless steels.

Used mainly in permanent magnets, in super-high-speed steels and corrosion-resistant 'super-alloys'.

Now used mainly where very high electrical conductivity is required; also in brass and bronzes.

Of little use as an engineering metal, because of softness and scarcity. Used mainly in jewellery, and as a system of exchange.

Quite soft when pure, but rarely used in engineering in the unalloyed form. As steel, our most important metal.

Not really the densest of metals, as the phrase 'as heavy as lead' suggests. Very resistant to corrosion – used in chemical engineering, but main uses in batteries and pigments.

Used in conjunction with aluminium in the lightest of engineering alloys.

Very similar to iron in many ways – used mainly as a deoxidant and desulphurizer in steels.

The only liquid metal at normal temperatures. Thermometers and scientific equipment.

A heavy metal, used mainly in alloy steels. One of the main constituents of modern high-speed steels. Also used in stainless steels.

A very adaptable metal, used in both ferrous and non-ferrous alloys. The metallurgist's main 'grain-refiner'. Principal uses – stainless steels and for electroplating.

Also known as 'columbium' in the USA. Used mainly in alloy steels, and in high-temperature alloys.

A precious white metal. Used in scientific apparatus, because of its high corrosion-resistance; also in some jewellery.

Has the highest electrical conductivity, but is used mainly in jewellery and, in a few countries, for coinage.

Widely used but increasingly expensive. 'Tin cans' carry only a very thin coating on mild steel. Very resistant to corrosion. A constituent of bronze.

A fairly light metal, which is becoming increasingly important as its price falls due to the development of its technology.

Used in electric lamp filaments, because of its high melting-point. Is also the main constituent of most high-speed steels and heat-resisting steels, but main use now is in cemented carbides.

Now used mainly in the production of atomic energy.

Used in some alloy steels as a hardener.

Used widely for galvanizing mild- and low-carbon steels. Also as a basis for some die-casting alloys. Brasses are copper-zinc alloys.

Used as a grain-refiner in steels. It is also used for atomic energy applications.

Index